Australia's Biodiversity and Climate Change

Will Steffen

Andrew A Burbidge • Lesley Hughes • Roger Kitching • David Lindenmayer
Warren Musgrave • Mark Stafford Smith • Patricia A Werner

Australian Government

National Library of Australia Cataloguing-in-Publication entry

Australia's biodiversity and climate change/Will Steffen ... [et al.].

9780643096059 (pbk.)

Includes index.
Bibliography.

Biodiversity – Climatic factors – Australia.
Biodiversity conservation – Australia.
Climatic changes – Environmental aspects – Australia.

Steffen, William L., 1947–

333.950994

Published by
CSIRO PUBLISHING
150 Oxford Street (PO Box 1139)
Collingwood VIC 3066
Australia

Telephone: +61 3 9662 7666
Local call: 1300 788 000 (Australia only)
Fax: +61 3 9662 7555
Email: publishing.sales@csiro.au
Web site: www.publish.csiro.au

Front cover (clockwise, from top left): Snow gums, *Eucalyptus pauciflora niphophila*, Kosciuszko National Park, NSW (DEWHA/Tim Bond); leaf beetles on a snow gum leaf, Namadgi National Park, ACT (DEWHA/Dionne Bond); southern leaf-tailed gecko, *Phyllurus platyurus*, Lane Cove National Park, NSW (DCC/Erika Alacs); water-lilies, *Nymphaea macrosperma*, Mornington, WA (DEWHA/Nick Rains); hummock grasslands, *Triodia* sp., Gawler Ranges National Park, SA (DEWHA/Tim Bond); red-eyed tree frog, *Litoria chloris*, QLD (Wiki Commons/Muhammad Mahdi Karim); Gouldian finch, *Chloebia gouldiae*, Wongalara NT (DEWHA/Steve Murphy).

Back cover (left to right): Coastal sub-tropical rainforest, Sea Acres Nature Reserve, NSW (DEWHA/Tim Bond); scentless rosewood in Murramarang National Park, NSW (DEWHA/Tim Bond); hummock grassland on Newhaven Sanctuary, NT (DEWHA/Nick Rains).

Set in 10/13 Adobe Minion and ITC Stone Sans
Cover and text design by James Kelly
Typeset by Desktop Concepts Pty Ltd, Melbourne
Printed in China by 1010 Printing International Ltd

The paper this book is printed on is certified by the Forest Stewardship Council (FSC) © 1996 FSC A.C. The FSC promotes environmentally responsible, socially beneficial and economically viable management of the world's forests.

CSIRO PUBLISHING publishes and distributes scientific, technical and health science books, magazines and journals from Australia to a worldwide audience and conducts these activities autonomously from the research activities of the Commonwealth Scientific and Industrial Research Organisation (CSIRO).

Preface

Australia has a rich natural biodiversity, with a high proportion of species found nowhere else in the world. This biodiversity underpins much of our country's economy, as well as contributing to our special national identity. In most parts of Australia, biodiversity is in decline from the pressure of threats such as habitat loss and invasive species. The Australian Government recognises climate change as a key additional threat to the conservation of the country's biodiversity, and the assessment presented in this book is part of a national response to this threat.

The Australian Government's climate change policy is built on three pillars: reducing Australia's greenhouse gas emissions; helping to shape a global solution; and adapting to unavoidable climate change. The Australian Government ratified the Kyoto Protocol in December 2007 and is working towards a post-2012 multilateral agreement for addressing climate change that is equitable and effective, and includes agreement on a long-term global goal for emissions reductions.

This book assesses the vulnerability and potential for adaptation of Australia's biodiversity, as well as consideration of the societal (governmental, policy, institutional) changes that might sustain Australia's biodiversity in a climate changing world. Whilst focussed on actions under the third pillar, adaptation, it has relevance to all three pillars.

Government focus on climate change adaptation has significantly increased in recent years. This is reflected, for example, in the endorsement by the Australian Natural Resource Management (NRM) Ministerial Council of the National Biodiversity and Climate Change Action Plan in 2005. Subsequently a series of vulnerability assessments have been commissioned, including for Australia's National Reserve System, for fisheries and aquaculture, for fire regimes and biodiversity, and for coasts. It is also reflected in the terrestrial and marine biodiversity decline reports prepared in 2005 and 2008 respectively, and in a range of national activities. Examples include development of climate change action plans for the Great Barrier Reef, for fisheries and for commercial forestry, the dissemination of information for managers of Natural Resource Management regions, and the review of the National Agriculture and Climate Change Action Plan. An assessment of the implications of climate change for Australia's World Heritage properties has also recently been completed under the auspices of the Environment Protection and Heritage Council, and a climate change strategy for Australia's botanic gardens endorsed by the Council of the Heads of the Botanic Gardens, has been published. Adaptation research planning for a range of key vulnerable sectors, including terrestrial, freshwater and marine biodiversity, is being nationally coordinated by the National Climate Change Adaptation Research Facility and associated National Adaptation Research Networks.

In late 2006, the NRM Ministerial Council commissioned a strategic assessment of the vulnerability of Australia's biodiversity to the impacts of climate change. To assist with this assessment, the Australian Government established a Biodiversity and Climate Change Expert Advisory Group. The Expert Advisory Group comprised Professor Will Steffen (The Australian National University, Canberra, Chair), Dr Andrew A Burbidge (WA Department of Environment and Conservation, Perth), Professor Lesley Hughes (Macquarie University, Sydney), Professor Roger L Kitching (Griffith University, Brisbane), Professor David Lindenmayer (The Australian National University), Professor Warren Musgrave (University of New England, Armidale), Dr Mark Stafford Smith (CSIRO Climate Adaptation Flagship, Canberra) and Professor Patricia A Werner (The Australian National University).

The Expert Advisory Group has assessed the vulnerability of Australia's terrestrial, freshwater and marine biodiversity to climate change, and suggested policy and management options to reduce this vulnerability. They have drawn on international and Australian published research, as well as results of current research programs and unpublished information provided by Australian experts. This research includes the results of other biodiversity-related NRM Ministerial Council actions such as the assessments of the implications of climate change for fire and biodiversity, and for the National Reserve System, and results from a series of workshops held with researchers, policy makers and biodiversity managers that examined the implications of Australia's changing climate for biodiversity and our management strategies. The main report of the assessment has been peer-reviewed and is published here. It is also available electronically on the Department of Climate Change's website (http://www.climatechange.gov.au), together with related products.

The insights gained through the assessment will provide additional guidance, information and ideas for biodiversity practitioners in developing climate change adaptation strategies to protect Australia's biodiversity. To further facilitate this, the Expert Advisory Group also developed two other documents: the Summary for Policy Makers distils the key messages and policy directions identified through the assessment process and a concise Technical Synthesis summarises the evidence base that underpins the assessment. The NRM Ministerial Council has acknowledged the significant contribution of the reports to understanding of the challenges for Australian biodiversity conservation in a changing climate.

The Council is continuing its effort to confront the challenges of climate change and has identified a broad-ranging suite of climate change priorities to be addressed over the period 2009 to 2012. These priorities include additional national actions to further develop climate change adaptation policy across and within land and marine sectors, including for biodiversity. Council has also encouraged integration of the findings into policy. The focus of the new national strategy – *Australia's Biodiversity Conservation Strategy 2010–2020* – will be on increasing resilience to climate change whilst also tackling the range of other stressors on Australia's biodiversity. This biodiversity vulnerability assessment also provides the basis for the terrestrial biodiversity National Adaptation Research Plan, one of the set of plans developed by the National Climate Change Adaptation Research Facility.

Department of Climate Change
September 2009

Acknowledgments

The Expert Advisory Group acknowledges the many ways in which a very large number of people have contributed to this project – through provision of written material, discussions and consultations, reviews and critiques of our efforts, and in managing the process itself. We especially thank Ross Bradstock, Barry Brook, Margaret Byrne, Lynda Chambers, Sarah Comer, Alan Danks, Peter Dann, Ken Green, Mark Hovenden, Tim Low, Graham Marshall, Paul Marshall, Leanne Renwick, Richard Williams, Steve Williams, Colin Yates and Bill Young for their direct contributions in writing expert boxes.

Discussions and consultations with scores of researchers, experts, managers, policy makers, and institutional representatives played a valuable role in the dynamic process of evaluation, debate, and innovative thinking involving a vast array of complex issues. We thank our colleagues for sharing their time, expertise, insights and their latest research findings. We thank Michael Dunlop (CSIRO DSE), Joern Fischer (The Australian National University), Mark Hovenden (University of Tasmania), Graham Marshall (University of New England), Robert Nadeau (George Mason University), Craig Nitschke (VIC Forestry) and Terry Root (Stanford University) as well as scientists at the Australian Antarctic Division: Ian Allison, Dana Bergstom, Andrew Constable, Tas van Ommen, Tony Press, Martin Riddle and Michael Stoddart; and marine biologists at the CSIRO Division of Marine and Atmospheric Research in Hobart: Alistair Hobday, Nic Box, Alan Butler and Elvira Poloczanska.

Many contributions to our efforts were made by means of five Australian Department of Climate Change (DCC)-sponsored workshops: 'Climate Change, Species and Ecosystems: Identifying Key Science Questions for Australia' (13–19 October 2007); 'States and Territories and Climate Change' (18–19 December 2007); 'Climate Change Impacts on Biodiversity: Honours Presentations' (5 February 2008); 'Institutional and Governance Issues in a Climate Changing World' (12 February 2008); and 'The Effects of Climate Change on Fire Regimes in Areas Managed for Biodiversity' (26 March 2008). Organisers and participants in these workshops, other than DCC staff and the authors, included the following: Will Allen (NSW Southern Rivers Catchment Management Authority); Alan Andersen (CSIRO DSE); Graeme Barden (DEWHA); Melanie Bishop (University of Technology Sydney), David Bowman (University of Tasmania); Ross Bradstock (University of Wollongong); Natalie Briscoe (University of Melbourne); Barry Brook (University of Adelaide); Linda Broome (NSW DECC); Alan Butler (CSIRO DMAR); Margaret Byrne (WA DEC); Lynda Chambers (BOM); Jane Carder (ACT Government); Geoff Cary (The Australian National University); Peter Clarke (University of New England); Hal Cogger (Australian Museum); Sonia Colville (DEWHA: Biodiversity Conservation Branch); Mark Conlon (NSW DECC); Garry Cook (CSIRO DSE); Brooke Craven (Tas DPIW); Ian Cresswell (DEWHA: National Oceans Office); Bruce Cummings (Parks Australia); Laura Dakuna (DEWHA); Gwendolyn David (University of Queensland); Jocelyn Davies (CSIRO/Desert Knowledge CRC); Kimberley Dripps (Vic DSE); Michael Dunlop (CSIRO DSE); Brendan Edgar (private consultant); Neal Enright (Murdoch University); Adam Felton (The Australian National University); Andrew Fisher (SA DWLBC); Gordon Friend (Vic DSE); Stephen Garnett (Charles Darwin University); Louise Gilfedder (Tas DPIW); Malcolm Gill (The Australian National University); Andreas Glanznig (Invasive Animals CRC); Ken Green (NPWS, NSW); Cate Gustavson (WA DPI); Meredith Henderson (SA DEH); Kevin Hennessy (CSIRO DMAR); David Hilbert (CSIRO TFRC); Richard Hobbs (Murdoch University); Paul Houlder (NSW DECC); Mark Howden (CSIRO DSE); David Keith (NSW DECC); Andrew Kennedy (NSW DPI); Darren Kriticos (CSIRO DSE); Bruce Leaver (DEWHA/Parks Australia); Adam Liedloff (CSIRO DSE); Rosie Lohrisch (DEWHA); Kevin Love (Vic DSE); Tim Low (University of Queensland); Andrew Lowe (SA DEH/State Herbarium); Chris Lucas (BOM); Ian Mansergh (Vic DSE); Graham Marshall (University of New England); Paul Marshall

(GBRMPA); Lachie McCaw (WA DEC); Belinda McGrath-Steer (SA DEH); Richard McKellar (WA DEC); Stacey McLean (Brisbane City Council); Astrida Mednis (DEWR); Fiona Melvin (Monash University); Rhonda Melzer (Qld EPA); Dale Mesaric (LaTrobe University); Alex Milward (Qld OCC); Laurence McCook (GBRMPA); Robert McDougall (Macquarie University); James Moore (James Cook University); Chris Morony (SA DEH); John Neldner (Queensland Herbarium, EPA); Hong Dao Nguyen (University of Sydney); Katherine O'Connor (University of Tasmania); Jessica O'Donnell (Macquarie University); Stuart Pearson (Land and Water Australia); Nicole Pitt (University of Tasmania); Stephen Platt (Vic DSE); Kathy Preece (Vic DSE); Alex Rankin (DEWHA); Michael Ross (DAFF); Stephen Roxburgh (University of New South Wales/CSIRO); Barry Russell (NT Government/Museum and Art Gallery of the NT); Gary Saunders (NSW DECC); Sarah Sharp (Environment ACT); James Shirley (private consultant); Jessica Stella (James Cook University); Steve Sutton (Bushfires NT); Michelle Swan (University of Western Australia); Liz Tasker (NSW DEC); Brian Walker (CSIRO DSE); Nicola Ward (Vic DSE); Judy West (CSIRO Centre for Plant Biodiversity Research); Rob Whelan (University of Wollongong); Peter Wilcox (Bushcare/Australian Natural Resource Management Team); Joel Williams (Flinders University); Mike Williams (SA Department for Environment and Heritage); Paul Williams (Qld Department of the Environment); Richard Williams (CSIRO DSE); Sara Williams (Qld EPA); Steve Williams (James Cook University); Joab Wilson (RMIT University); Brendan Wintle (University of Melbourne); Colin Yates (WA DEC); David Yeates (CSIRO Entomology); Alan York (University of Melbourne); Andrew Young (CSIRO Plant Industry); Andrew Zacharek (DEWHA/National Oceans Office); and Charlie Zammit (DEWHA/Biodiversity).

All Australian states and territories provided helpful advice on various drafts of the report through relevant agencies. We thank them for their efforts in making the assessment more regionally and policy relevant. We greatly appreciate the expert peer reviews of the entire near-final report provided by Professor Richard Hobbs (Murdoch University) and Professor David Karoly (University of Melbourne).

We thank the Department of Climate Change for the opportunity to undertake this strategic assessment of biodiversity vulnerability in a climate changing world. In particular, we are grateful for the support, cooperation, advice and encouragement of Jo Mummery and Anne-Marie Wilson. Lalage Cherry and Liz Dovey provided tireless energy, many hours of hard work, much valuable intellectual feedback and input, and day-to-day management of materials and people, all of which were essential for both the assessment itself and the production of this book. We thank them and their colleagues who assisted at various stages of the assessment including Cristina Davey, Anna van Dugteren, Brendan Edgar, Angas Hopkins, Brendan Kelly and Stefanie Pidcock.

We also thank those who helped with the production of the book itself – layout, design, figures, editing and indexing. They include Dave Gardiner, Clive Hilliker, John Manger, Tracey Millen, James Coffey and Sherrey Quinn.

Contents

List of authors

Professor Will Steffen, Executive Director, Climate Change Institute, The Australian National University (Chair)

Dr Andrew A Burbidge, Wildlife Research Centre, Western Australian Department of Environment and Conservation and Consulting Conservation Biologist

Professor Lesley Hughes, Department of Biological Sciences, Macquarie University

Professor Roger Kitching, Griffith School of the Environment, Griffith University

Professor David Lindenmayer, The Fenner School of Environment and Society, The Australian National University

Professor Warren Musgrave, Consulting Economist and Emeritus Professor, University of New England

Dr Mark Stafford Smith, Science Director, CSIRO Climate Adaptation Flagship and Research Fellow, Desert Knowledge Cooperative Research Centre

Professor Patricia A Werner, The Fenner School of Environment and Society, The Australian National University

1 The climate change challenge

This assessment of the vulnerability of Australia's biodiversity to climate change was begun in early 2007 in response to a request from the Natural Resource Management Ministerial Council. This introductory chapter outlines the scope of the assessment and the approach to it taken by the Expert Advisory Group (EAG). The structure of the assessment is then presented in the form of a chapter-by-chapter synopsis, which lays out the flow of logic in the assessment and the major topics addressed by the EAG. Finally, the chapter describes the aims and characteristics of the assessment's key messages and policy directions.

1.1 SCOPE OF AND APPROACH TO THE ASSESSMENT

Human-driven climate change is now widely acknowledged to be a reality, with impacts discernible for a large number of sectors. One of the most vulnerable sectors is biodiversity, which is already under pressure from a wide range of existing stressors. Climate change presents an additional challenge, on top of and interacting with existing stressors.

In 2006 the Natural Resource Management Ministerial Council adopted as a priority action the preparation of a strategic assessment of the vulnerability of Australia's biodiversity to climate change. The Australian Greenhouse Office (now part of the Department of Climate Change, DCC) of the former Department of the Environment and Heritage commissioned the assessment and, in early 2007, formed the EAG to conduct it. The terms of reference for the assessment were to: (i) cover terrestrial, freshwater and marine environments; (ii) be strategic in nature and provide policy directions for future adaptation planning (i.e. it will not be a systematic, region-by-region, community-by-community assessment); (iii) include an assessment of the scientific observations and predictions around impacts/responses to climate change; and (iv) provide comments on ways biodiversity management can adapt to enhance the resilience of Australian biodiversity to the impacts of climate change.

The terms of reference for the EAG thus take a broad view of the assessment, and explicitly include the need to provide advice on policy directions and management strategies to enhance resilience of biodiversity to climate change, and to reduce the impacts arising from its interaction with existing stressors. In addressing this latter term of reference, the focus has been on strategic advice rather than on a large number of specific policy recommendations. However, where appropriate, specific policy options are suggested as examples of the types of actions that might be required to put the strategic advice into action.

This is the first such national assessment of the vulnerability of Australia's biodiversity, in its entirety, to climate change. However, the focus is primarily on terrestrial biodiversity, for two reasons. Firstly, there has recently been a thorough analysis of the impacts of climate change on marine biodiversity generally

Figure 1.1 Billabong with paperbarks *Melaleuca* spp. and lotus lilies *Nelumbo nucifera*, Yellow Water, Kakadu National Park, Northern Territory. Source: AUSCAPE. Photo by Jean-Paul Ferrero.

(Hobday *et al.* 2006) and the Great Barrier Reef in particular (Johnson and Marshall 2007). Secondly, there has been relatively little research on the consequences of climate change for freshwater biodiversity, so there is relatively little literature to draw upon. However, to the extent possible, freshwater biodiversity is included in the assessment.

The primary audience for the assessment is the biodiversity conservation policy and management sectors at all levels of government, and in the rapidly increasing private sector community that contributes to biodiversity conservation. In addition, many scientists from the biodiversity conservation, ecology, resilience, institutional/governance and climate change research communities will find useful information and insights in the assessment. Finally, interested members of the general public may find some of the more general analyses and strategic discussions to be accessible and informative.

The EAG has taken a broad, long-term perspective to its task. In terms of the projected magnitude and rate of the climate change, Australia's (and the world's) biodiversity is facing a threat equivalent to those of the abrupt geological events that triggered the great waves of extinction in the past. Thus, our experience in biodiversity conservation over the past century may provide only partial or limited guidance for dealing with the climate change threat. This, coupled with the considerable uncertainties in the precise nature of future climate change at local and regional scales, presents daunting challenges to the assessment. We have thus stepped back from the usual climate scenario-driven approach to vulnerability assessments and gone back to fundamental ecological principles as the basis for analysis and synthesis. This approach is reflected in the structure of the book, and in the nature of the key messages and suggested policy directions.

1.2 STRUCTURE OF THE ASSESSMENT

Two features of the book are important to understand its structure and flow. Firstly, we have based our analysis and recommended ways forward on a small set of ecological principles that characterise the ways in which: (i) individual species interact with their surrounding environment; (ii) species interact with each other in communities and ecosystems; (iii) ecosystems and landscapes are structured; and (iv) environmental change affects the structure and functioning of ecosystems. These principles underpin the analyses of current biodiversity change and those projected under further climate change, as well as the policy and management principles required to deal with these challenges. Thus, they underpin the analyses throughout the book.

Secondly, climate change – although having some very important and unique characteristics in terms of its consequences for biodiversity – is considered to be another stressor that adds to and interacts with a range of existing stressors that have already significantly changed and diminished Australia's biodiversity. Thus, viewing climate change in isolation from other stressors, particularly now and for the next few decades at least, is misleading and counterproductive in terms of policy and management. However, without early and vigorous mitigation actions, climate change has the potential by the second half of the century to become an overwhelmingly profound and pervasive driver of change in Australia's biotic fabric, resulting in many extinctions and the formation of novel ecosystems that may not provide the essential ecosystem services on which humans depend.

The book begins in *Chapter 2: The nature of Australia's biodiversity* with a long-time perspective on the evolution of Australia's biota – why Australia is so species-rich, why our biodiversity is unique, and why the conservation of our biodiversity is so important. The chapter sets the stage for the rest of the book by describing the environmental conditions under which our biota has slowly evolved, and the nature of the interactions between the continent's two waves of human colonisers and the rest of its environment. This background is essential for understanding the profound implications of the European settlement of the continent and the looming challenge of climate change, both of which drive change of such magnitude and rate that Australia's biosphere is undergoing rapid and continuing transformation.

Chapter 3: Australia's biodiversity today describes in much more detail the two centuries of acute change since European settlement. The chapter briefly discusses the proximate and ultimate drivers of current change in Australia's biodiversity, and then focuses on the recorded changes at the genetic, species and ecosystem levels. Much of the emphasis has historically been at the species level, dealing with extinctions, threatened species, and changes in the relative abundance and distribution of species driven by land clearing and other stressors. The role of introduced species is particularly important, as are the higher-level changes occurring at community and ecosystem levels. Although many will already be familiar with the historic record of change in Australia's biodiversity, it is an integral part of this assessment as these changes continue to unfold and their drivers continue to operate. As noted previously, without an effective integration of present stressors and their consequences into climate-oriented research, policy making and management, any efforts at climate change adaptation to enhance the conservation of Australia's biodiversity are virtually certain to fail.

The next two chapters review and synthesise available information on recent climate change. *Chapter 4: The rate and magnitude of climate change* provides a succinct overview of the current state of the science in climate change. Organised around the global and the Australian scales, it describes the climate changes that have already been observed over the past one to two centuries and outlines the range of projections for Australia for the rest of this century. However, given the very high uncertainty associated with many aspects of climate change projections, they do not play a strong role in the assessment in terms of providing reliable local- and regional-scale information around which specific adaptation actions can be confidently taken.

The ways in which climate change is already affecting Australia's biota, and will potentially affect it, are described in *Chapter 5: Responses of Australia's biodiversity to climate change*. It begins by describing the different nature of climate change as a stressor, in terms of: (i) its rate compared to geological timescales; and (ii) the major changes in the basic physical and chemical environment underpinning all life. Drawing on the basic ecological principles introduced in Chapter 3, this chapter outlines the ways in which Australia's biodiversity is already responding to climate change, ranging from physiological to ecosystem levels. It then considers predicted trends in biodiversity change, and focuses on the probable nature and direction of changes in species and ecosystems in response to general climatic trends rather than to specific scenarios. An important feature of the chapter is a treatment of several difficult but important issues – uncertainties, non-linearities, time lags, thresholds, feedbacks, rapid transformations, synergistic interactions and surprises.

Chapter 5 is different from other chapters in its much greater level of detail, with thorough referencing. The chapter gives the reader an excellent overview of the current state of knowledge on climate change impacts on Australia's biodiversity, and thus provides a valuable reference source. However, our knowledge base is highly focused on the species level and on climate change as an isolated stressor, as this reflects the current state of the science. Given that the most important impacts of climate change on biodiversity will undoubtedly be the indirect ones at the community and ecosystem levels along with the interactive effects with existing stressors, the current state of knowledge is not adequate for anything but the most general guidance on adaptation approaches. Much more research, focusing on the knowledge gaps outlined in Chapter 5, is required before direct climate science-driven approaches to biodiversity adaptation can be undertaken successfully in all but a few cases.

However, the situation for policy and management is not hopeless. The ecological principles introduced in Chapter 3 provide the underpinning for re-evaluating the current approaches to biodiversity conservation, many of which will still be very important in a world of changing climate but others of which will need to change. Concepts such as resilience and transformation provide positive, proactive avenues to reduce the vulnerability of biodiversity to climate change despite the lack of knowledge about system-level responses and the large uncertainties associated with the projections of future climate change. The emphasis is on making space and opportunities for ecosystems to self-adapt and reorganise, and on the maintenance of fundamental ecosystem processes that underpin vital ecosystem services. Such approaches require transformation of the institutional architecture to implement the revised and new policy and management tools. These issues are dealt with in the next two chapters.

Current policy and management for biodiversity conservation is considered in *Chapter 6: Current biodiversity management under a changing climate*. Beginning with a description of current management principles, the chapter then analyses the current set of conservation strategies and tools, and the current policy and institutional landscape in the context of the existing threats to biodiversity. Current policy and management are then discussed as a platform on which to build effective climate adaptation, and focus on the aspects of the current approaches that will be especially effective in the context of climate change. The chapter concludes with an evaluation of the current constraints – such as resource limitations and knowledge gaps – that provide challenges to policy and management in meeting their objectives.

Turning towards the future, *Chapter 7: Securing Australia's biotic heritage* focuses on ways in which the adaptive capacity of Australia's biodiversity can be enhanced. Two background analyses – one on the nature of the future climate change threat from the perspective of biodiversity and the other on the major socio-economic trends sweeping across Australia that could affect biodiversity – set the stage for an exploration of what can be done to build adaptive capacity. We again return to the guiding principles of Chapter 3 to inform management approaches, which then provide

Figure 1.2 Animal trapping as part of a post-fire fauna survey. Source: CSIRO Sustainable Ecosystems.

Figure 1.3 The tiny mountain pygmy possum *Burramys parvus*, which faces extinction as the Earth warms. Source: Linda Broome.

the basis for suggesting strategies and tools to meet the climate change challenge. The nature and magnitude of the climate change challenge imply that innovative governance systems and a much larger resource base will be required to implement the new and revised strategies. The chapter also presents a conceptual framework for a more systematic regional approach to biodiversity conservation, which builds on the existing and projected socio-economic trends as a platform for improving biodiversity outcomes, and includes consideration of three stylised climate change scenarios for the 21st century.

Finally, *Chapter 8: Responding to the climate change challenge* concludes with a set of key messages and policy directions, based on the analyses in the preceding chapters.

1.3 KEY MESSAGES AND POLICY DIRECTIONS

Our messages have the following characteristics:

- The key messages and policy directions are not recommendations in the formal sense that they present specific policy proposals that have undergone a thorough analysis of policy options. Our aim is to present the broad directions in which policy, management tools and strategies, governance, and public perceptions of the importance of biodiversity conservation must go if the vulnerability of Australia's biodiversity to climate change is to be reduced significantly.
- The set of messages and policy directions should be viewed as an integrated package – the key messages are linked, and often one relies on one or more of the others. They are not designed as a list that can be prioritised but rather as a mutually reinforcing set that provides a powerful framework for turning around the current trend of biodiversity decline even in the face of climate change.
- Some of the key messages and policy directions are not derived from specific pieces of analysis found in the body of the book. Rather, they are derived from a broad-based synthesis of more detailed information and analyses presented in several chapters. An example is the message that without early, vigorous and ongoing mitigation measures by the nations of the world, there is a high probability of more severe

climate change and the associated risk of unavoidable, much higher rates of biodiversity loss in the coming decades and centuries.

- Mitigation features prominently in the key messages. Although adaptation is often conceptualised and treated as separate from mitigation, this is a fundamental mistake that can easily lead to counterproductive outcomes for both mitigation and adaptation. Where appropriate, we have integrated the two towards building synergistic outcomes for both.

The set of key messages and policy directions aim for a fundamental rethink of how biodiversity conservation is approached – from the basic underpinning science through to the management tools needed to enhance adaptive capacity; and the policy instruments, resource base and institutional architecture needed to respond effectively to the climate change challenge. A vital prerequisite to adopting the new policy directions is a greater public understanding of the role of biodiversity and ecosystem services in human well-being, and a greater public participation in biodiversity policy development and resourcing. Although much has been achieved in biodiversity conservation in recent decades, the nature of the climate change challenge demands a greatly strengthened conservation effort for Australia to secure its biotic heritage for the 21st century and beyond.

2 The nature of Australia's biodiversity

This chapter outlines why Australia is so species-rich, why so many species occur on this continent and nowhere else, and why the conservation of biodiversity is important. The chapter is a prelude to Chapter 3, which deals with the significant changes that have occurred in Australia's biodiversity since European settlement in 1788. This review also sets the scene for the commentary in Chapters 4 and 5 that discusses how climate change may further affect biodiversity in Australia, and for Chapters 6 and 7, which discuss current management strategies for biodiversity conservation and how these may need to change to enhance the resilience of Australian biodiversity in a changing climate.

2.1 WHY IS AUSTRALIA SO RICH IN BIODIVERSITY?

'Biodiversity' is a term used to encompass the variety of life on Earth. It refers to the variety of all life forms: the different plants, animals and micro-organisms, their genes, and the communities and ecosystems of which they are part. Biodiversity is usually recognised at three levels: genetic diversity, species diversity and ecosystem diversity. It is most often used with reference to the variety and number of species and subspecies (e.g. the diversity of birds is approximately 10 000 species), and it also applies to genetic variation within species, differences between populations, and variation at higher levels of organisation such as the community or ecosystem, or even variation within a biome.

Between 7 and 10% of all species on Earth occur in Australia. More than 4500 species of marine fishes – and the greatest number of species of red and brown algae, crustaceans, sea squirts and bryozoans in the world – live in Australian inshore waters (Chapman 2005; ABRS 2008). Fifty-seven per cent of all mangrove species are found in Australian intercoastal zones. There are more than twice as many species of reptiles in Australia as there are in the United States, and Australian deserts support more lizard species than any other comparable environment (Chapman 2005; James and Shine 2000; Morton and James 1988). The distribution and abundance of biodiversity is not even across the continent. Some groups are more species-rich in the tropics and high rainfall areas (e.g. birds and butterflies), while others such as lizards, scorpions and grasshoppers are more species-rich in deserts.

Australia has been isolated from other land masses for many millions of years. This isolation – together with a range of other factors such as its flatness and its nutrient-poor soils – has led to the continent supporting large numbers of species found nowhere else, and to combinations of key ecological processes that are different from those occurring elsewhere in the world. The continent has a rich diversity of biogeographic regions on land (Fig. 2.1) and in coastal seas (Fig. 2.2).

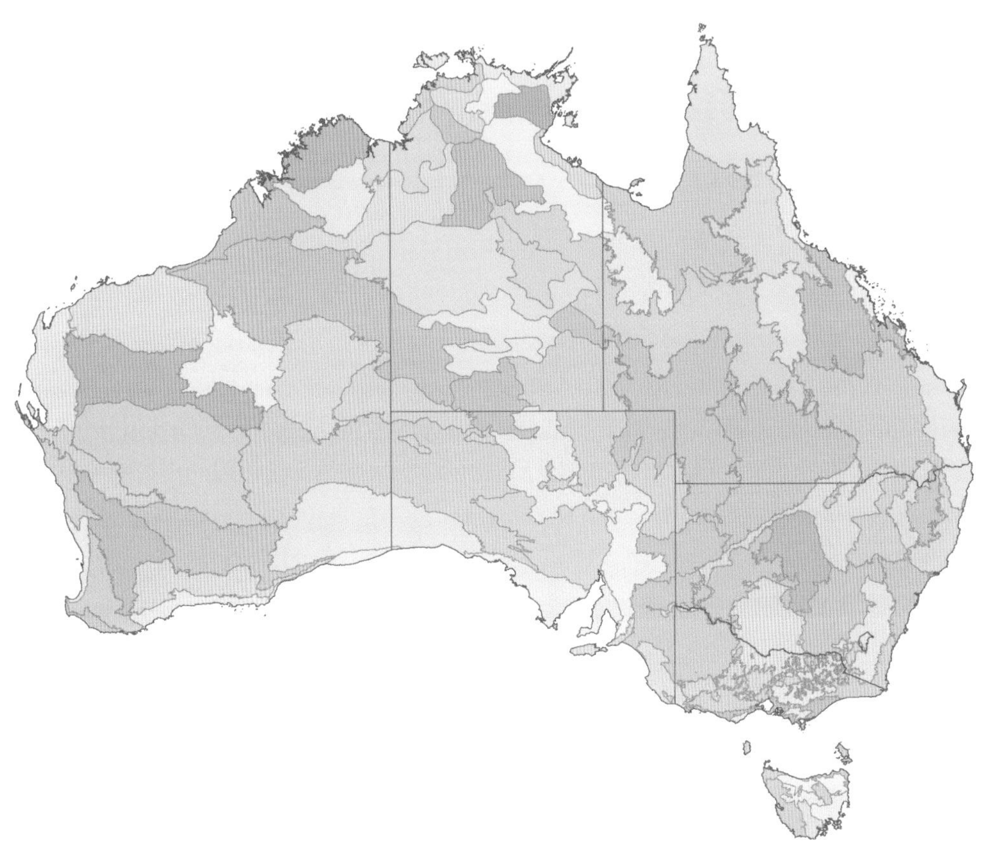

Figure 2.1 Australia's diverse terrestrial bioregions. Some 85 bioregions and 403 subregions have been identified. A bioregion is a distinct ecologically and geographically defined area. Source: Department of the Environment, Water, Heritage and the Arts.

Many of Australia's species, and even whole groups of species or taxonomic families, are endemic to this continent (Table 2.1). About 85% of terrestrial mammals, 91% of flowering plants, 90% of reptiles and frogs, and 95% of ectomycorrhizal fungi are found nowhere else (Castellano and Bougher 1994; Chapman 2005; Lindenmayer 2007a). More than 50% of the world's marsupial species occur only in Australia (Dickman and Woodford Ganf 2007). In addition, most Australian groups of plants and animals have particular features differentiating them from counterpart groups on other continents. Several reviews have offered explanations for the unique features and unusually high level of biodiversity found on this continent (e.g. Orians and Milewski 2007).

The world's biodiversity is not distributed evenly. In a landmark study, Mittermeier *et al.* (1998) identified 25 'biodiversity hotspots' comprising only 1.4% of the land surface of the Earth, where as many as 44% of all species of vascular plants and 35% of all species in four vertebrate groups are confined (see http://www.biodiversityhotspots.org/). One of the 25 hotspots occurs within Australia – Southwest Australia. The World Conservation Monitoring Centre has developed a list of 17 megadiverse countries – a group of countries that harbour more than 70% of the Earth's species and are therefore considered extremely biodiverse.

Australia's biodiversity is not evenly distributed across the continent or around its shores. A study by the Commonwealth Department of Climate Change has identified a number of terrestrial areas with particularly high species diversity (Fig. 2.3). No comprehensive study of Australia's marine hotspots is available, but some areas are known to be particularly biodiverse, e.g. the Great Barrier Reef, the Coral Sea, the South Tasman

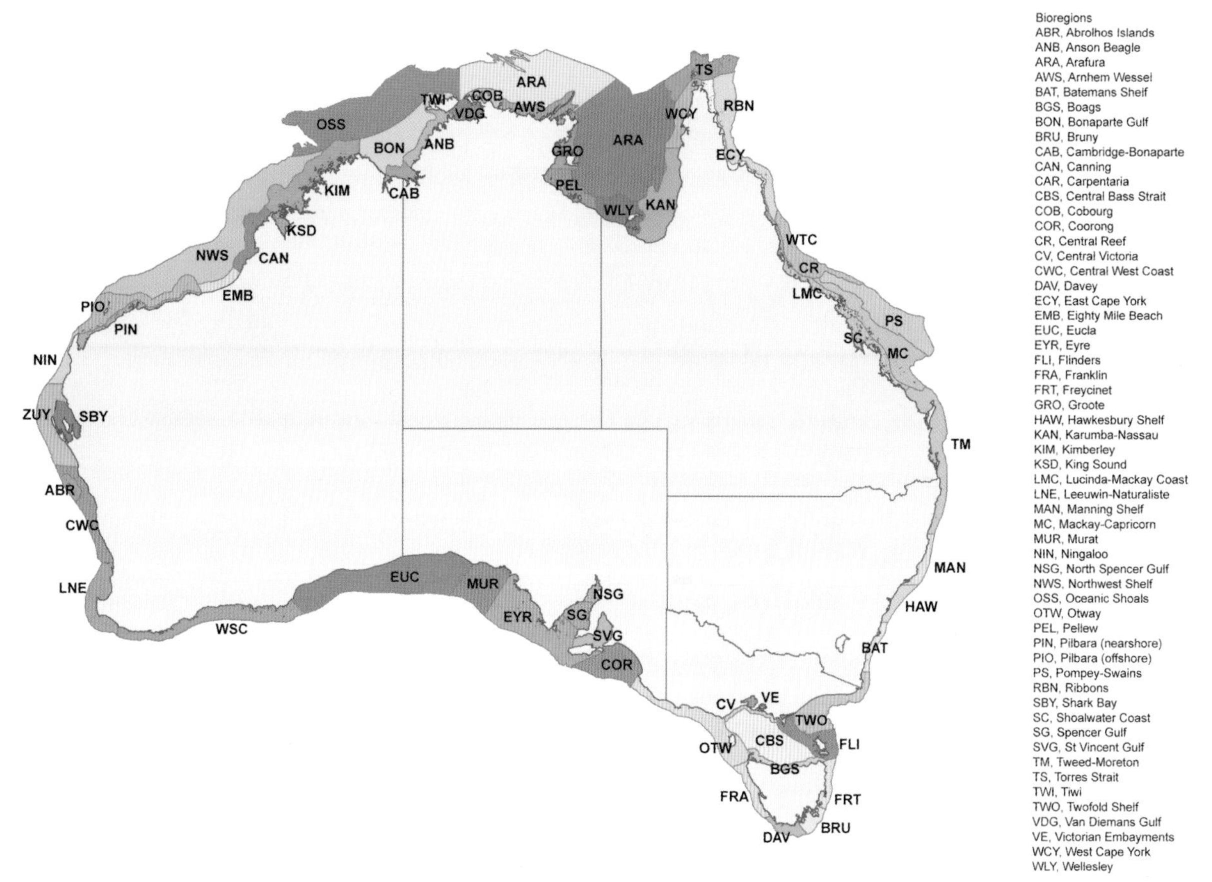

Figure 2.2 Australia's meso-scale marine bioregions. Source: Department of the Environment, Water, Heritage and the Arts.

rise and other southern sea mounts, the Ningaloo fringing reef, the waters of the Dampier Archipelago, Scott Reef, and much of the Kimberley coast.

To understand the current and future challenges to Australia's biodiversity associated with climate change, it is important to firstly understand Australia's environmental and ecological history. This chapter contains an overview of the origins and nature of Australia's unique biodiversity.

2.2 AUSTRALIA: THE ISOLATED CONTINENT

2.2.1 Breaking free from Gondwana

The super-continent of Gondwana was initially composed of the landmasses that subsequently fragmented into the continents and subcontinents of South America, India, Antarctica, South Africa and Australia. The Gondwanan landmass started to disperse between 170 million and 180 million years ago. India was the first to break away, followed by Africa, and then New Zealand, which started to move north. By the end of the Cretaceous period (65 million years ago), South America and Australia were still joined to Antarctica. The continent of Australia eventually broke free of Antarctica approximately 45 million years ago (late Eocene epoch) and moved northwards (Table 2.2). Almost all of the biota on the Australian landmass was then isolated from the biota of other continents for millions of years. Relict elements of the biota that arose during this period and still flourish in Australia include the plant genera *Banksia* (Fig. 2.4) and *Nothofagus* (southern beech), the monotremes and the parrots (family Psittacidae).

Table 2.1 Australia's rich species diversity: global status for major animal, fungus and plant groups.

Species	Global diversity
Marine fish	One of the most diverse fish faunas in the world, with more than 4500 species
Sharks and rays	54% of the entire chondrichthyan fauna is endemic to Australia
Ectomycorrhizal fungi	95% endemic (22 genera and three endemic families)
Terrestrial vertebrates	1350 endemic terrestrial vertebrates, far more than the next highest country (Indonesia, with 850 species)
Terrestrial mammals	305 species, of which 258 (85%) are endemic; more than 50% of the world's marsupial taxa occur only in Australia
Birds	17% of the world's parrots occur in Australia – more than 50 species (second-highest level of endemism after Brazil and the same as Colombia)
Reptiles	89% endemic; some groups such as front-fanged snakes (family Elapidae), pythons and goannas are more diverse than elsewhere in the world Australian deserts have the world's highest lizard species diversity
Frogs	93% endemic (highest level of endemism of any vertebrate group in Australia); 220 species
Marine invertebrates	17.8% of the world's crustaceans, 22% of bryozoans and 29.4% of sea squirts occur in Australian waters
Vascular plants	91% of flowering plants are endemic; 17 580 species of flowering plants, 16 endemic plant families (the highest in the world) and 57% of the world's mangrove species
Butterflies and moths	Many groups are unique to Australia

Source: Modified and updated from Lindenmayer (2007a, p. 33). See also Chapman (2005) and Crisp *et al.* (1999).

The structure and functioning of many Australian ecological communities differ markedly from analogous communities on other continents (Keast *et al.* 1959; Krebs 2008; Orians and Milewski 2007). For example, extreme events and disturbances have a greater influence on most Australian communities than in North American, South American, European, northern Asian and southern African communities, where biotic interactions such as competition and predation are more likely to control species abundances (Krebs 2008; Orians and Milewski 2007). In addition, whole trophic levels within Australian communities seem to be occupied by different types of species compared with other continents. For example, many of the ecological niches occupied by placental mammals on other continents are filled by marsupials, birds, reptiles or insects in Australia.

Studies of biodiversity in Australian systems have provided critical tests of ecological theories developed in other areas of the world. For example, two of the major environmental features of Australian deserts – highly variable rainfall and soil infertility – are actually characteristic of most parts of the continent as a whole, with peculiar expressions of these two features in various biogeographical regions. These have informed syntheses of ecological work at a global level (e.g. Andrewartha and Birch 1954; Keast 1959; Orians and Milewski 2007; White 2005).

2.2.2 Continental drift to the dry mid-30 degree latitudes

The Australian continent is centred at about latitude 30° south. Most of the world's great deserts are also found at approximately this latitude (north and south of the equator) and Australia is no exception (Table 2.2). The continent experiences a very low average annual rainfall (Fig. 2.5) and a very low percentage of rainfall runoff to surface waters (Fig. 2.6).

Natural selection and genetic drift in isolation produced very high degrees of endemism (Table 2.2), including entire families of species found only in Australia, as well as the evolution of large numbers of species within many plant genera (e.g. *Eucalyptus* and *Acacia*), vertebrate genera (e.g. skinks *Ctenotus*;

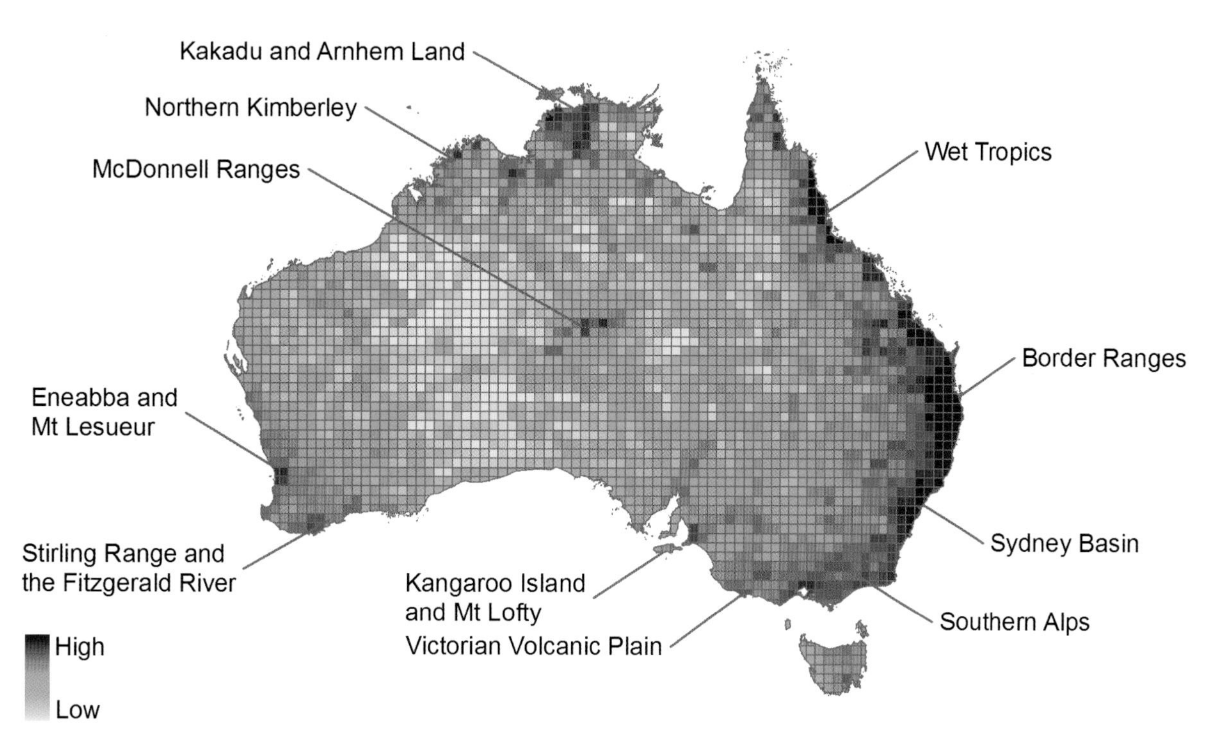

Figure 2.3 Australia's terrestrial areas of highest species richness. Source: Australian Natural Heritage Assessment Tool.

Fig. 2.7), ant genera (e.g. *Camponotus*), butterfly genera (e.g. *Trapezites*), beetle genera (e.g. jewel beetles *Castarina*) and crane flies (*Gymnoplisia* spp.).

Figure 2.4 *Banksia hookerana*, one of more than 100 species of *Banksia*, occurs near Eneabba, Western Australia. Source: Andrew Burbidge.

The increasing aridity of the Australian continent over the past 20 million years has favoured organisms that could adapt to dry conditions, resulting in a high level of diversity of groups such as reptiles and xerophyllous plant families (Table 2.2). Freshwater species that could survive long periods of drought were also favoured; for example, the ephemeral crustaceans of salt lakes (Williams 1984).

If climate change results in an even drier Australian climate, the fact that much of the nation's flora and fauna is pre-adapted to high aridity environments could give a degree of resilience not found in other parts of the world suffering similar drying conditions. However, the degree to which this might be true is highly dependent on a large and complex set of factors, which are outlined in later chapters of this publication, especially Chapter 5.

2.2.3 Missing the ice ages

Most of mainland Australia and Tasmania escaped continental ice sheeting during the series of ice ages of the Pleistocene epoch, which began about 2.4 million

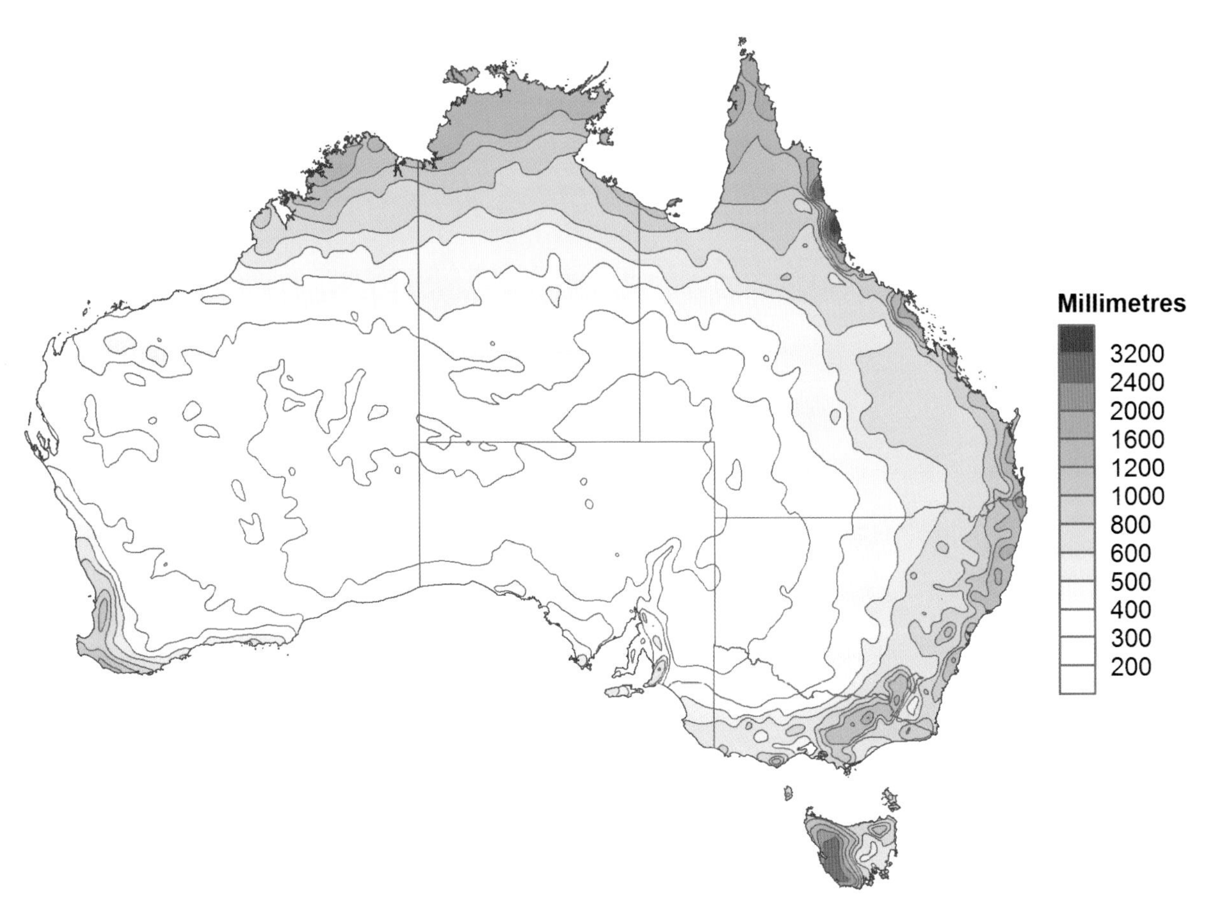

Figure 2.5 Average annual rainfall across Australia (in mm). Source: Australian Bureau of Meteorology.

years ago (Table 2.2). In contrast, continental ice sheets at the mid to high latitudes of the Americas, Europe and Asia not only restructured geological features, but also renewed substrates, grinding rock into nutrient-rich soils. Australia missed out on this process of soil formation and replenishment. As a result, Australian soils are some of the oldest and most nutrient-poor in the world (Lindsay 1985). Furthermore, water bodies are also relatively nutrient-poor (except when nutrient-enhanced by human activities or erosion) when compared with other regions of the globe (Williams 1984).

Adaptation to nutrient-poor soils is a distinctive characteristic of the Australian biota (Table 2.2). For example, the Australian flora has a high degree of ever-greenness and sclerophylly, both of which help plants retain hard-won nutrients (Bowman and Prior 2005). Nitrogen-fixing plants are major components of most vegetation types in Australia, compared to elsewhere in the world where soil nitrogen occurs in much higher concentrations (e.g. Orians and Milewski 2007). Many of the anti-herbivore defences of plants involve compounds such as resins and gums in leaves and stems, and mechanical structures such as hard seed coats. These carbon-based mechanisms in Australia are in contrast to the nitrogen-rich defence compounds used in most other floras (see Westoby *et al.* 2002, for low-nutrient implications for leaves).

Low-nutrient vegetation has important consequences for herbivores. Most of Australia's herbivorous native vertebrates have slow metabolic rates, especially the arboreal herbivores (Braithwaite *et al.* 1993), an advantage when vegetation is nutrient-poor. There are no native ruminant grazers or browsers, and insects play a large role both as leaf-eaters and seed-eaters. Many reptiles in the Australian arid zone prey on

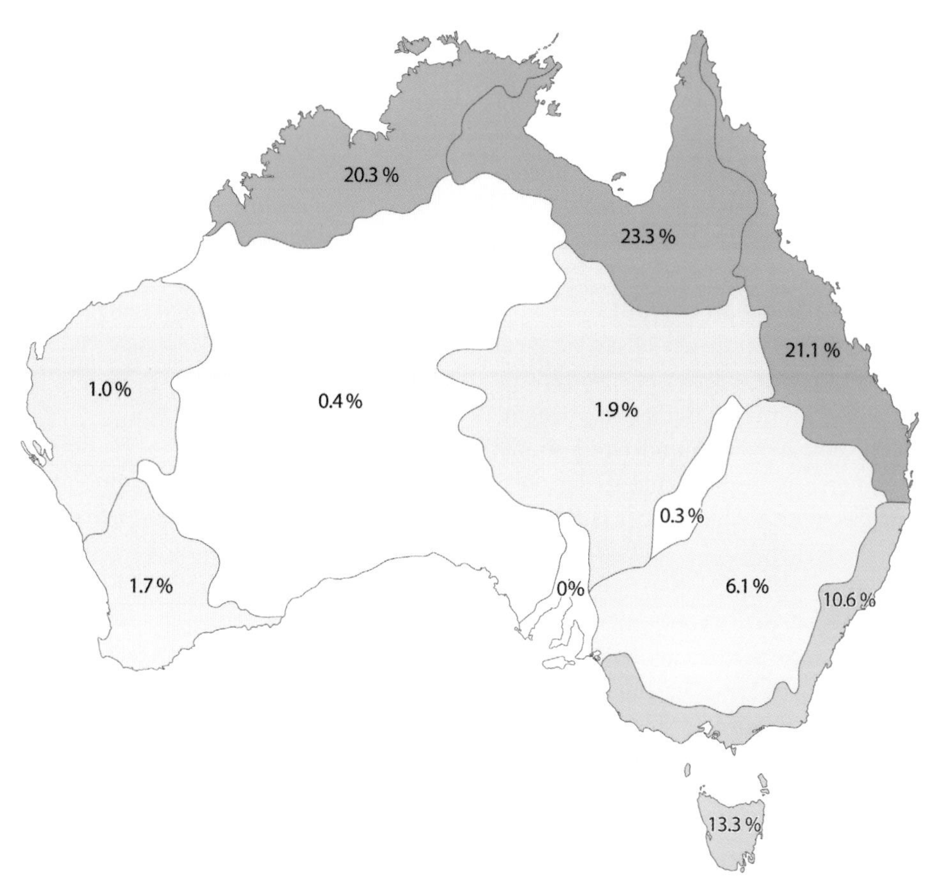

Figure 2.6 Rainfall runoff patterns across Australia (annual average percentages). Source: Lindenmayer (2007a), modified from data of Australian Bureau of Meteorology.

invertebrates such as termites, which are particularly adept at surviving on nutrient-poor vegetation (Morton and James 1988). Australia also has a higher proportion of myrmechochorous plant species whose seeds are adapted for dispersal by ants (Berg 1975).

Figure 2.7 Leopard skink *Ctenotus pantherinus*. Nearly 100 species of *Ctenotus* occur in Australia. Source: © Bert & Babs Wells/DEC.

Fire plays a profound role in Australian landscapes today (Bradstock *et al.* 2002). Low-nutrient vegetation has direct and indirect consequences for fire regimes. High levels of resins and similar carbon-rich compounds in the flora, coupled with slow decomposition rates, produce large, highly flammable fuel stocks. Frequent fires may further reduce nutrient status so that the low-nutrient/fire relationship becomes self-reinforcing (Orians and Milewski 2007). The inverse relationship between grazing by large herbivores and frequency of fires across the world's savannas has led some workers to regard fire as an 'alternative grazer' (Bond and Keeley 2005).

Among higher plants, the traits conferring an advantage under very dry or very nutrient-poor conditions (e.g. resprouting, well-protected seeds) also provide advantages for survival after fires (Orians and Milewski 2007; Whelan 1995). There are likely to be some pre-adapted components of the Australian flora that will survive, or even benefit, if expected climate changes result in more frequent and/or more intense fires in many Australian ecosystems (Pittock 2005). See also Chapters 4 and 5.

2.2.4 In the middle of ocean currents

Australia is located at the confluence of several great oceans. The strongest regional driver of climate variability is the El Niño–Southern Oscillation phenomenon (ENSO). The climate is also under monsoonal influences in the northern regions, and under polar influences from the south (Table 2.2). It is the only continent where the major currents on both east and west coasts run in the same direction – from the equator to the pole, from a warm to a cool climate.

The Australian climate is characterised by a high degree of variability, with extremes in temperature and precipitation (droughts, floods and storms), due primarily to large oceanic influences from tropical to sub-Antarctic latitudes. The incidence and intensity of rainfall in many localities is unpredictable, apart from some regional areas with reliable annual seasonal rainfall and/or temperature patterns (Table 2.2). As a result, there has been strong selection for organisms and traits that increase survival probabilities during extreme drought, and heat or cold. Examples include aestivation (e.g. water-holding frogs *Cyclorana*, *Notaden*; Fig. 2.8); large, long-lived seed banks; opportunistic life histories tied to rainfall events; and special physiologies, morphologies and life histories among animals (e.g. rapid post-disturbance population recovery among marsupials) and plants (e.g. acacias such as mulga *Acacia aneura*; many eucalypt species) that can withstand great variations in rainfall or temperature. Some freshwater fish have evolved to cope with great variations in water flow and temperature (e.g. spawning of Murray cod *Maccullochella peelii peelii* is prompted by flood events).

Figure 2.8 Desert spade-foot frog *Notaden nichollsi* is well adapted to desert living, where it spends most of its time underground. Source: © Bert & Babs Wells/DEC.

2.2.5 A flat continent

Australia has limited topographic relief (Table 2.2). Less than 5% of the land is more than 600 m above sea level (Fig. 2.9). Topographic barriers such as mountain ranges may therefore not have been as great a barrier for species dispersal as on some other continents (Heatwole 1987). However, a lack of topographic variability may limit the ability of many species to migrate upslope as temperatures increase as a result of rapid climate change. If species need to migrate, elevational gradient is important as it can mean less distance to travel to find suitable climate; for example, a couple of kilometres uphill may be equivalent to hundreds of kilometres across flat landscapes.

2.3 THE FIRST HUMANS COLONISE AUSTRALIA

A range of animals and plants colonised Australia as the continent moved towards Asia over the past 20 million years. These included species of bats, rodents, birds and Indo-Malaysian plants (Breed and Ford 2007; Churchill 1998; Specht and Specht 1999). *Homo sapiens* was one of the colonisers during the most recent ice age. These colonisations were facilitated by sea levels that were significantly lower than they are today.

The first humans – Indigenous Australians – arrived on the continent between 40 000 and 65 000 years ago (Table 2.2). They came, and remained, as hunters and gatherers, but they changed the landscape through hunting and by their use of fire (Bowman 2003). At approximately the same time, many large species of mammals became extinct. Natural selection favoured plants and animals able to adapt to

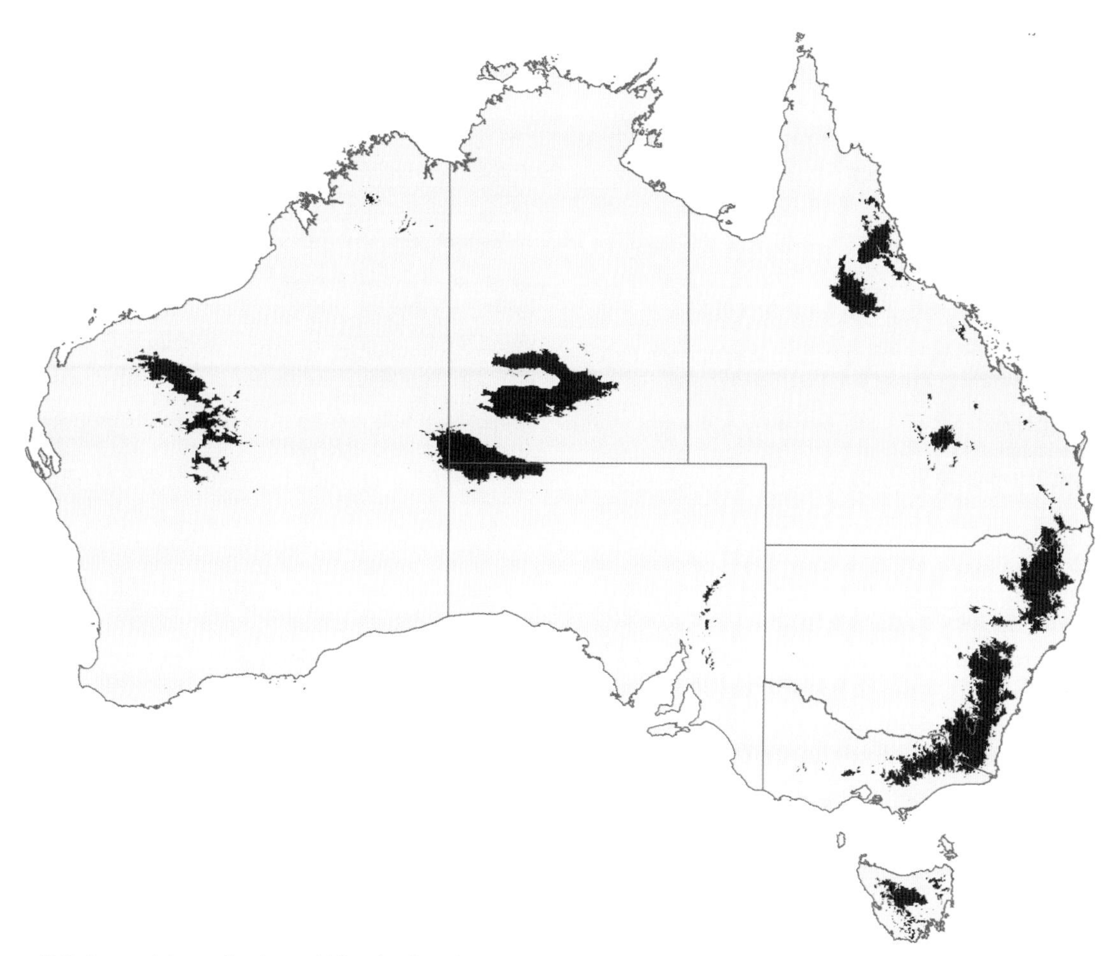

Figure 2.9 Areas of Australia above 600 m in elevation. Source: James W Shirley, using 2.5 m digital topographic data of Geosciences Australia.

human-derived changes in fire regimes. Plants and animals reacted individualistically to the intensity, timing and frequency of new fire regimes (Whelan 1995).

Subsequent human colonists and visitors came to Australia, especially to the north, starting about 6000 years ago (Table 2.2). These later peoples introduced the dingo *Canis lupus dingo*, the largest terrestrial placental mammal in Australia (other than *Homo sapiens*) at the time. The introduction of this species is thought to have contributed to the extinction of at least two species of large carnivorous marsupials on mainland Australia – the thylacine *Thylacinus cynocephalus* and the Tasmanian devil *Sarcophilus harrisii*. Both of these species survived only in Tasmania, where the dingo failed to colonise. The dingo also probably affected the distribution and relative abundance of its prey, and is known to maintain such an effect today (see Chapter 3). Human colonists of northern Australia over the past 6000 years also introduced several species of plants. The distribution of most of these species has remained localised (e.g. the tamarind tree *Tamarindus indica*), but other species have become subdominant or minor elements in the local flora (e.g. bamboo *Bambusa* spp.).

2.4 THE NEXT WAVE OF HUMAN COLONISTS: THE EUROPEANS

Europeans arriving in Australia over 200 years ago found landscapes that were shaped, in large part, by Aboriginal burning (Bowman 2003). These new

Table 2.2 Environmental and evolutionary consequences for Australia's biodiversity of (A) geology and (B) human colonisation.

A. Geology	Environmental consequences	Evolutionary consequences	Reflected in current biodiversity
Break up of Gondwana and isolation of Australia	Continent moved northwards	Selection and genetic drift occurs in isolation from biota of other continents over millions of years	Endemism extremely high; unique flora and fauna on a global scale
Plate tectonics moves Australia into 30 degrees latitude zone	Aridity: the world's great deserts occur at these latitudes	Selection of terrestrial biota under very dry, and often very variable, conditions	Large evolutionary speciation within endemic genera
Located at confluence of several great oceans with strong ENSO and monsoonal influences, and polar influence to the south	Variable environments driven by extremes in temperature and precipitation (e.g. drought, storms)	Selection for organisms and traits that respond favourably to highly variable climates (e.g. marsupial vs placental mammals)	Many opportunistic physiologies (e.g. eucalypts, marsupials, water-holding frogs); large annual plant seed banks, desiccation resistance of many aquatic species
Pleistocene ice ages: continental ice sheets do not form on Australian mainland	Soils were not rejuvenated by glacial action; soils remain very nutrient poor	Selection for organisms and traits appropriate to low soil nutrient and water availability	High degree of sclerophylly in terrestrial flora
Volcanic activity and uplift have not occurred in recent geological time; old mountains and plateaus highly eroded	Topography of continent is essentially flat; soils are nutrient poor	General lack of selection for dispersal distance	Some species of native biota have limited capacity to disperse rapidly over long distances
B. Human colonisation of Australia	**Environmental consequences**	**Evolutionary consequences**	**Reflected in current biodiversity**
First colonisers from the north (40 000–65 000+ years b.p.)	Increased incidence of fire; modified fire regimes	Selection of traits under modified fire regimes; vegetation changes availability of habitat and food for animals; extinctions of some large mammals	Community composition a product of the shifts in relative abundance of plant species (e.g. high degree of fire-resistant and fire-enhanced species)
Incidental visitors, especially in the north (6000+ years b.p.)		Introduction of dingo to mainland Australia and some plants in the north	Extinction of two large carnivorous marsupials on mainland Australia; introduced plants in north remain locally abundant or widespread but subdominant only
European settlement (220 years b.p.)	Land clearing; drainage of wetlands; pollution and redirection of freshwater flows; nutrient additions to land and runoff to water bodies; introduction of both carnivores and hard-hoofed herbivores; introduction of exotic plants and pathogens; introduction of exotic freshwater (e.g. European carp *Cyprinus carpio*) and marine species; habitat modification and destruction; overfishing	Selection for colonising and rapidly growing species; changes in competition and predator–prey relationships in community structure and functioning create changes in selection pressures on species morphologies, physiologies, phenologies and life histories	Shifts in abundances and distributions of species (some to extinction); loss of community integrity and hence ecosystem function. See Chapter 3.

colonisers further altered the landscapes (Table 2.2). They cleared forests, woodlands, mallee, brigalow, and other vegetation such as salt marsh and mangroves; drained wetlands; and redirected watercourses. They added fertiliser, crops and new pasture plants to the land. An unintended result was increased nutrient runoff to freshwater bodies and marine ecosystems. Other key modifications to landscapes included wind- and water-induced soil erosion after the removal or alteration of vegetation cover. Irrigation for crops, land clearing and other farming practices led to soil degradation including increased salinity. Many native species were targeted as pests, including the dingo, or exploited for various uses. Introductions of both carnivorous and herbivorous vertebrates as well as numerous invertebrates, plants (some becoming invasive weeds), fungi and pathogens also occurred. These landscape-level changes are ongoing (see Chapter 3).

European settlement and landscape changes were rapid and extensive. As a consequence, many elements of Australia's biota experienced rapid shifts in distribution and abundance within a few decades of colonisation and some species became extinct (e.g. the Darling Downs hopping-mouse *Notomys mordax* and Gould's mouse *Pseudomys gouldii*). In some cases, the functioning of entire ecosystems was dramatically altered; for example, temperate native grasslands and grassy box-gum temperate woodlands in the southern wheat–sheep belt (Table 2.2). The nature and magnitude of these ecological changes and losses of biodiversity are described in more detail in Chapter 3.

2.5 AUSTRALIA'S MAJOR BIOMES

'Biomes' are large, biologically distinctive parts of the Earth, the spatial occurrence of which is correlated with climate. The life forms of the biota, not the individual species, are the biological units used for the analysis, description and classification of biomes (Eblen and Eblen 1994). Biomes are usually described by their major life forms (e.g. temperate forest, tropical rainforest, coral reef) or other commonly used terms (e.g. alpine, savanna). The term 'biome' encompasses the sum of the plant and animal communities within a given climatic zone, successional stages of those communities, and any embedded geological features such as rivers, lakes, glaciers and bogs. Shifts in the relative proportions of a biome's components can occur due to land use changes, but an entire biome is rarely significantly altered other than by a change in climate.

Despite massive changes in land use since European settlement, Australia's major biomes are still clearly defined. The IBRA classification (Thackway and Cresswell 1995 and later revisions) couples biomes with geographical regions to create 85 'bioregions' for Australia (Fig. 2.1). Major transformations in Australia's bioregions may occur as the climate changes. This is discussed further in Chapter 5.

2.6 WHY IS THE CONSERVATION OF AUSTRALIA'S BIODIVERSITY IMPORTANT?

Over the past two decades, biodiversity conservation has become a major objective of most countries worldwide, and it was adopted as an aim of the many nations (including Australia) that ratified the Convention on Biological Diversity (the Rio Convention) developed at the 1992 'Earth Summit' in Rio de Janeiro, Brazil. There are four broad groups of reasons for conserving living things and the ecosystems that they form. These are briefly outlined below.

2.6.1 Ecosystem services

Living organisms contribute to the life support system of Earth on which humans depend. Plants provide the oxygen we breathe, maintain the quality of the atmosphere, moderate climate and its impacts, regulate freshwater supplies, and generate and maintain the topsoil. Plants, animals and microbes dispose of wastes, generate and recycle nutrients, control pests and diseases, pollinate crops, and provide a genetic store from which we can benefit in the future. These values of biodiversity are often termed 'ecosystem services' (Daily 1997; Millennium Ecosystem Assessment 2005) and they can have significant impacts on human well-being and health (Box 2.1). Some ecosystems services embrace or overlap utilitarian values of individual species (section 2.6.2), heritage values (section 2.6.3), and existence and ethical values (section 2.6.4) described later.

The concept of 'ecosystem services' is relatively recent, being first used in the late 1960s. Research on

Box 2.1 Ecosystem services: an example of monetary valuation. (David Lindenmayer)

It has been estimated that the environment returns an estimated US$33 trillion in goods and services to human society each year, about 1.8 times global gross domestic product (Costanza *et al.* 1997). However, the true value of ecosystem services are often not well accounted for until they are gone (McIntyre *et al.* 2000). This was illustrated by the 2004 Boxing Day tsunami that killed 200 000 people, made two million people homeless and led to economic losses of US$6 billion (Mooney 2007). The areas where the greatest tsunami damage occurred were also those places where extensive clearing of mangrove forests had taken place (Kathiresan and Rajendran 2006). For example, more than half of the mangrove forests along the coast of Thailand had been replaced by shrimp farms between 1975 and 1993 (Barbier and Cox 2002). In the late 1990s, the value of shrimp exports was US$1–2 billion, or returns of US$194–209 per ha. This figure seems significant until compared with financial losses and losses of human life associated with the 2004 Boxing Day tsunami. This figure further pales into insignificance when other values of mangrove forests to local human communities are estimated – up to US$35 000 per ha from timber and non-timber products alone.

ecosystem services has grown dramatically in the past decade in response to a growing need to place a value on biodiversity. The ecosystem services concept focuses on human dependence on the environment and reduces the complexity of natural systems to a manageable (comprehensible) number of services that people receive from ecosystems. This can help engage stakeholders and the community in dialogue about what services are needed where, when and by whom (Cork *et al.* 2008). Understanding and categorising ecosystem services also aids in making decisions about management priorities (Wallace 2007). For example, the IUCN estimated that 46% of the total value of ecosystems services to humans originates in wetlands (IUCN 1999), a statistic that could help set priorities in a particular region. Table 2.3 categorises and provides examples of ecosystem services from Australian natural biodiversity. This table follows the Millennium Ecosystem Assessment (2005) in categorising ecosystem services, although there is debate about classification systems. For example, Wallace (2007) proposed separating processes for achieving services (means) from the services themselves (ends).

Some economists have attempted to place monetary values on ecosystem services with results that are, at least, consistent with the belief that such values are very high (Costanza *et al.* 1997). For example, the rivers, wetlands and floodplains of the Murray–Darling Basin are estimated to provide A$187 billion in ecosystem services annually (Thoms and Seddon 2000). The equivalent estimates for Australian terrestrial ecosystems are A$325 billion per year (Jones and Pittock 1997). By comparison, the entire Australian gross domestic product in 2005–06 was about A$965 billion (Australian Bureau of Statistics 2008).

2.6.2 Utilitarian values of individual species

Another important reason for the conservation of biodiversity is its economic contribution or monetary value of individual species or groups derived for particular regions. Australia's biota and its landscapes, for example, attract tourists to Australia. The koala *Phascolarctos cinereus*, alone, is estimated to be worth over a billion dollars annually to the Australian tourism industry (Australian State of the Environment Committee 2001). The Great Barrier Reef was estimated to be worth A$6.9 billion to the Australian economy in 2005–06 (Access Economics 2007).

Plants, animals and micro-organisms provide nearly all of our food, medicines and drugs. Many commercial products are made from naturally occurring plants and animals found in the Australian landscape, including renewable resources like paper, leather, fuel and building materials. Indeed, the role of trees and vertebrates in providing human shelter, fuel and food are well known. This is in contrast to the economic uses and products of relatively inconspicuous invertebrates, which are relatively unknown. Examples of the economic use of invertebrates are shown in Table 2.4. Products derived from native plant and animal species range from cosmetics, adhesives

Table 2.3 Examples of ecosystem services from Australia's natural biodiversity.

Type of service	Service	Examples from Australia's natural biodiversity
Provisioning services	Food	Macadamia nut *Macadamia integrifolia*; bush food; meat and hides from kangaroos; honey; fisheries
	Fibre	Eucalypts provide timber and pulp
	Fuel	Wood; biofuels (e.g. ethanol from sugar cane)
	Genetic resources	Use of Australia's marine and terrestrial organisms for medical drug development; tree breeding for plantations
	Biochemicals, natural medicines, etc.	Oil mallees; Kakadu plum *Terminalia ferdinandiana* has 50 times the concentration of vitamin C in oranges; anti-cancer drugs are being developed from wild plants
	Ornamental resources	Wildflower industry; shell collecting
	Fresh water	Naturally vegetated catchments provide millions of dollars worth of potable water for Australian cities and for irrigation each year; vegetation controls groundwater tables – its loss can lead to soil salinisation
Regulating services	Air quality regulation	Microbial immobilisation of sulphur compounds; plants absorb a variety of pollutants
	Climate regulation	Phytoplankton in the ocean have absorbed about a quarter of the CO_2 produced by human activities since the industrial revolution; changes in land cover can influence local temperature and rainfall; sequestration of greenhouse gases in trees, etc.
	Water regulation and purification	Vegetation and micro-organisms remove pollutants from water; naturally vegetated watercourses limit flooding; wetlands purify water
	Erosion control	Naturally vegetated watercourses show minimal erosion compared with cleared watercourses; mangroves prevent coastal erosion
	Disease regulation	Insects, birds and other predators naturally control disease-carrying insects
	Pest regulation	Predation of pest insects by native animals; predation of house mouse by native carnivores; biological control
	Pollination	Insects, birds and mammals pollinate crops and native plants species that are utilised by people
Cultural Services	Cultural diversity and heritage	Plants and animals are an integral part of Aboriginal culture
	Spiritual and religious values	The close association of Aboriginal Australians with country, including the Dreaming, in which ancestral beings, including animals, created the land and its biodiversity; many non-Aboriginal Australians feel connected to plants, animals and the wild places they form
	Recreation and ecotourism	Australia's natural scenery and biodiversity are the major attraction for tourists visiting our country; amateur fishing; bushwalking; study of natural history
	Aesthetic values	Australia's natural landscapes and biodiversity provide us with a sense of place
	Knowledge systems	Use of traditional and scientific knowledge systems for managing land and biodiversity
	Inspirational values	Australia's cultural identity is closely associated with the bush and its plants and animals; some Australian species are national icons; appreciation of the wonder of living things
	Educational values	Biodiversity helps us learn about the world and its complexity
Supporting services	Soils formation	The organic component of soil comes from biodiversity
	Photosynthesis; primary production	Primary production via photosynthesis provides the basic building blocks for all carbon-based life forms on Earth, including *Homo sapiens*; natural vegetation and algae replenish the oxygen we breathe and require for energy utilisation
	Nutrient cycling	Symbiotic bacteria in many Australian plants fix nitrogen from the atmosphere, thus fertilising the soil
	Water cycling	Transpiration transfers groundwater to the atmosphere

Source: Millennium Ecosystem Assessment (2005); Wallace (2007).

Table 2.4 Examples of economic uses of invertebrates.

Economic uses and products	Animal groups
Adhesives	Velvet worms, earthworms and marine worms
Ant repellents	Ants and wasps
Antibiotics research and development	Ants
Anticoagulants	Leeches
Art materials	Butterflies
Biological control agents	Nematodes, mites, insects
Biological control of weeds	Leaf hoppers
Brain research	Sea slugs, nematodes
Bird repellents	Leaf hoppers
Control of crystallisation	Molluscs
Cryoprotectants	Springtails, mites
Ecotourism foci	Glow worms (flies)
Industrial products (concrete, car parts, ceramics, fibres)	Molluscs, spiders
Pigments	Cochineal beetles
Termiticides	Ants

Source: Modified from Lindenmayer (2007a).

and anticoagulants to industrial products used in concrete, car parts, ceramics and fibres.

Only a limited number of the Earth's species have been used by humans. Many biological resources, including those from species many would regard as useless today, will become valuable in the future (Beattie and Ehrlich 2001). This emphasises the need for the preservation of biodiversity to ensure opportunities for future generations of people, including their unknown or unexpected future needs.

2.6.3 Heritage values and national identity

Heritage values of a place are defined under Australia's *Environment Protection and Biodiversity Conservation Act 1999* (Cwlth) as: '[including] the place's natural and cultural environment having aesthetic, historic, scientific or social significance, or other significance, for current and future generations of Australians' (s. 528). Australia's biota is an important dimension of the national character, and particular groups of people, especially Aboriginal Australians, gain additional value from their association with significant aspects of our biotic heritage. Examples include red deserts, coral reefs and rainforests to koalas, kangaroos and wallabies, which are national emblems for various football teams, airlines and other national institutions.

2.6.4 Existence and ethical values

Many believe that biodiversity has an 'existence' value. That is, plants, animals and other organisms should be conserved because of their beauty, symbolic value or intrinsic interest, or on religious or philosophical grounds. Many people would feel a loss if the world's beautiful and interesting plants and animals, and the wild places they inhabit, disappeared. For many, biodiversity also has value in its own right, regardless of its utility to humans. Each species may be viewed as a unique and irreplaceable product of millions of years of evolution. Put simply, the rich diversity of other life forms is a core part of our own humanity.

2.7 BIODIVERSITY IN AN ECONOMIC CONTEXT

Much of the value of biodiversity as an ecosystem service is not captured in markets, and consequently is not included in the national accounts. Thus, both these

important measures fail to represent the true value of biodiversity to society. This failure means that popular perceptions of the value of biodiversity are less than they should be. Not only does this have important implications for the way we manage biodiversity by reducing the urgency we might otherwise feel to reverse its loss, but also it results in serious under-investment in biodiversity conservation. It is important to understand the difficulties that contemporary economic systems have in dealing with biodiversity, as these must be overcome if we are to significantly improve the effectiveness of biodiversity conservation in future (see Chapters 6 and 7).

Market and market-like institutions are appropriate when services can be established with low transaction costs (particularly the costs of exclusion) as private goods (e.g. defining private property by issuing tradeable permits). Otherwise, as for many ecosystem services associated with biodiversity, collective institutional arrangements will be required. Even where market institutions are appropriate, they need to be established, administered and adapted via collective arrangements.

The design and choice of collective institutional arrangements for sustaining ecosystem services associated with biodiversity are critical for strengthening our capacities to adapt under the pervasive complexity and uncertainty of climate change. In particular, the types of investments we make now in adaptation will set the framework for transforming institutional structures and building innovative solutions (see Chapters 6 and 7).

2.8 THE NATURE OF THE CHALLENGE

The prolonged isolation of the Australian continent, coupled with the lack of topographic relief and limited nutrient levels of most soils, are among the key factors that have resulted in the evolution of Australia's unique and highly diverse biota.

The conservation of Australia's species-rich and distinctive biota is critically important for heritage, ethical, intrinsic and utilitarian values, and for the roles that it plays in the maintenance of ecosystem services. At the planetary scale, biodiversity is crucially important for the maintenance of our own life support system.

Shifts in abundance and distributions of biota, including species extinctions and the drivers responsible for such changes, are addressed in Chapter 3. The challenge to live in a manner that is more ecologically sustainable will require changes in attitudes; innovative institutional arrangements, management skills and economic instruments; and far greater understanding of the biology of Australia's (and the Earth's) species and ecological communities.

3 Australia's biodiversity today

This chapter reviews the state of Australia's biodiversity at the present time and the changes that have occurred over the last 220 years. Ten key ecological principles are introduced, which are relevant for interpreting past and future changes in biodiversity from the population to the landscape levels. This chapter identifies key stressors, other than climate change, which have operated and continue to operate to produce the biodiversity and landscapes that we experience today. Proximate stressors – such as direct exploitation, land clearing, water and fire management, and exotic species – reflect ultimate drivers such as human population size, individual ecological footprints, agriculture, urbanisation, mining and 'perverse' incentives. Global drivers, such as globalisation and climate change, are superimposed upon and exacerbate the continental-level pressures.

3.1 A TRANSFORMED BIOSPHERE

Drivers of biodiversity changes continue to operate in Australian ecosystems today. They include habitat fragmentation, spread of invasive species, reduced water availability to freshwater systems and marine pollution. This confounds our ability to unequivocally attribute the ongoing changes in Australian biodiversity to climate change or some other factor. More importantly, it means that climate change is affecting a heavily perturbed biosphere. It makes planning, management and policy formulation to reduce the vulnerability of Australia's biodiversity to climate change much more difficult. There is no doubt, however, that dealing effectively with current drivers of biodiversity change will be critically important to decrease the vulnerability of Australia's biota to climate change.

Understanding the impacts of historic and present drivers of changes, apart from climate change, is a fundamental first step for scientists and managers. This chapter provides an overview of these impacts, based on underlying ecological principles that determine how species and ecosystems respond to changes in the environment around them. This knowledge, especially the fact that Australia's biosphere has already been modified extensively by human actions, is the necessary background and context for an understanding of the potential further impacts of climate change, including indirect impacts through impacts on other drivers (Chapter 5), and the policy and management responses required to deal effectively with this additional challenge (Chapters 6 and 7).

3.2 UNDERLYING ECOLOGICAL PRINCIPLES

The science of ecology addresses the structure and functioning of communities and ecosystems, including the distribution and abundance of individuals and populations of biological species that form ecosystems (Krebs 2008). Key principles regarding responses of biota and ecosystems to environmental change provide the basis to understand and manage ecological change. The most important relevant principles are summarised briefly in Table 3.1, and are central to the analysis presented throughout the rest of this assessment.

Returning to fundamental ecological principles is necessary because Australia's environment, and the Earth's environment as a whole, is now operating under conditions not experienced previously (Steffen *et al.* 2004). The nature of changes occurring simultaneously across the continent, their magnitudes and rates of change are unprecedented. The past is no longer a reliable guide to the future. Thus, understanding and applying ecological principles is essential to maintain or enhance Australia's biodiversity in this century of rapid environmental change. There are no shortcuts to sound policy and effective management.

Table 3.1 presents 10 key ecological principles relevant to environmental change, with examples from the past (pre-climate change). They fall into four clusters:

- Principles 1–3 deal with the relationships between individual species and the surrounding environment. In particular, these principles address the 'environmental envelope' that broadly defines the locations where species are found on the land or in the sea, and the characteristics of species that determine their ability to respond to environmental change.
- Principles 4 and 5 deal with the role of individual species in communities or ecosystems.
- Principle 6 looks at ecosystem structure, and Principles 9 and 10 can be applied to all levels of organisation, but especially to entire landscapes.
- Principles 7 and 8 describe phenomena that are applicable to all ecological systems, at all levels.

These 10 principles (Table 3.1) are described further below.

- *Species differences.* Regardless of how species are defined, each represents a uniquely different evolutionary solution to the survival problems posed by past and present environments. Each species has a fundamental niche defined by its physiological tolerances to the physico-chemical environment. Within this envelope of tolerance, the species occupies a 'realised' niche – the area actually occupied because of limitations due to other organisms such as competitors, predators, etc.
- *Scale (time and space).* The distribution and abundance of species vary in space and time. These spatial and temporal drivers operate on a variety of scales so that the observed occurrence and abundance of a particular species may reflect processes that occur from continental to local spatial scales, and on time scales from the geological to the immediate.
- *Life histories and population genetics.* The ability of a species to respond to environmental changes imposed upon it will reflect the 'life history' attributes of that species. These include reproductive rates, longevity, dispersal ability, genetic variability and phenotypic plasticity. Although every species is unique, life history attributes occur in repeated sets so, practically, species can be organised into a manageable number of functional groups. Species that adapt well to change generally have high reproductive rates, short longevity and high mobility. Any long-term mismatch between life history attributes and environmental pressures will result either in local extinction or, on a longer time scale, evolutionary adaptation.
- *Species' interactions.* Species do not occur in isolation. A range of interspecific interactions occur including competition, predation, parasitism and mutualism. Interacting sets make up ecological communities; when considered with their abiotic components, these form ecosystems.
- *Species' roles.* Within communities and ecosystems, some species have wider impacts than others. Small numbers of species may determine the state of entire communities, and play a major role in the self-organisation of a community. All of them have effects that cascade through entire food webs in either a bottom-up or top-down fashion. These 'special' or 'key' species include key structural species (sessile, usually large, organisms that set the physical structure of a community), foundational species (the base of the food web, especially in open-water systems), ecological

engineers (modifiers of the physical environment or water courses) and keystone species (having a disproportionate effect on the entire community relative to their size or biomass, such as top predators) (Box 3.1).

- *Trophic structures and ecosystems.* Communities are structured into so-called trophic levels on the basis of what eats what. These trophic levels organise entire communities into food webs. The addition of species and community-level interactions with the abiotic environment produces the emergent properties of ecosystems. As environments have changed either through long-term natural processes or by shorter-term human activities, novel ecosystems have arisen. Impacts on particular species can cascade through trophic levels to produce changes in the entire system.
- *Multiple drivers of ecological change.* There are many drivers that impact upon communities and ecosystems, and these drivers frequently display interactions among themselves, leading to complex and often unpredictable outcomes. Drivers such as land clearing, exotic species introductions, urbanisation, pollution and fire regimes change Australian ecosystems and will interact with any future climate-driven changes.
- *Non-linearity.* Ecological changes driven by environmental changes may be proportional to changes in the drivers (that is, they are linear), but in many instances there will be thresholds across which rates of change alter or even jump to different levels. These non-linear changes add unavoidable uncertainty to predictions (Box 3.2).
- *Heterogeneity.* Environments are essentially heterogeneous in space and time relative to any individual, species or community. The levels of heterogeneity differ from ecosystem type to ecosystem type but, in general, greater levels of heterogeneity will sustain more biotic diversity. Heterogeneity within and between ecosystems occurs at multiple scales as do the drivers that effect change.
- *Connectivity.* Connectivity is a relative term that reflects the individual species or community type under consideration. It refers to the location of resources and habitat that are conducive to survival and reproduction of the species in the landscape. In other words, these resources must be in close enough proximity to be available through dispersal of individuals, offspring, or reproductive units (e.g. seeds, pollen). This is not the same as landscape diversity, although higher landscape diversity may well enhance connectivity for species or communities (Box 3.3).

The relationships among individual species (which are often the focus of biodiversity conservation) and the provision of ecosystem services (which rely on well-functioning ecosystems) are especially important in periods of rapid environmental change. At larger scales, the interactions among ecosystems, particularly the nature and strength of the connections among them, become very important as rapid environmental change drives differential responses of individual species and groups of species. These 10 principles will be used as a foundation for some of the discussion in Chapters 6 and 7.

Principles 7 and 8 are particularly important to consider when assessing changes to biodiversity in the face of climate change. Firstly, at the ecosystem level, change is almost always driven by multiple, interacting drivers. In fact, historical and existing drivers of change to biodiversity remain substantially more important, even today, than climate change. The impacts of climate change will undoubtedly grow in importance rapidly as the century progresses. Climate change should not be viewed as a new, independent driver, but rather as the latest in a number of significant, human-related drivers of change to Australia's biodiversity that will interact with other drivers in complex ways. Secondly, non-linear responses to drivers of any kind of environmental change are common (Box 3.2). Statements such as 'expect the unexpected' and 'prepare for surprises' are well-grounded in ecological science.

Human beings have been part of Australian ecosystems for at least 40 000 years, and more likely 60 000–65 000 years (Chapter 2). Indigenous Australians modified the continent's landscape primarily through hunting (several species of large mammals disappeared; Principles 4 and 6 – changing species–species interactions and consequently altering trophic

Table 3.1 Ecological principles relevant to environmental change.

Subject	Ecological principles	Explanation	Relevance to environmental change excluding climate change	Australian examples
1. Species differences	Every species is different. Species distributions and abundances reflect individual responses to their environments.	• Environmental determinants include the physico-chemical environment and the biological environment. • These define their fundamental and realised niches. • Species specific features such as dispersal ability, reproductive rate and competitive ability affect their spatial and temporal dynamics.	Responses to human-induced changes on terrestrial native species differ widely in response to urbanisation and land clearing. Native marine species responses to overfishing or water pollution also differ widely. Dispersal ability affects the ability of species to respond to landscape or seascape modification.	Among birds, noisy miners *Manorina melanocephala* succeed under urbanisation whereas most other honeyeaters (family Meliphagidae) do not. Eucalyptus species in northern savannas are well adapted to fire, whereas adjacent monsoon forest species are killed by fires.
2. Scale (time and space)	Species abundances and distributions are responses to drivers at different scales.	• Species respond differentially to drivers at different scales. • Key scales are both spatial and temporal, including local to continental, and immediate to geological.	Some exotic predators operate on widely different scales; other impacts are location-specific.	Feral cats *Felis catus* are ubiquitous – from rainforests to deserts; European carp *Cyprinus carpio* are generalists, surviving in pristine and highly degraded habitats; fire ants *Solenopsis invicta* are pests of urban ecosystems; and pasture weeds are important in agricultural landscapes.
3. Life histories and population genetics	Life history attributes determine the ability of species to respond to change. Population genetic variability and breeding systems also determine the ability to respond.	• Attributes include reproductive rates, longevity, dispersability, genetic variability and phenotypic plasticity. • These occur in repeated sets organising species into a manageable number of functional groups. • Species that adapt well to change have high reproductive rates, short longevities and high mobilities.	Native species that become agricultural 'pests' are pre-adapted to rapid responses. Some native species cope with radical change better than others because of life history and/or genetic plasticity.	Eastern and western grey kangaroos *Macropus giganteus* and *M. fuliginosus* have greatly expanded on grazing lands with water points. Native bushflies *Musca vetustissima* are massively abundant in cattle regions; rainbow lorikeets *Trichoglossus haematodus*, and ringtail and brushtail possums *Pseudocheirus peregrinus* and *Trichosurus vulpecula*, have thrived and expanded in urban areas

Subject	Ecological principles	Explanation	Relevance to environmental change excluding climate change	Australian examples
4. Species interactions	No species exists in isolation from other species.	• Species interact in pairs, trios, etc. • Interactions include competition, predation and mutualism. • Interacting sets make up ecological communities; when considered with their abiotic components, they are called ecosystems.	Introduced food plants change distributions of native herbivores. Introduced animals change whole communities.	Lantana *Lantana camara* and camphor laurel *Cinnamomum camphora* sustain the Wonga pigeon *Leucosarcia malenoleuca* and Emerald dove *Chalcophaps indica*. Citrus trees in non-rainforest regions have allowed citrus swallowtail butterflies *Papilio fuscus*, *P. anactus* and *P. aegeus* to expand their ranges hugely.
5. Species roles	Some species affect ecological structure and processes more than others within communities and ecosystems.	• Single species may determine the state of entire communities. • These include, e.g. key structural species, foundational species, keystone species and ecological engineers.	Forestry management practices affect whole local communities of plants and animals. Marine communities can be vastly changed by the alteration of single species, especially the primary producers or top predators. Some exotics have far more impact than others. Changes in relative abundances of species with changed fire regimes.	Removal of 'over-mature' trees and stags affect hole-nesting mammals and birds. Exotic seaweed (e.g. *Caulerpa taxifolia*) shades and out-competes native seagrass species. Reduction of sea urchin predators (e.g. abalone *Haliotis* spp.) and lobsters (family Nephropidae) result in reductions in kelp forest, resulting in 'urchin barrens'. Introduced cane toads *Bufo marinus* may have widespread impacts, whereas house geckoes *Hemidactylus frenatus* are restricted in distribution and impact.
6. Trophic structures and ecosystems	Species are structured by their means of obtaining food into a larger trophic structure, or food web. The interaction of biota and physico-chemical environment yields ecosystem-level processes.	• Changes at the species level can cascade through trophic levels to produce a change in the entire system. • Changes in geological time and those due to human activities have produced novel ecosystems.	Changes in plants through fertilisation or grazers have cascaded through ecosystems to produce tree dieback or changes in primary productivity, or fire regimes with further consequences.	Eucalypt dieback results, in part, from a cascade of interactions associated with superphosphate application and clearing. Northern savanna changes in grazers or exotic grasses produce cascading effects on fire regimes, nutrient cycling and tree cover in Kakadu National Park.

Table 3.1 Ecological principles relevant to environmental change. (Continued)

Subject	Ecological principles	Explanation	Relevance to environmental change excluding climate change	Australian examples
7. Multiple drivers of ecological change	Communities and ecosystems change in response to many drivers, and these drivers themselves may interact.	• Communities and ecosystems respond to drivers of change in different ways. The timing and relationships among the drivers can also affect outcomes. Hence, when there is more than one driver present, the responses can be complex, sometimes leading to unpredictable outcomes.	Major drivers have included land clearing, exotic species, urbanisation and pollution. Some regional ecosystems have changed due to land clearing, others due to exotic predators, urbanisation, etc. These changes are also evident at the individual species levels (above).	Reduction of brigalow *Acacia harpophylla* woodland due to clearing of vegetation for agriculture. Reduction of limestone grasslands due to urbanisation. Change in desert communities due to exotic predators. Reduction in mangrove, seagrass and salt marsh communities due to coastal development. Shift in coral reef communities dominated by hard corals to those dominated by algae as a result of overfishing of herbivorous fish species and increased nutrients.
8. Non-linearity	Changes can be non-linear.	• Changes in species abundances and distributions, community structure, or ecosystem function may be proportional to changes in the drivers (even those that are linear). • However, there may be thresholds where rates of change alter or even jump to different levels (i.e. non-linear response). • As a consequence, this inherently increases uncertainty in predictions.	Native communities can absorb small numbers of exotics, beyond which radical change occurs. Vegetation clearing can gradually reduce native populations to a point where mating encounters become infrequent and/or offspring survival plummets – then the species collapses rapidly (non-linearly).	Impact of fishing leading to the collapse of stocks. Canopy clearing beyond some basic levels destroys the canopy assemblage of plants and animals. Horticultural escapes have sat 'dormant' for decades but then suddenly became weeds, e.g. *Mimosa pigra*, *Opuntia* spp., *Lantana* spp.

Subject	Ecological principles	Explanation	Relevance to environmental change excluding climate change	Australian examples
9. Heterogeneity	Variations in time and space (heterogeneity) enhance biotic diversity.	Because different species are different sizes, have different home ranges and dispersal abilities, and interact with other species at different times and distances, then the greater the variability of resources and habitat, the greater the number of species.	Variability of rainfall and resources – as well as disturbances by storm, drought, insect outbreaks, and traditional fire practices – have been a fundamental feature of Australian landscapes. Recent changes to these, especially fire regimes in terrestrial systems, have reduced biodiversity.	Homogenisation of vegetation on northern floodplains caused by introduced *Mimosa pigra* has reduced total biodiversity. The application of a single fire regime to large landscapes has reduced total biodiversity in northern savannas and central desert grasslands.
10. Connectivity	Connectivity among resources and habitat required by species determines the longer-term patterns of species distributions and abundance, and ultimately, longer-term landscape-scale biodiversity.	Connectivity is an entirely relative term to the individual species or community type under consideration. It refers to the location of resources and habitat that are conducive to survival and reproduction, in close enough proximity to be available through dispersal of individuals, offspring or reproductive units (e.g. seeds, pollen). What might appear as a highly fragmented landscape to one species might also be one of high connectivity to another species.	Changes in land use and land management have been the main causes of changes in connectivity in terrestrial systems. Fragmentation of previously continuous landscapes, sea bottom types, or vegetation may have changed connectivity for many species and communities.	Many species are under threat by the extreme fragmentation of the Western Australian wheat belt region. The East Australian Current is the main mechanism connecting coral reefs along the east coast of Australia, from the tropics to the marginal subtropical reefs such as those at the Solitary Islands and Lord Howe Island.

Box 3.1 Cassowaries as a keystone species. (David Lindenmayer)

The southern cassowary *Casuarius casuarius* is a keystone species in the tropical rainforests of north Queensland because it is the only disperser of more than 100 plant species that have very large fruits (Crome 1994). It excretes the seeds of fruits intact, often with parts of the original fleshy covering adhering to the seed, making it a particularly effective dispersal vector for rainforest plants. The role of the southern cassowary is filled by many species of vertebrates (mammals, birds and reptiles) in rainforest communities elsewhere in the world (Jones and Crome 1990). Without the southern cassowary, the dynamics of rainforest communities in north Queensland would be changed markedly and the status of many plants (and the animals that are, in turn, associated with them) would change. Reductions in numbers of the southern cassowary could have important consequences for ecosystem functioning.

Southern cassowary *Casuarius casuarius*. An endangered species due to loss of habitat. North Queensland, Australia.
Source: AUSCAPE. Photo by Jean-Paul Ferrero.

structures) and through the modification of fire regimes (Principle 7 – the alteration of an existing disturbance regime). However, the ecosystem services that Indigenous Australians extracted from the continent's biota – primarily food – were very modest, leaving the structure and functioning of Australian ecosystems largely intact compared with their pre-human state.

Following European settlement, the loss of a large number of native species over much of their previous range and the introduction of many alien organisms have vastly changed the interactions among species in many ecosystems (Principle 4) and altered the structural and functional roles that many species play (Principle 5). The complete conversion of natural ecosystems into heavily managed production systems, especially in the south-eastern and south-western parts of the continent, has placed new ecosystems with different trophic structures and altered chemical environments on the landscape (Principle 6). Conversion and modification of ecosystems over most parts of the continent has transformed the nature and pattern of connectivity (Principle 10): the biotic fabric of Australia is significantly different from what it was 200 years ago. Most importantly, these ongoing changes to Australia's biodiversity are the result of multiple, interacting drivers (Principle 7), most of which continue to operate into the 21st century.

The next two sections address the most fundamental questions related to the many changes that have occurred in Australia's biodiversity over the past two centuries. What features of contemporary Australian culture and society drive such widespread changes to our biota? What is the current status of our biodiversity and where is it headed?

3.3 DRIVERS OF CHANGES IN AUSTRALIA'S BIODIVERSITY

As Chapter 2 described, the natural drivers that have shaped Australia's unique biota operate on geological timescales; that is, over millions and hundreds of millions of years. At those timescales, the pace of change is slow, allowing time for organisms to migrate, to respond slowly to changing disturbance regimes, and

Box 3.2 Non-linearity leads to thresholds and tipping points. (David Lindenmayer)

Some kinds of disturbances can induce non-linear or threshold changes in ecological processes, species interactions and population sizes in which there is a sudden switch from one state to a markedly different one (Walker *et al.* 2004). Crossing such process-related thresholds may produce changes that are either irreversible or extremely difficult to reverse (Walker and Salt 2006; Zhang *et al.* 2003).

Particular ecological examples of non-linearities include the build-up of nutrients in freshwater ecosystems. These impact upon the freshwater communities in a linear fashion until a threshold is reached at which point there is a massive outbreak of, for example, Cyanobacteria leading to eutrophication and a fundamental change in the food web dynamics of the whole ecosystem.

Thresholds related to patterns of landscape structure occur in some Australian landscapes (Radford *et al.* 2005) and those overseas (Homan *et al.* 2004). For example, thresholds may exist for the amounts of particular kinds of habitat or land cover types in a landscape. When these thresholds are transgressed, sudden changes in species abundance or ecosystem processes may occur, leading to changes in system state or 'regime shifts' (Folke *et al.* 2004). Hypothetically, thresholds are more likely to be crossed and regime shifts more likely to occur when levels of particular kinds of habitat or types of land cover are low (e.g. 10–30%; Radford *et al.* 2005).

The impacts of thresholds and their potential irreversibility mean that a better understanding of thresholds is critical to ensure that management practices do not inadvertently drive ecosystems, species and ecological processes close to critical change points.

- It is difficult to identify critical change points or thresholds and to anticipate (and prevent) regime shifts before these occur (Groffman *et al.* 2006).
- Thresholds may not exist for some biodiversity measures (e.g. total numbers of species) because of contrasting responses of individual species (Lindenmayer *et al.* 2005).
- Not all trajectories are characterised by critical breakpoints (Groffman *et al.* 2006; Parker and MacNally 2002).

Thus, differentiating the kinds of species, landscapes, ecosystems and ecological processes prone to threshold responses from those that exhibit other kinds of responses is very important (Lindenmayer *et al.* 2005).

The key point to recognise, despite these and other caveats, is the very real possibility of non-linear responses in ecosystem processes and landscape modification, and the existence of critical zones in which rapid change occurs.

---

to adapt genetically. Australia's biota faces a very different situation today. Arguably the most important feature of the drivers of change associated with the arrival of Europeans 220 years ago is the unprecedented rate at which they operate – several orders of magnitude greater than the rate of change with which Australia's biota has evolved.

This section identifies:

- those factors associated with contemporary Australian society that directly affect organisms ('proximate' drivers)
- those factors that act indirectly on organisms, mainly through socio-economic forces and institutional arrangements ('ultimate' drivers)
- those more recent factors of external origin that are now influencing Australia's biota ('global' drivers).

3.3.1 Proximate drivers

Direct removal through hunting and trapping

The most direct driver of biodiversity change is human activity that deliberately or accidentally kills organisms. The historical and continued harvesting of target fish has the potential for massive impacts on native marine species. In 2006, 19 stocks out of a total of 97 were being overfished, with a further 51 classed as 'uncertain' (Larcombe and McLoughlin 2007). Overall, the 2006 data agree with a long-term trend of

**Box 3.3 Connectivity.
(David Lindenmayer)**

Connectivity relates to the ability of species, ecological resources and processes to move through landscapes – not only in the terrestrial domain, but also in aquatic systems and between the two. Connectivity, and in particular the value of corridors, has been much debated. Some debates about connectivity stem from the term being too broadly conceived, rendering it difficult to use in practice, and from different interpretations of terms. Lindenmayer and Fischer (2007) suggested that some of the controversy might be avoided by making a careful distinction between:

- habitat connectivity or the connectedness of habitat patches for a given taxon
- landscape connectivity or the physical connectedness of patches of a particular land cover type as perceived by humans
- ecological connectivity or connectedness of ecological processes at multiple spatial scales.

Lindenmayer and Fischer (2007) further noted that although the three connectivity concepts are interrelated, they are not synonymous. In some circumstances, habitat connectivity and landscape connectivity will be similar (Levey *et al.* 2005). In others, habitat connectivity for a given species will be different from the human perspective of landscape connectivity. Furthermore, connectivity is always relative to the individual species that move or exchange genetic material (e.g. pollen), which is a function of distance, timing and other species that might be present. A single habitat that appears to have high connectivity to one species would be highly fragmented to another. For these reasons, putting the important principle of connectivity into practical terms for biodiversity conservation can be difficult (Chapters 6 and 7).

increasing numbers of stocks that are overfished and/or subject to overfishing. Stocks that continue to be exploited unsustainably include deepwater sharks (chondrichthyans), orange roughy *Hoplostethus atlanticus*, gemfish *Rexea solandri*, several species of tuna – southern bluefin *Thunnus maccoyi* and bigeye *Thunnus obesus* – swordfish *Xiphias gladius* and commercial scallops *Pecten fumatus*. Many of these species are the top predators (Principle 5; Table 3.1) within their ecosystems, and their declines have cascading impacts on ecosystem structure and functioning (Principle 6).

Land clearing

Vast tracts of Australia's landscape have been transformed through vegetation clearing for agriculture and settlement over the past two centuries (Fig. 3.1). As recently as 2000, some 600 000 ha per year were being cleared, particularly in the semi-arid regions of the eastern states. Forests on the mainland east coast and in Tasmania have been reduced by about 30–50% of their original extent, particularly those on more fertile soils. Clearing has occurred in the east from Victoria to central Queensland, tracking the westward slopes along the Great Dividing Range. In the south-west of the continent, much of the woodland has disappeared due to clearing for arable farming. Today, the Bassian/Eyrean-Bassian zones have the most fragmented natural biomes on the continent. Rainforest has been reduced in area by more than half, and other specialised ecosystems are formally described as endangered (Castles 1992).

Clearing has a direct impact through loss and fragmentation of habitat, and is often followed by:

- grazing by stock
- planting of crops
- changes in water, fire and nutrient regimes
- invasion by weeds and exotic animals (Principle 7).

These events underpin major environmental problems such as erosion, salinisation and weed invasion. Large-scale land clearing can also affect regional climate directly through, for example, changes in reflectivity and evapotranspiration (McAlpine *et al.* 2007).

Redistribution of water resources and changes in nutrient capital

Major changes in water and nutrient availability due to agriculture and urban development are the:

- widespread application of fertilisers – especially superphosphate – on crop and pasture lands, and nitrification of water bodies

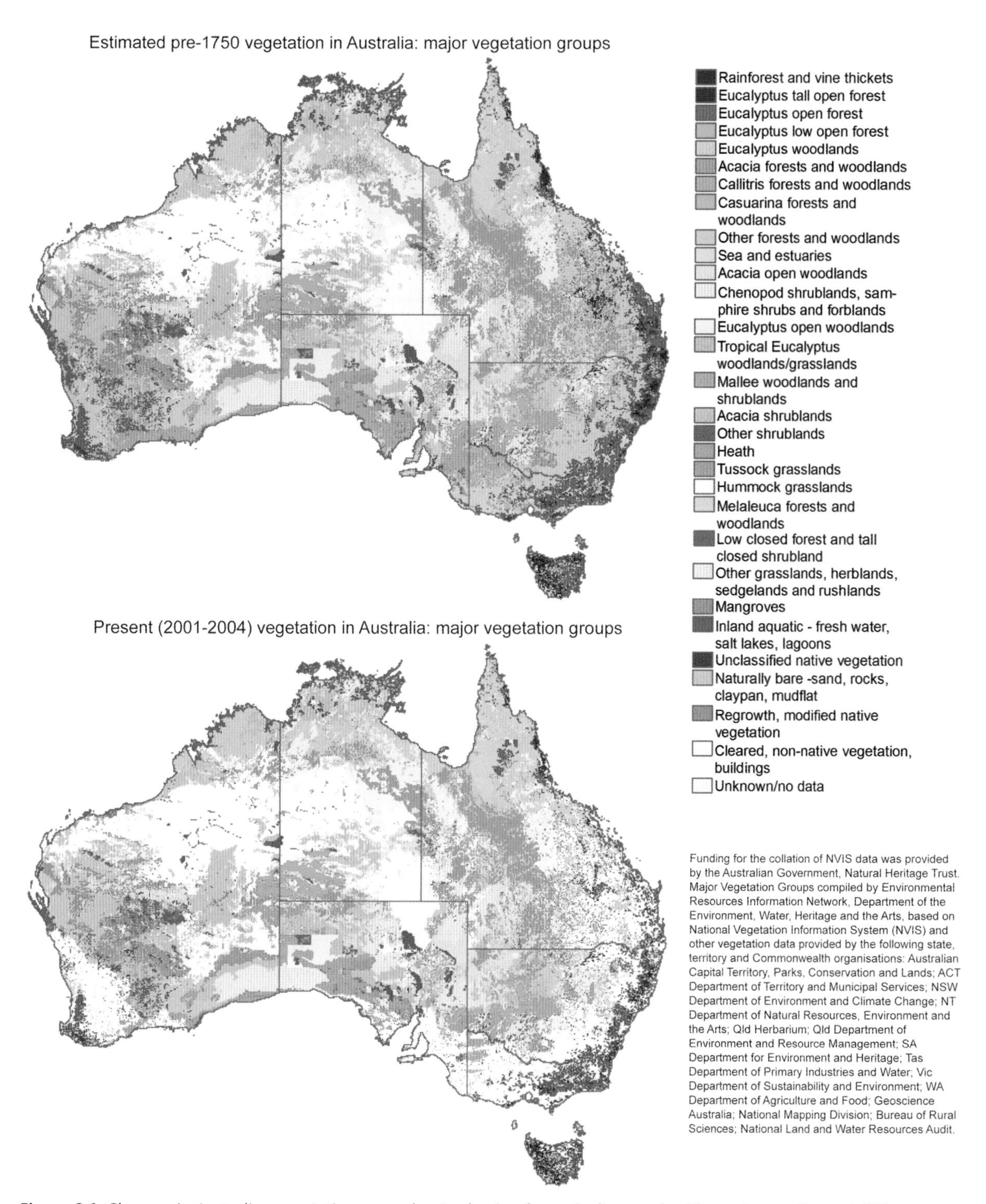

Figure 3.1 Changes in Australian vegetation cover due to clearing for agriculture and settlements over the past 220 years. Source: Environmental Resources Information Network.

- construction of watering points in grazing areas
- diversion of water for crop irrigation.

Increasing the nutrient capital on Australia's infertile soils, almost always in order to grow crops and fodder, has many unintended consequences for native biodiversity. The addition of superphosphate to pastures in Australia in conjunction with widespread tree clearing, and what we now know was overstocking, leads to a cascade of biological interactions, such as increasing pasture tree dieback. Figure 3.2 shows how the change in one factor – increasing soil nutrients by fertilisation – resulted in both an increase in understorey competition and the loss of certain native insects. This then, through the massive outbreaks of other leaf-eating insects, eventually made the trees more susceptible to the pathogen *Phytophthora cinnamomi*, killing the host trees (Landsberg and Wylie 1983), and providing an example of Principles 6 and 7. Nutrients added to soil can also make their way into water bodies as pollutants, and change entire communities and ecosystems (McKergow *et al.* 2004).

Nutrient runoff from human settlements has also eutrophied some Australian groundwaters and water bodies, including coral reefs and other marine systems affected by river outflows (Bell 1992; Bell and Elmetri 1995; Brodie 1997). Globally, hypoxia in the coastal regions of the world has increased over the past 60 years, leading to more than 400 'dead zones' affecting a total area of >245 000 km^2 where sea bed organisms

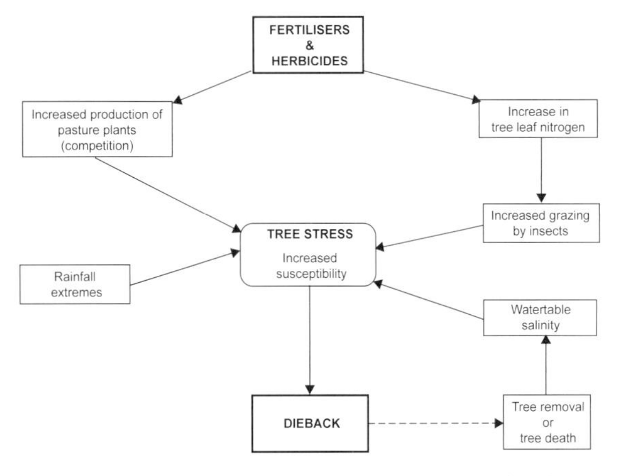

Figure 3.2 Dieback of pasture trees as an example of an ecological cascade. Source: Modified from Landsberg and Wylie (1983).

and fish can no longer survive (Diaz and Rosenberg 2008); the extent of these zones in Australian waters is not large, but trends are in the same direction as the rest of the world.

Redistributing scarce water resources, primarily for cropping and other agricultural enterprises, has had significant impacts on biodiversity. The most dramatic example is the massive diversion of water in the Murray–Darling Basin. Similar impacts on biodiversity occur in river basins throughout Australia. Urban settlements require significant sources of fresh water, necessitating the construction of dams, diversion of water bodies, extraction of groundwater for human consumption, watering of gardens and recreation. Dams and impoundments, by their very nature, redistribute water, with knock-on effects on local and downstream biodiversity.

Some forms of agricultural practice have compromised the inherent ability of the underlying ecosystem to recover from change – its resilience – so that restoration is virtually impossible. Extensive long-term use of fertilisers has changed some soils to such an extent that the original naturally nutrient-poor ecosystem will most likely never return to its previous state. In some northern pastures, long-term cattle grazing has irreversibly changed native vegetation (Sharp and Whittaker 2003). In an even more extreme case, long-term irrigation leading to surface salinisation has alienated those areas in perpetuity. In some instances, of course, 'new' ecosystems of conservation value can be established in place of the original types; in others, such as newly established salt pans, changes may be irreversible.

Introduction of new species

The introduction of new species to the continent has been a major driver of change to Australian biodiversity. The knock-on effects of such exotic species as the red fox, rabbits, prickly pear *Opuntia*, domestic cats and pigs is well documented (Figs 3.3, 3.4), but several other species are also having significant impact on biodiversity, for example, camels, feral cattle, donkeys, goats, cane toads, some species of invertebrate animals such as ants, and pathogens such as *Phytophthora* and the chytrid fungus. Weeds, many of which were brought in for fodder or for horticultural use, are an

Figure 3.3 A pair of feral cats feeding on prey in grass. Feral cats feed on native species and can pose a serious threat to native wildlife. Source: AUSCAPE. Photo by Kathie Atkinson.

increasing problem (Fig. 3.5). Undoubtedly 'sleeper' species – introduced some years ago but with limited, local distributions currently – could become major problems in the future, especially as the environment changes. Fortunately, Australia now has some of the strongest quarantine laws in the world, and in general, new species introductions have declined in recent years. Despite past history, however, current policies continue to permit introductions of new species to improve agricultural production (especially pastures) and for horticultural purposes, angling and the pet trade which, despite quarantine assessment, may still have potential to cause problems in future.

Figure 3.5 *Mimosa pigra* young plant in wild rice *Oryza rufipogon* (native). A sole Mimosa plant can produce over 200 000 seeds a year, and the seed pods float, enabling them to spread rapidly during the Wet. The plant has already taken over more than 800 sq km of wetlands in the NT Adelaide River floodplains, Northern Territory, Australia. Source: AUSCAPE. Photo by Frank Woerle.

Changes in fire regimes

Australia's plants and animals have become variously adapted to particular fire regimes, depending on biome and region (Dixon *et al.* 1995; Flematti *et al.* 2004; Whelan 1995). The three aspects of a fire regime – intensity, frequency and seasonality – have been greatly modified over much of the continent with the arrival of Europeans (Gill 1975), often to minimise the potential threats to persons and property (Cary *et al.* 2003). Before European presence in Australia, fire regimes in most of Australia consisted mainly of frequent fires lit by Aboriginal people as an aid to their hunting, gathering and cultural activities (Fig. 3.6); and occasional wildfires that originated from lightning strikes (Bowman 2000; Rose 1995). With the natural distributions of some plant and animal species already reduced (due to clearing, exploitation, competition with exotics

Figure 3.4 Rabbits *Oryctolagus cuniculus*, introduced pest to Australia. Source: AUSCAPE. Photo by Jean-Paul Ferrero.

Figure 3.6 A fire lit for hunting purposes by Aboriginal people, southern Great Sandy Desert, Western Australia. Source: Andrew Burbidge.

and so forth), the chances of a single fire having a disproportionate impact on the well-being of a whole species has increased. This is especially the case for rare and endangered invertebrates or herbaceous plants for which whole populations may cover only a few hundred square metres.

Mineral extraction

The extraction of mineral resources affects localised land areas, so in many cases mining has minimal direct impacts on biodiversity. However, in areas where highly endemic species and rare ecosystems occur in small, spatially restricted patches, mining operations potentially can have direct and disproportionate impacts on biodiversity (Box 3.4). The effects on biodiversity of the extraction of petroleum and natural gas, especially from sea beds, are largely unknown. Similarly, the deposition of waste products from mining – currently about one billion tonnes of waste rock annually (Mudd 2008) – into landfills and water bodies has potentially damaging but more or less unknown impacts on local biodiversity. Because of its radioactivity, waste from uranium mining may present additional hazards for biota in its vicinity.

3.3.2 Ultimate (indirect) drivers

Drivers of biodiversity change in Australia are ultimately linked to human activities. Some are associated with production/consumption (e.g. food production, marketing and use), or with individual lifestyle and societal expectations, whereas others underpin development (e.g. human population numbers and growth, roads, minerals and energy production), or are associated with institutions (subsidies). A common, integrative method to estimate the impact of a society on the environment is to combine its population level with the ecological impact associated with the goods and services that each person in the society consumes.

Population growth

Australia's rate of population growth (about 1.2% per year) is one of the highest in the developed world.

Box 3.4 An example of biodiversity vulnerability to mining in Western Australia. (Andrew Burbidge)

The banded ironstone formation (BIF) ranges of the mid-western region of Western Australia provide an interesting example of mining impact on biodiversity. Strong international demand for iron ore has encouraged several companies to explore these ranges. In November 2007, the Western Australian Environmental Protection Authority (EPA) had completed formal assessments of three BIF mining proposals and was assessing another three. There were, in addition, a further three at the feasibility stage of evaluation and some 25 prospects that were under exploration.

Although forming a very small proportion of each bioregion, the BIF ranges are of very significant biodiversity value due to their unique geology, soils and relative isolation. The ranges are important due to the presence of endemic plant species – some of which are listed as threatened under Australian and Western Australian legislation (*Environment Protection and Biodiversity Conservation Act 1999* (Cwlth) and *Wildlife Conservation Act 1950* (WA)) – rare and restricted plant species, and highly restricted and distinct plant communities. Based on survey information to date, every range is distinctly different from the other sampled ranges from an ecological perspective.

There is no clear understanding of the level of loss that can be inflicted on the identified range communities without permanently compromising the sustainability of local ecosystems and their component species. Of the BIF ranges of the Yilgarn Craton, only two are within established conservation reserves, and even those are within categories where policy and legislation allow exploration and mining subject to recommendations (not concurrence) of the Minister for the Environment (notwithstanding any requirements for assessment under the *Environmental Protection Act 1986* (WA)). Accordingly, no BIF ranges are adequately protected to fully protect rare flora, vegetation and fauna against development.

In order for the Western Australian State Government to balance expectations for economic development with biodiversity conservation requirements, a series of strategically planned 'compromise' conservation reserves would need to be identified, where mining cannot proceed.

Paynter's tetratheca *Tetratheca paynterae* subsp. *paynterae* growing on weathered banded ironstone. The subspecies is restricted to 4 ha on Windarling Range where it is being impacted by mining of iron ore.
Source: Andrew Brown/DEC.

Depending on assumptions about fertility, lifespan and net immigration rates, the Australian population is projected to grow from the current 21 million to 25–33 million by 2050 and to 26–51 million by 2100 (Foran and Poldy 2002). Most of the projected increase is due to immigration; in fact, if there were zero net immigration, the population would drop to a projected 17 million by 2100. In the absence of any national population policy that is an unlikely scenario, so we may assume that the human population of Australia will continue to increase.

Increases will be accompanied by increased demand for food and other resources from the land. Unless innovative and appropriate steps are taken, pressures

Box 3.5 Human population growth as a driver of biodiversity loss. (David Lindenmayer)

Australia is a large country with relatively few people. However, even our comparatively small population size and low population density has had major direct and indirect impacts on the continent's biota.

There is not necessarily a simple cause and effect relationship between human population size and species loss. For example, many species of native mammals have been lost from arid areas of Australia where humans are scarce. In these cases, feral predators and changing fire regimes may be the primary causes of species loss. Furthermore, Luck *et al.* (2004) have reported strong positive correlations between human population density and species richness for birds, mammals, amphibians and butterflies in Australia, perhaps reflecting the fact that humans tended to build cities where basic productivity (and species diversity) were naturally high. Thus, there is considerable potential for future spatial 'conflict' between the expansion of the nation's population and the loss of biodiversity. Indeed, a large proportion of Australia's threatened species occur in peri-urban areas at high risk from the further expansion of cities and towns.

The current rate of population increase in Australia is relatively high; in fact it is one of the highest in the developed world. In summary, it is impossible to divorce biodiversity loss and environmental degradation from the controversial issue of Australia's population size (Foran and Poldy 2002).

on biodiversity will increase, regardless of climate change (Box 3.5). Because immigration is the projected source of population increase, the rate of population growth can be modulated through the political decision-making process.

Ecological footprint

The total resources required to support an individual or a population is said to be the 'ecological footprint' for that individual or population (Rees 1992; Wackernagel and Rees 1996). The resources include not only those used directly (e.g. food, fibre, water, housing, appliances, schools, etc.), but also the amount of resources required to deliver goods and services to the individual (e.g. transport, infrastructure, fuel, minerals, electricity), and to absorb the wastes produced by human activities (e.g. absorption of carbon dioxide by oceans). The ecological footprints of countries, cities or settlements can be estimated by scaling up the hectares required to sustain the resource demands of one person (Global Footprints Network 2007). The range in per capita footprint ranges from 0.5 ha to 10.5 ha, with a worldwide average of 2.0 ha. (Fig. 3.7). Australia has a per capita ecological footprint of about 6.8 ha.

An important insight from calculating ecological footprints is the relationship between urban dwellers and their impact on the environment, including biodiversity. An inevitable consequence of urbanisation is that urban dwellers become disconnected from the ecosystem services that sustain them, both in fact and in awareness. Their footprints are found in places far distant from the cities in which they live. Even some environmentally aware urban dwellers may not fully comprehend the ultimate impacts of their lifestyles and consumption patterns on biodiversity in Australia and elsewhere on the planet.

Primary industries

In terms of pressure on biodiversity, one of the most important components of an ecological footprint is associated with primary industries. Agriculture, fishing, aquaculture and forestry are especially important to Australia, and support human populations here and elsewhere with food, fibre and other valuable products. These services obtained from highly modified ecosystems are essential for human well-being, and thus enjoy high priority in the economic, social and cultural life of the country. However, provision of these valuable services often has unintended and unwanted consequences for Australia's biodiversity. For example, farming is largely responsible for current habitat loss and fragmentation in rural areas, dramatically reducing both suitable 'living space' and connectivity for many native species. There is a growing recognition of this problem, leading to more emphasis on land stewardship,

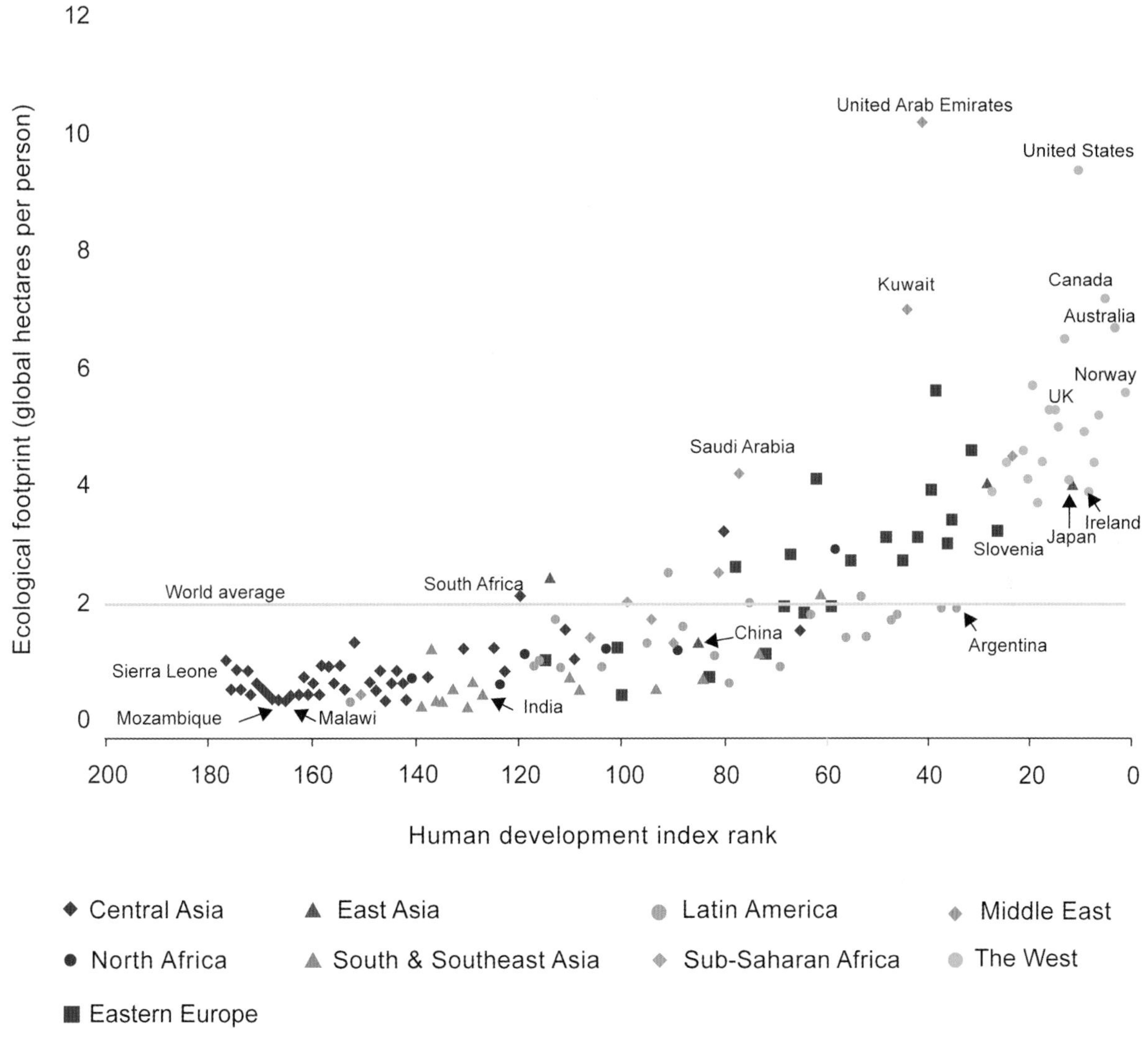

Figure 3.7 Ecological footprints (hectares per person) of various countries. Also shown is the ranking in the Human Development Index (HDI), an aggregate indicator of human well-being. Australia has achieved a very high ranking in HDI, but with a large impact on the natural environment. Source: EPA (2007) based on data from Global Footprint Network (2006), United Nations Development Programme (2006).

biodiversity conservation and, more broadly, sustainability (see Chapters 6 and 7).

Perverse incentives

Subsidies of various forms (e.g. direct payments, compensation, tax incentives) from Australian institutions are common, indirect drivers of biodiversity change. Subsidies for fisheries, forestry, land clearing, agriculture and grazing are driven by desired economic outcomes, but often have unintended and deleterious impacts on biodiversity. For example, the use of millions of tonnes of superphosphate that ultimately led to massive dieback of pasture eucalypts (Fig. 3.2) was propelled by large tax incentives and other subsidies for pasture 'improvement', as has been the introduction of many of the worst weeds in the north such as gamba grass *Andropogon gayanus*. The term 'perverse incentives' has been applied to these types of counterproductive subsidies (Myers and Kent 2001). Other examples include:

- direct subsidies for land clearing, which was once a condition on property leases
- financial encouragement by governments of industrial-style agricultural developments such as cotton and sugar
- payments of drought relief to maintain businesses on sub-marginal land that can no longer support them.

As landscapes move towards management for profit under new carbon markets, or for the production of biofuels, new types of perverse incentives could arise.

3.3.3 Global drivers

Over the rest of this century and beyond, the most prominent of the global drivers will be human-induced climate change. This is discussed in detail in the remaining chapters. There are, however, non-climatic drivers that are global in nature and extent.

The global marketplace, within which Australia's industries operate, influences land use and production decisions that have flow-on impacts for biodiversity. These pressures have become acute under economic globalisation and industry pressures for a global 'level playing field'. External factors such as loss of stratospheric ozone, airborne nitrogen deposition and intense international fishing pressure on marine fish stocks also have actual or potential impacts on Australian biodiversity.

Globalisation is also occurring in law through international treaties and conventions, and in public debate as driven by modern information technology such as the internet. Australia is already prominent in international discussions on a range of environmental mitigation issues, including biodiversity conservation. The 'international citizen' role will probably become more important for Australia in coming decades.

3.4 THREATENED SPECIES AND VULNERABLE ECOSYSTEMS

The drivers described in the previous section have led to profound changes to the biotic fabric of the continent. This section describes changes in diversity over the past 220 years at the genetic, species and ecosystem levels. Much of the emphasis is on the species level, as this is the scale at which biodiversity conservation is often focused. The ecological community and ecosystem levels are also very important, as ecosystem functioning is largely dependent on the interactions among organisms within ecosystems, superimposed on the larger spatial scales of topography, soils, nutrients, hydrology and climate. Ecosystems provide many 'free' but critical services to humans (Chapter 2).

3.4.1 Genetic-level changes

Habitat fragmentation due to land clearing and modification is the most important proximate driver of change at the genetic level. In some cases fragmentation has entirely eliminated sub-populations of native species (e.g. long-nosed bandicoots *Perameles nasuta* in the Sydney metropolitan area (Australian Museum 2008); southern hairy-nosed wombats *Lasiorhinus latifrons* in South Australia), and in others has split large interbreeding populations into smaller, isolated sub-populations (e.g. heathland species such as one-sided bottlebrush *Calothamnus quadrifidus*, and the wandoo *Eucalyptus wandoo* and salmon gum *E. salmonophloia* woodlands in the agricultural region of Western Australia) (Byrne *et al.* 2008). This diminishes total genetic variability within the species and reduces the potential for biological adaptation (e.g. Yates *et al.* 2007). This reduction in adaptive capacity is of special concern in the context of climate change, in which species will come under intensifying pressure to adapt or disperse to new locations (Chapter 5).

3.4.2 Species-level changes

Species-level changes include local or continental extinctions, changes in the relative abundance and distribution of species, and the addition of species through intentional and unintentional introductions.

Extinction, functional extinction and threatened species

Extinction usually occurs when a change in a species' environment is beyond its ability to adapt. Over the past 220 years, Australia has lost many species, probably at a rate unprecedented since the Last Glacial Maximum 20 000 years b.p. More than 50 vertebrate species, and a comparable number of plant species, have been lost. Island species are particularly vulnerable. Extinctions

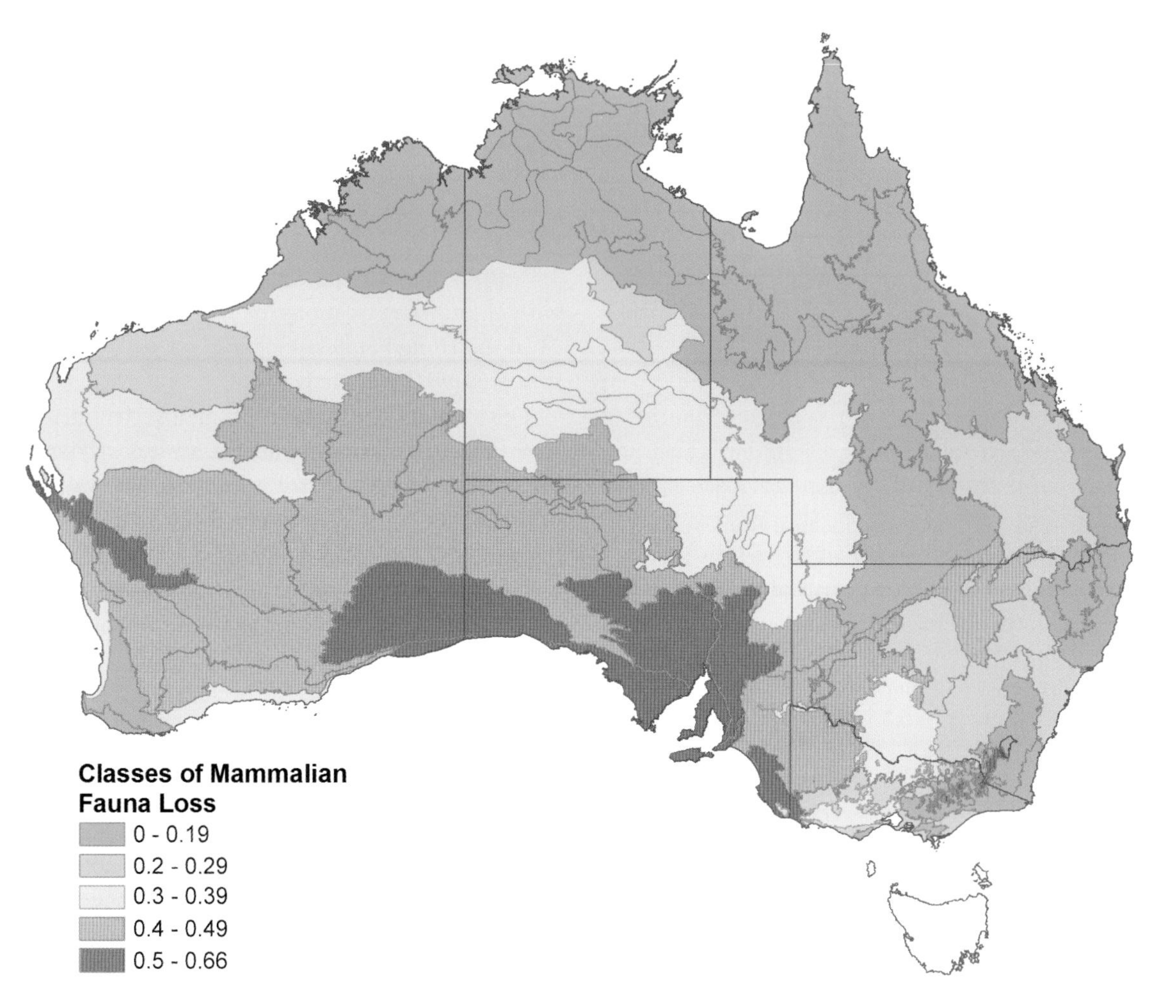

Figure 3.8 Proportion of loss of mammalian fauna in 76 bioregions of mainland Australia. The proportion is calculated by adding the number of extinct species to one half of the species in decline. Shown are five classes: 0–0.19, 0.2–0.29, 0.3–0.39, 0.4–0.49 and 0.5–0.66; where 0 = all species persisting, through to 1 = all extinct. Source: McKenzie *et al.* (2007).

and massive declines are best documented among birds, mammals, frogs and higher plants, where there is sufficient information to quantify the losses. However, similar patterns are also found among other, less well-known, groups such as butterflies and other invertebrates. Unfortunately, these numbers represent large percentages of endemic species in every group (see further on). The drivers of these extinctions have almost always been landscape transformations and/or the direct impact of introduced species (e.g. Lindenmayer 2007a; Primack 2001; Ward 2004).

Mammals. Australia has the worst record for mammal extinction and decline of any country (Box 3.6). In south-central Australia, the proportion of loss of mammalian fauna is greater than 50%, and more than half the area of the country has more than 30% loss (Fig. 3.4). Of 72 bioregions, only two small ones, North Kimberley and Tiwi-Cobourg, have had no extinctions or significant declines (McKenzie *et al.* 2007) (Fig. 3.8).

Modelling suggests that several factors are significantly related to mammal loss:

- rainfall (more loss in arid areas)
- environmental change (more loss in areas where habitat has been lost or degraded)
- mammal body weight (medium-sized mammals from 35–5500 g in body weight are most affected)

Box 3.6 Australia's extinction record for terrestrial mammals. (Andrew Burbidge)

Some 305 non-marine mammal species are known to have been present in Australia at European settlement. Eighty-four per cent (257) of these are endemic to continental Australia. The remaining 48 species also occur in New Guinea and/or nearby islands. Of the non-endemic species, 30 species are bats. Among Australian bioregions there was an average of 51 endemic mammal species (range: 28 to 89).

Twenty-two mammal species are now extinct. Eight more became restricted to continental islands, and 100 have disappeared from at least one of Australia's 85 bioregions. In parts of south-central Australia, more than half of the mammal species known to be alive when Europeans first settled Australia are now extinct or threatened with extinction, and one third of Australia's desert mammals are now extinct (Burbidge *et al.* 1988; Lindenmayer 2007a; McKenzie *et al.* 2007). Australia has accounted for 30% of the world's known mammal extinctions since the year 1600.

- whether a species shelters on the ground or in trees, burrows or rock piles (on-ground species are most affected)
- the density of introduced herbivores such as sheep and cattle (areas with high densities of herbivores have lost more mammal species)
- the length of time since foxes and cats first arrived in the region (the longer since arrival the more species lost) (McKenzie *et al.* 2007).

Birds. Only one bird species, the paradise parrot *Psephotus pulcherrimus*, is known to have become extinct on the Australian mainland since European settlement. In contrast, bird extinctions on oceanic islands, mainly on Lord Howe and Norfolk islands, have been very common. Island species of emus have disappeared from King Island (*Dromaius ater*) and Kangaroo Island (*Dromaius baudinianus*) off Australia's southern coastline (Ford 1979). On the mainland, local extinctions of populations have occurred as species distributions have shrunk. Examples are many, including the grey-crowned babbler *Pomatostomus temporalis* and the partridge pigeon *Geophaps smithii* (Barrett *et al.* 2003).

Higher plants. To the best of our knowledge, 49 species of vascular plants have become extinct in Australia over the past 220 years, according to the current *Environment Protection and Biodiversity Conservation Act 1999* (EPBC) List of Threatened Flora. This compares with 27 known extinctions in the whole of Europe and 74 in the United States since European settlement of that country (Lindenmayer 2007a).

Other groups. Extinctions of small vertebrates, and freshwater and marine organisms, are less well known, but undoubtedly unrecorded extinctions have occurred. Four well-documented extinctions of frogs have occurred in eastern Australia. Up to another 14 frog species are classified as endangered and two more as critically endangered, according to the current EPBC list and the Global Amphibian Assessment (GAA 2007). Extinctions of species within taxonomically less well-known groups such as fungi, mosses or insects are poorly documented. Indeed many species within these groups are simply undescribed. Marine extinctions have been less dramatic and considerably more difficult to establish beyond reasonable doubt, although the extinctions of two species of marine algae have been established (Millar 2003a, b; Monte-Luna *et al.* 2007). The nature of the marine environment means that species are generally more mobile and have more extended ranges than terrestrial species.

Many Australian species have already reached such low numbers that they no longer have an effective role in the functioning of ecosystems, and are therefore considered 'functionally extinct'. These include some of the mammals listed in Table 3.3, some birds such as the night parrot *Pezoporus occidentalis* (Garnett and Crowley 2000) and the many critically endangered plants where only one or two very small populations survive (Atkins 2006). Functional extinctions often have flow-on effects to other species. A higher proportion of Australian plant species are pollinated by vertebrates, compared with plants on other continents (Orians and Milewski 2007), and many of these vertebrates are threatened species. Some native orchids are pollinated by only a single species of wasp (Hoffman and Brown 1998). Hence, if the wasp cannot complete its life cycle because of the disappearance of the food resources on which it depends, the orchid also will fail to reproduce.

Functional extinctions of species can have dramatic effects on entire communities or ecosystems. This is especially true if the functional extinction concerns a key species (Table 3.1, Principle 5). Two examples follow:

- Some coral species (e.g. *Acropora* spp.) have declined to such an extent in inshore waters of the Great Barrier Reef region due to eutrophication that they can no longer function as key structural species, being replaced by less complex and less diverse algal-dominated inshore reef communities (Table 3.1, Principle 5). The functional extinction of the corals affects hundreds of other species that depend upon their presence.
- Most non-arboreal Australian mammals (including the burrowing bettong *Bettongia lesueur* and brush-tailed bettong *B. penicillata*), the bilby *Macrotis lagotis*, the potoroos (*Potorus* spp.) and many bandicoots (*Perameles* and *Isoodon* spp.), some lizards and many ground-dwelling birds dig burrows and/or turn over soil when searching for food. They are functionally ecological engineers (Table 3.1, Principle 5) and provide an important ecosystem function that enhances soil water penetration and reduces runoff; this has far-reaching consequences for many plants and soil organisms. Some plants, known as disturbance opportunists, require this kind of soil turnover to germinate. Many of these animals, even those not officially threatened with extinction, have experienced severe reductions in their ranges (Table 3.3) or declines in abundance; accordingly the important ecosystem services to which they contribute are, in turn, threatened.

Compared with the high extinction rates already experienced by the Australian biota, an even higher number of species are considered at risk of extinction. The perceived extinction threat to these species is generally classified in Australia by state and national programs using international criteria agreed upon by the IUCN (Box 3.7). Most of the listed species have declined dramatically in numbers and range, and are on distinct 'extinction trajectories' (Lindenmayer 2007a).

Warning signals that a species may be threatened with extinction include:

- massive and/or rapid declines in distribution and/or abundance
- declines in the species' functional position within a community

Box 3.7 What does 'threatened' mean? (Andrew Burbidge)

In Australia, most jurisdictions and specialists follow terminology established by the IUCN when describing species and ecological communities that are threatened with extinction or destruction. 'Critically endangered' (CR), 'endangered' (EN) and 'vulnerable' (VU) are used for species or communities that meet certain criteria, either those developed by the IUCN or as prescribed in legislation or policy, with CR denoting the highest level of risk of extinction. 'Threatened' is a collective term used to describe any of these levels of threat to the survival of a species or an ecological community.

Providing figures from threatened species lists greatly understates the degree of species declines because:

- the legalistic nature of such lists means that species will be listed only when there is overwhelming evidence that they are threatened with extinction
- insufficient information about most of our biodiversity prevents a proper assessment of status. Invertebrate animals, for example, comprise over 95% of all species, but most Australian invertebrates are not scientifically described, and even for described species, distribution and conservation status are rarely known.

Nevertheless, it is clear from available information on vascular plants and vertebrate animals that many species have disappeared from large parts of their former ranges and that if current trends continue, the rate of extinctions in Australia will increase.

Table 3.2 The number of species and subspecies declared as threatened, under the *Environment Protection and Biodiversity Conservation Act 1999*, as at March 2009. Additional nationally threatened species and subspecies are listed by states and territories, and are yet to find their way onto the EPBC Act list.

	Extinct	Critically endangered	Endangered	Vulnerable	Total
Total plant species and subspecies	**49**	**81**	**523**	**665**	**1318**
Mammals	27	4	33	55	119
Birds	23	6	41	61	131
Reptiles	0	2	14	37	53
Frogs	4	2	14	12	32
Fishes	1*	3	16	25	45
Invertebrates	0	16	12	6	22
Total animal species and subspecies	**54**	**23**	**130**	**192**	**402⁺**

* Listed as 'extinct in the wild', i.e. remains in captivity.
⁺ Plus one species of fish listed as 'conservation dependent'.
Compiled from data at: http://www.environment.gov.au/biodiversity/threatened/index.html.

- declines in reproduction and recruitment to the next generation to a level below that necessary to sustain a population.

In long-lived species, such as cockatoos, marine mammals and trees, effective monitoring may identify some of these warning signals decades before the declines appear in the adult populations (e.g. lack of regeneration in tree populations (Saunders *et al.* 2003)).

The number of Australian species classified as 'critically endangered', 'endangered' and 'vulnerable' is of great concern. Beyond the species already extinct there is a total of 1269 vascular plants, 92 mammals, 108 birds, 53 reptiles, 28 frogs and 44 fishes including sharks listed as threatened species, at some level, according to the EPBC Act list (Table 3.2). Thirty-one per cent of Australia's mammals, 14.5% of frogs and 15.8% of birds are classified as extinct, critically endangered, endangered or vulnerable, using the current EPBC Act list based on the numbers of total species from Chapman (2005). On a per capita basis, Australia's standing in terms of threatened species is substantially worse than the next two worst-performing countries (United States and Mexico) (Fig. 3.9), although different criteria in different countries make comparisons difficult on a worldwide basis (Lindenmayer 2007a).

The formal listing of communities, species and populations under threat occurs in Australia under both Commonwealth and state legislation. Although generally effective, this continually generates jurisdictional issues and, sometimes, undesirable outcomes. Species whose natural distributions encompass just a small fraction of a particular state may be listed as of concern in that state despite being common elsewhere.

Changes in the relative abundance and distribution of species

Apart from threats to species survival, many species' ranges have decreased at an increasing rate since European settlement. These shrinking ranges are unprecedented, compared with historical and palaeo-records (e.g. Burbidge 2004). The best-known changes in species' ranges are within the mammal, bird and frog populations, many of which have suffered massive reductions, often to the point of species collapse. Unfortunately, much less is known among other groups of organisms, although they too have undoubtedly suffered reduced distributions and abundances, given the interdependency of species groups (Table 2.1, Principle 4).

Mammals. Many of Australia's mammal species now have distributions covering less than 20% of their original range (Table 3.3). Some of the greatest reductions in range have occurred with numbats *Myrmecobius fasciatus* (fig. 3.10), which now occupy less than 5% of their original range, and burrowing bettongs

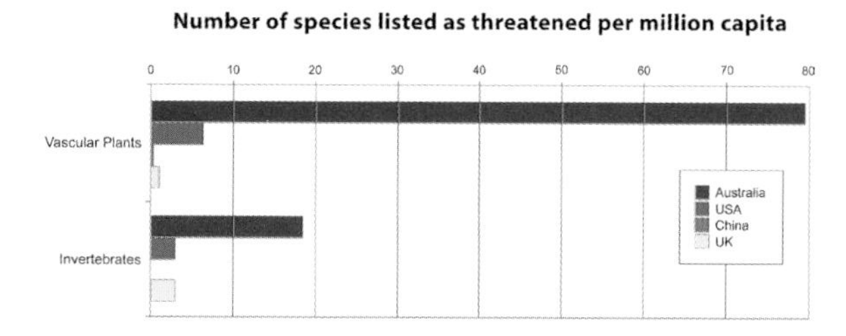

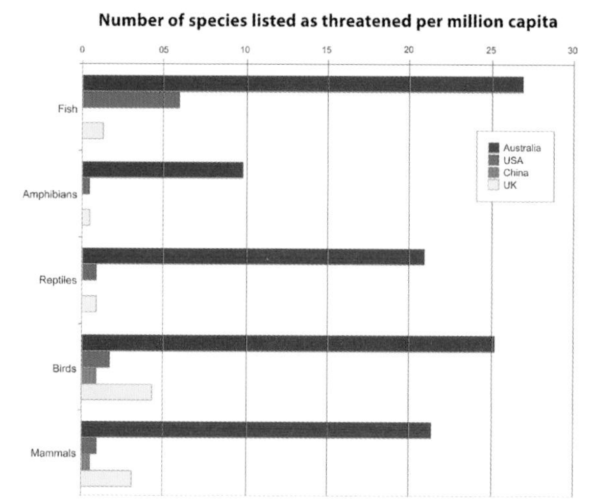

Figure 3.9 Number of threatened species of vascular plants, invertebrates and several groups of vertebrates per capita for Australia, United States, China and the United Kingdom. Numbers are calculated on the basis of definitions used by the *Environment Protection and Biodiversity Conservation Act 1999*. Source: Lindenmayer (2007a, p. 37).

Figure 3.10 The numbat *Myrmecobius fasciatus* once occurred though much of southern arid and semi-arid Australia but became restricted to small parts of the south-west. Source: © Bert & Babs Wells/DEC.

Bettongia lesueur (Fig. 3.11) and rufous hare-wallabies *Lagorchestes hirsutus*, which now occupy less than 1% of their original range (van Dyke and Strahan 2008).

Birds. The reporting rates of many species of Australian woodland bird species (that is, the number of sightings relative to the number of observers as monitored by Birds Australia; Barrett *et al*. 2003) have declined 11–51% over the past two decades, reflecting a decline in abundance as well as range (Table 3.4). Birds that originally had an Australia-wide distribution such as the Australian bustard *Ardeotis australis* and the brolga *Grus rubicunda* have suffered massive declines and, although still widely distributed, occur in much smaller numbers and have largely disappeared from their southern and eastern ranges. In contrast, other large and charismatic birds such as the

Figure 3.11 The burrowing bettong *Bettongia lesueur* once had one of the widest distributions of any Australian mammal but survived only on three islands off the west coast. Source: © Bert & Babs Wells/DEC.

Table 3.3 Examples of species collapse among Australian mammals, comparing their historic and current ranges.

Common name (scientific name)	Historic range (sq km)[1]	Current range (sq km)	Collapse (%)
Banded hare-wallaby (*Lagostrophus fasciatus*)	489 868	607	>99
Burrowing bettong (*Bettongia lesueur*)	4 371 154	607	>99
Greater stick-nest rat (*Leporillus conditor*)	1 325 043	607	>99
Rufous hare-wallaby (*Lagorchestes hirsutus*)	1 961 902	1215	>99
Bridled nailtail wallaby (*Onychogalea fraenata*)	1 097 876	10 022	99
Northern hairy-nosed wombat (*Lasiorhinus krefftii*)	105 991	1519	99
Brush-tailed bettong (*Bettongia penicillata*)	1 771 786	53 451	97
Hastings river mouse (*Pseudomys oralis*)	269 078	7593	97
Numbat (*Myrmecobius fasciatus*)	1 924 243	58 918	97
Dusky hopping mouse (*Notomys fuscus*)	902 900	42 518	95
Heath rat (*Pseudomys shortridgei*)	235 975	14 881	94
Smoky mouse (*Pseudomys fumeus*)	150 939	12 755	92
Tasmanian bettong (*Bettongia gaimardi*)	512 342	47 681	91
Dibbler (*Parantechinus apicalis*)	99 006	10 326	90
Leadbeater's possum (*Gymnobelideus leadbeateri*)	43 733	5063	88
Red-tailed phascogale (*Phascogale calura*)	176 450	28 852	84
Bilby (*Macrotis lagotis*)	5 295 921	946 026	82

[1] Historic range at the time of European settlement.
Source: Modified from Lindenmayer (2007a, p. 41).

magpie goose *Anseranas semipalmata*, although occupying a reduced range (about 50% of their historical range; Blakers *et al.* 1984), are still abundant and may even be re-establishing themselves in areas such as south-east Queensland.

Reptiles. Declines among reptiles are less dramatic than among mammals or birds. Nevertheless, 204 of Australia's 765 species require some level of remedial management to halt their declines (Cogger *et al.* 1993). The main drivers of decline are land clearing and the introduction of predators. For example, juvenile carpet pythons *Morelia spilota* have disappeared from many parts of their range due to fox predation (Michael and Lindenmayer 2008), and goannas across northern Australia are being severely impacted by the westward spread of the cane toad *Bufo marinus* (Ujvari and Madsen 2008). Species that have naturally small ranges, such as the western swamp tortoise *Pseudemydura umbrina* and the pygmy blue-tongue *Tiliqua adelaidensis* have been especially sensitive to environmental changes. Significant numbers of sea snakes are killed as by-catch in prawn fisheries (Ward 2000).

Amphibians. Abundances of amphibian species are in serious decline (Alford and Richards 1999), perhaps due in part to their requirements for at least two different habitats to complete their life cycle. The decline in frog species has been variously attributed to land clearing, water pollution and the introduction of predatory fish. There seems to be an emerging consensus that the fungal disease, chytridiomycosis, has played a major role in the decline and extinction of numerous species worldwide, including frog species mainly at high altitudes along the east coast of Australia. The increase in this disease has multiple drivers (Table 3.1, Principle 7), depending on location. Nevertheless, there is growing evidence that this global phenomenon is driven by both human disturbance and rising temperatures (Carey and Alexander 2003).

Fish. Similarly, the abundances of several Australian fish species are in decline. The Australian native freshwater fish fauna are particularly vulnerable, with only 144 species compared with 600 in North America, 2000 species in South America and 1500 in Africa. Most Australian native species of freshwater fish are

Table 3.4 Change in reporting rates of some Australian woodland birds during the past two decades.

Common name (scientific name)	% decline nationally
Flame robin (*Petroica phoenicea*)	51
Scarlet robin (*Petroica boodang*)	34
White-fronted chat (*Epthianura albifrons*)	34
Dusky woodswallow (*Artamus cyanopterus*)	33
Jacky winter (*Microeca fascinans*)	32
Diamond firetail (*Stagonopleura guttata*)	28
Crested shrike-tit (*Falcunculus frontatus*)	24
Varied sittella (*Daphoenositta chrysoptera*)	20
White-winged triller (*Lalage sueurii*)	17
Southern whiteface (*Aphelocephala leucopsis*)	13
Red-capped robin (*Petroica goodenovii*)	11

Source: Lindenmayer (2007a, p. 64).

endemic to coastal drainage areas. In the Murray–Darling Basin there are only 46 species, which possibly reflects the harsh, variable arid conditions that have prevailed within inland Australia. Unfortunately, these native fish populations are at less than 10% of pre-European levels, based on catch data over the past 100 years (Murray–Darling Native Fish Strategy 2003). Climate change could be an additional stressor to an already highly modified aquatic environment, particularly given the Murray–Darling Basin's prominence in agriculture with strong competition for water.

In contrast to declining trends in the above species, several other native species show increases in abundance and distribution over the past two centuries, probably as an unintended outcome of human activities. In agricultural and pastoral areas, increased food supply and/or watering points have produced increased numbers of red kangaroo *Macropus rufus* and western grey kangaroo *Macropus fuliginosus*, as well as some birds (e.g. the cattle egret *Ardea ibis*, galah *Eolophus roseicapillus*, little corella *Cacatua sanguinea* and crested pigeon *Ocyphaps lophotes*; James *et al.* 1999; Landsberg *et al.* 1999).

In peri-urban and urban areas, human developments have changed continuous native vegetation into extensive 'edge environments', favouring a suite of larger aggressive species that originally were restricted to previously relatively scarce 'edge' habitats. This is especially evident where forests have been cleared or fragmented, producing increases in bird species such as noisy miners *Manorina melanocephala*, bell miners *M. melanophrys*, blue-faced honeyeaters *Entomyzon cyanotis*, ibis (*Threskiornis* spp.), butcherbirds (*Cracticus* spp.), currawongs (*Strepera* spp.), rainbow lorikeets *Trichoglossus haematodus* and corellas (*Cacatua* spp.). Some species of butterflies have benefited by extensive street plantings of suitable food plants or even weed species (Braby 2000). Reptiles such as goannas (*Varanus* spp.), water dragons (*Physignathus lesueurii*) and some of the smaller skinks (e.g. the garden skink *Lampropholis delicata*) have thrived in the resource-rich suburbs of some cities. Similarly, the brushtail possum *Trichosurus vulpecula* and ringtail possum *Pseudocheirus peregrinus* have adapted well to human settlement.

In the arid zone of Australia, some native woody plant species such as *Acacia* spp. have increased in distribution and/or abundance over the past few decades (Gifford and Howden 2001). It is not clear to what extent this 'woody thickening' is related to more recent changes in fire regimes, changes in grazing pressure, long-term trends in precipitation, increased atmospheric CO_2 concentration or other factors (Chapter 5).

In addition to these semi-natural changes in the range of native species, the ranges of many species of Australian animals and plants have been expanded artificially through deliberate or accidental introductions outside their natural range. Celebrated examples include koalas *Phascolarctos cinereus* on Kangaroo Island, blue-winged kookaburras *Dacelo leachii*, rainbow lorikeets *Trichoglossus haematodus* and yabbies *Cherax destructor* in Western Australia, and superb lyrebirds *Menura novaehollandiae* in Tasmania. Less known are many instances where Australian native plants exotic to an immediate location have escaped from gardens and invaded the local landscapes (Low 2002). Tropical species such as the umbrella tree *Schefflera actinophylla* have been transported deliberately to subtropical areas. In general, native species that have moved into new areas have been classed as undesirable by local management agencies and treated accordingly (Chapter 6).

Introduced species

Technically, local biodiversity is increased when a new species enters an area in which it has not previously occurred or when an exotic species is introduced into an existing community. This includes the native species introduced deliberately. However, very rarely does a species enter a new area without significant effects on the biota already present. Introduced species compete for resources with indigenous flora and fauna. In general, a successfully introduced species is usually common and not at risk of extinction, as its abundance and distribution show a tendency for increase. In contrast, the displaced species are more likely to become increasingly uncommon and at risk, at least locally.

Over the past 220 years, Australians have deliberately introduced many plants and animals to the continent and released them into the wild, with good intentions at the time (Hobbs and Humphries 1995; Rolls 1969), but most often to the detriment of native biodiversity. Other species have arrived accidentally through inadequate quarantine. Even more have been introduced for horticultural or household purposes, or for agricultural purposes such as pasture improvement (Lonsdale 1994), but have become environmental weeds well beyond the region of their first introduction (Low 1999).

The most successful species introductions to Australia (e.g. foxes *Vulpes vulpes*, cats *Felis catus*, rabbits *Oryctolagus cuniculus*, goats *Capra hircus*, pigs *Sus scrofa*, rats (*Rattus* spp.), mice (e.g. *Mus musculus)*, water buffalo *Bubalus bubalis*, horses *Equus caballus*, cane toads *Bufo marinus*, trout (e.g. *Oncorhynchus mykiss* and *Salmo trutta)*, carp *Cyprinus carpio*, mosquito fish (*Gambusia* sp.) and other freshwater fishes, honeybees *Apis mellifera*, and a host of plants including prickly pear (*Opuntia* spp.), mimosa *Mimosa pigra*, gamba grass *Andropogon gayanus* and mission grass *Pennisetum polystachion*) have interacted negatively with native species by competing successfully for food and habitat or by direct predation, causing an overall decline in Australian biodiversity. Some of the introduced, naturalised species have direct and immediate effects on other biota by means of toxins and other chemicals (e.g. cane toads, *Parthenium* weed, cotton bush (*Asclepias* spp.)) (Everist 1974) or by causing disease (e.g. the water mould *Phytophthora cinnamomi*, which affects many plants, and the chytrid fungus *Batrachochytrium dendrobatidis*, which affects frogs).

Some introduced species, now naturalised, have changed physical environments to such an extent that the entire ecosystem is affected, including altering water flow and fire regimes, or penetration of light in water columns or to lower layers of vegetation in forests. These have significant secondary effects on native flora and fauna. Examples include lianas in the tropics; for example, rubber vine *Cryptostegia grandiflora*, gamba grass *Andropogon gayanus*, Asian water buffalo *Bubalus bubalis* and mimosa *Mimosa pigra* on the floodplains of the north; and water hyacinth *Eichhornia crassipes*, alligator weed *Alternanthera philoxeroides* and salvinia *Salvinia molesta* in freshwater systems.

Australian coastal waters have also received large numbers of exotic species. Outbreaks of the introduced aquatic aquarium plant caulerpa *Caulerpa taxifolia* in estuaries in New South Wales have successfully competed with native seagrass species, resulting in shading and dieback of seagrass meadows. Among the classic examples that have established over wide areas are the starfish *Asterias amurensis* from the north Pacific, the European shore-crab *Carcinus maenas*, the isopod *Synidotea laevidorsalis* and the invasive green alga *Codium fragile* subsp. *tomentosoides* (Byrne *et al.* 1997; Chapman and Carlton 1994; Thresher *et al.* 2003; Trowbridge 1996).

Although not often considered as introduced or invasive species, the domestic animals of agriculture, most notably cattle and sheep, have had the most profound impacts on the Australian landscape of any introduced species. Vast areas of the woodland, native grassland and rangelands of the continent have been cleared, fertilised and grazed. Pastoral activities change soil structure, water flows, fire regimes, soil compaction, nutrient regimes and light penetration. Impacts are dramatic and stretch from the below-ground microbial communities to many native mammals and woody plants. In addition, a range of exotic grasses introduced to improve pastures now dominate some landscapes. These transformed areas provide important services to Australians as well as export income to the economy, but have also driven many deleterious changes in the native biota.

Crops are also almost entirely introduced species. Arable lands have been modified to grow food crops and plantation forests, with changes in landscapes similar to those resulting from pastoral activities – with the addition of changes in topography and drainage, water use and salinisation, especially under

Box 3.8 The cane toad: an introduced species that has direct and indirect effects on Australian biota. (David Lindenmayer)

The cane toad *Bufo marinus* was introduced to Queensland in 1935 from Hawaii, where they were also an introduced species from South America, to control the beetle pests of sugar cane such as the greyback cane beetle *Dermolepida albohirtum*. From the initial release sites near Cairns, the cane toad is now established in large areas of Queensland, northern New South Wales and the Northern Territory, and is likely to invade the whole of tropical northern Australia (Amos *et al.* 1993). The rate of spread is about 25 km per year, but it is possible that it might reach other areas faster because of accidental translocations in shipments of food such as bananas (O'Dwyer *et al.* 2000). Although it is an example of a biological control program gone badly wrong, it also yields insights into how an introduced species affects whole ecological communities, as scientists monitor the cane toad's entry into new territory.

The cane toad was introduced without *any* research as to either its effectiveness as a control agent, or its potential impacts on native fauna (Low 1999). The cane toad has toxic skin secretions that can kill vertebrate predators including goannas, quolls and snakes. Phillips *et al.* (2003) estimated that up to 30% of Australia's terrestrial snake species may be at risk from the cane toad, as well as other reptile predators such as goannas. In addition, the early life history stages of the cane toad (eggs, tadpoles, hatchlings) are toxic to some aquatic predators (Crossland 2001; Williamson 1999). The cane toad failed to control the targeted greyback cane beetle, instead relying on a range of other prey, including native invertebrates and small vertebrates, and competing for food with predators such as snakes and lizards. It also eats bird's eggs and nestlings, such as those of the tunnel-nesting rainbow bee-eater *Merops ornatus* (Boland 2004).

Some of the native species affected by cane toads have shown remarkable and rapid adaptive responses. For example, Phillips and Shine (2004) reported that the mean body sizes of two snakes eaten by cane toads – the red-bellied black snake *Pseudechis porphyriacus* and the common tree snake *Dendrelaphis punctulata* – have increased in the 80 years since cane toads have begun spreading across Queensland, thus decreasing their vulnerability to cane toad predation. Snakes not eaten by cane toads did not show such a change in mean size. The capacity for rapid adaptive response will be paramount to survival of any population in the face of any rapid environmental change.

Cane toad.

Source: Lindenmayer (2007a).

irrigation. Many areas have been degraded beyond any reasonable hope of restoration or regeneration, with negative consequences for the actual or potential biodiversity. Again, these transformed areas provide important ecosystem services to Australians and provide income to the economy, but usually to the detriment of many native species.

The few notable exceptions to the negative effects of species' introductions on Australia's native biodiversity have been the release of some insects and microorganisms introduced as biological control agents against other introduced species; for example, cactus moth *Cactoblastis cactorum* against prickly pear (*Opuntia* spp.), myxomatosis against rabbits, and dung beetles *Onitis viridulus* and *Onthophagus taurus* against cattle ordure (Fenner and Fantini 1999; Waterhouse and Sands 2001). Almost every case of successful biocontrol has involved target organisms that are themselves exotic species occurring in highly modified environments. In a few of the unsuccessful cases, the intended control agents have became widespread and even more destructive pests than the target species; an example is the cane toad *Bufo marinus* (Box 3.8).

Most of Australia's introduced species are here to stay. Although many are under control in defined areas, full eradication has proved virtually impossible once the organisms have become established. Eradications have been possible on some offshore islands (Burbidge and Morris 2002) and with some insect species in highly circumscribed mainland areas. Examples of mainland eradications include the pawpaw fruitfly *Bactrocera papayae* in north Queensland (Cantrell *et al.* 2002) and isolated outbreaks of pest ant species in Kakadu (Hoffmann and O'Connor 2004). Major control programs are underway for the red imported fire ant *Solenopsis invicta* in the Brisbane area (Molony and Vanderwoude 2008) and for the yellow crazy ant *Anoplolepis gracilipes* on Christmas Island (Threatened Species Scientific Committee 2005), both of which threaten indigenous wildlife. In all cases these eradication and extensive control attempts have been, and continue to be, very expensive.

3.4.3 Community- and ecosystem-level changes

Changes occur in species interactions as responses to environmental change, and usually have knock-on effects to whole communities and ecosystems. These higher-order changes range from direct species–species interactions – such as mutualism, competition and predation – to changes in the ways in which species influence the structure and functioning of ecosystems, including cascading impacts through ecosystems, and the formation of novel communities and ecosystems (Table 3.1, Principles 4, 5 and 6).

Species–species interactions

Mutualism. Introduced species can affect mutualistic relationships among native species. For example, the introduction of honeybees *Apis mellifera* into Australia many decades ago interfered with the pollination of some native plants, and may have further led to a reduction in the availability of nectar and pollen for native animals such as native bees (Pyke 1990). One consequence for some plants may be an increase in inbreeding due to honeybee pollinators negatively affecting bird pollination which, relative to bee pollination, promotes outcrossing (Stankowski *et al.* in review). Honeybees have also increased competition for tree hollows with native mammals and birds, as bee nests are constructed and defended (Emison 1996; Harper *et al.* 2005).

Competition. Successful invasive species usually displace previously thriving native species by outcompeting them for space, food, habitat or other limiting resources. Invading plant species are often highly productive and fast growing, readily out-competing a wide range of less productive, slower growing native species and reducing total community biodiversity in consequence (Hobbs and Humphries 1995). Some of the best documented examples in Australia have been in the north (Cowie and Werner 1993), where it has been possible to monitor the impacts of invading plant species during their expansion phases. Mimosa *Mimosa pigra* out-competed floodplain vegetation through rapid growth, vegetative reproduction and a persistent seed bank (Lonsdale *et al.* 1989). European carp *Cyprinus carpio* successfully competes with freshwater native species. Many introduced bird, mammal, reptile and amphibian species have successfully outcompeted their native Australian counterparts (Hobbs and Humphries 1995).

Predation. Predators play significant roles at several levels. They have direct effects on their prey populations, compete with other predators for prey, mediate competition and relative abundances among their prey, and can have significant effects on community structure and ecosystem functioning (Table 3.1, Principle 5). The most successful introduced predators to Australia are 'generalists' such as the fox *Vulpes vulpes* and the cat *Felis catus*, which eat a wide variety of prey species and do not have specific habitat requirements. Foxes and cats are the main cause of most species' collapses in the arid zone of Australia (Jarman 1986; Kinnear *et al.* 2002).

Dingos *Canis lupus dingo*, a subspecies of the wolf, consume macropods, birds, pigs, wombats, buffalo, domestic stock, rabbits, reptiles and rodents (Corbett 2001; Newsome 2001), and probably play a role in limiting kangaroo numbers (Caughley *et al.* 1980). Dingos (Fig. 3.12) came to Australia from the north, never reached Tasmania, and most probably displaced through extinction thylacines *Thylacinus cynocephalus* and Tasmanian devils *Sarcophilus harrisii* from the mainland. Some philosophical debate remains as to whether dingos should be regarded as exotic animals (based on their relatively recent – within the past 6000 years – human-assisted arrival on the continent) or

Figure 3.12 Dingo *Canis lupus dingo*, Fraser Island, Queensland, Australia. Source: AUSCAPE. Photo by Dave Watts.

should be conserved as part of our threatened native fauna. Today they most likely play a role as a keystone predator, mediating competitive relationships among an array of prey species (Table 3.1, Principle 5.)

Interactions between dingos and other mammalian predators, including foxes and cats, are complex (reviewed by Glen and Dickman 2005). In some regions dingos compete with, and suppress, populations of these other two species (Glen and Dickman 2005), with indirect positive effects on populations of native species of prey, including large kangaroos (Daniels and Corbett 2003; Newsome 2001; Short and Smith 1994; Smith and Quin 1996). When dingos are removed from a region, foxes increase in number and, if both dingos and foxes are removed, then feral cat numbers increase (Johnson 2007; Linnell and Strand 2000).

Ecological cascades

The removal or addition of a key species has effects far beyond the other species with which they interact directly (Table 3.1, Principle 5). Second- and third-order effects create ecological cascades whereby entire food webs and ecosystems are substantially modified (Table 3.1, Principle 6). A classic example begins with the native red land crab *Gecarcoidea natalis* on Christmas Island, considered an ecological engineer species (Fig. 3.13). These crabs normally influence the entire ecosystem by modifying the soil, litter dynamics and the seedling bank in the plateau rainforests (O'Dowd and Lake 1989, 1990). However, supercolonies of the introduced species yellow crazy ant *Anoplolepis gracilipes* have wiped out red crab populations in large areas of the island. As a result, normally suppressed tree seedlings have burgeoned on the forest floor and furthermore, scale insects, mutualists of *Anoplolepis*, have infested the canopy trees in large numbers. The entire ecosystem, together with its characteristic endangered species such as Abbott's booby *Papasula abbotti* and the coconut crab *Birgus latro*, is undergoing a series of interrelated changes due to the invasion of a single species of exotic ant and the consequent increase in scale insects.

Ecological cascades are often centred on large introduced grazing mammals, including cattle and sheep, which compact soil, redistribute water flows, and preferentially feed on particular plants, thus changing the composition of plant assemblages and the amount of biomass within the ecosystem. These changes, in turn, have effects on a host of smaller native animals. Other physical changes such as reduced fuel loads in grazed areas affect fire regimes, which in turn have consequences for native tree and vertebrate populations. One such example, in a natural landscape, involves the introduced Asian water buffalo *Bubalus bubalis* in Kakadu National Park, wherein different ecosystems had different degrees of response and resilience, and which involved two ecological cascades (Box 3.9).

Novel communities and ecosystems

Over geological time scales, the constant mixing and matching of species dispersing into new areas

Figure 3.13 The red crab *Gecarcoidea natalis*, an ecological engineer species on Christmas Island, is declining because it is being killed by supercolonies of the introduced yellow crazy ant *Anoplolepis gracilipes*. Source: Andrew Burbidge.

Box 3.9 Ecological cascades in Kakadu National Park. (Patricia Werner)

Asian water buffalo *Bubalus bubalis* were introduced into northern Australia and dominated the natural landscape for 25 years, up to 1982. The buffalo were subsequently subjected to an eradication campaign, which rapidly removed more than 95% of the animals from the area within a few years (Skeat *et al.* 1996). The non-linear responses, cascades and potential for novel ecosystems described below are relevant to Principle 8 in Table 3.1.

The rapid increase in buffalo numbers to carrying capacity in the late 1970s produced effects at many levels, as reviewed by Skeat *et al.* (1996). These changes in fact made up an ecological cascade, or chain of events caused by the introduction of a single species (Petty *et al.* 2007). The impacts included:

- saltwater intrusion and death of freshwater floodplain species where levees were broken by buffalo activities
- changes in the abundance and relative composition of floodplain plant species throughout all the floodplain areas
- as a consequence of vegetation changes, a reduction in goose nest sites and decreases in some small mammal populations
- reduced total herbaceous vegetation and fuel loads in uplands wooded savanna
- a shift towards annual plant species, as the buffalo preferentially ate perennial species
- increased growth of established upland trees – as competitors were removed by the buffalo – but reduced recruitment into the canopy by juvenile trees
- as a consequence of vegetation changes and fuel loads, changes in size structure of the woody component of upland communities.

The eradication campaign later initiated a second ecological cascade of interacting responses to the withdrawal of the large grazer. Different ecosystems experienced the two cascades somewhat differently, especially regarding resistance to change with buffalo increases, and resilience or recovery after buffalo were removed. The floodplains had not been very resistant to change with the buffalo introduction, but they were quite resilient, recovering quickly to their previous state of productivity, species abundance and distributions. Comparatively, the dry upland savanna woodlands, which had been quite resistant to change when the buffalo were first introduced, experienced strong hysteresis effects, with time delays in the grassy understorey; this in turn affected fuel loads and fire regimes and, ultimately, tree regeneration. Due to long generation times of the woody component, it is not known whether these changes are reversible or if the system is on a new trajectory to an alternative community (Petty *et al.* 2007).

Water buffalo bull, Kakadu.

Source: Dave Lindner.

A photograph taken at Kapalga, within Kakadu National Park, in 1984 contrasting a buffalo-excluded area (right) with an area with buffalo.

Source: photo by Don Tulloch, provided by Patricia Werner.

generated novel communities and ecosystems (Davis 1984; Delcourt and Delcourt 1991; Wright 1984). The floodplain ecosystems of Kakadu National Park are geologically less than 6000 years old, and their plant associations less than 2000 years old (Chappell and Woodroffe 1985; Woodroffe *et al.* 1987). The Great

Box 3.10 Novel ecosystems. (David Lindenmayer)

Much of the current literature on biodiversity emphasises species extinctions resulting from landscape transformation and other factors (e.g. Primack 2001; Thomas *et al.* 2004; Ward 2004). However, some landscape transformations do not simply result in extinctions (Lindenmayer and Fischer 2006), but lead to the genesis of 'novel ecosystems', which are those containing 'new combinations of species that arise through human action, environmental change, and the impacts of the deliberate and inadvertent introduction of species from other regions'. Species may 'occur in combinations and relative abundances that have not occurred previously within a given biome' (Hobbs *et al.* 2006, p. 1; see figure).

Novel ecosystems present a challenge to some of the traditional thinking in conservation ecology, such as the focus on maintaining species abundances or ranges as the most appropriate response to landscape transformation (Table 6.1). They also challenge our values regarding novel ecosystems and raise scientific, moral and ethical dilemmas for resource management and policy making. For example:

- Where are novel changes acceptable or unacceptable?
- What can we do about changes that are deemed unacceptable?
- What kinds of ecosystem services might novel ecosystems provide?

Novel ecosystems also force us to examine concepts of resistance and resilience – what is resilience when the underlying environment is on a trajectory of change so that the system cannot be returned to a former state? New management tools and methods will be needed to address the concepts of resilience and transformation. This is discussed further in Chapter 7.

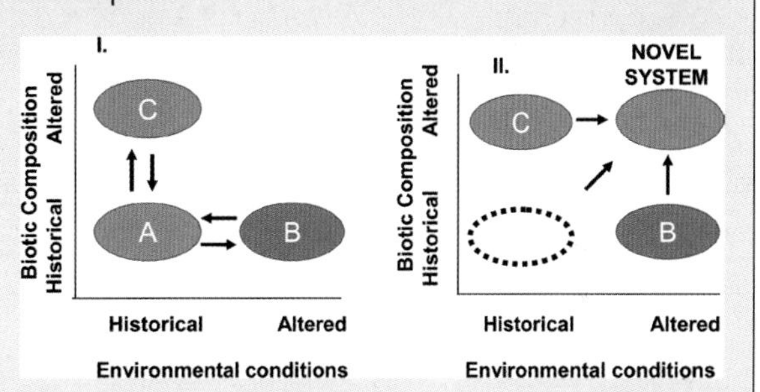

Formation of novel ecosystems. I. An ecosystem is altered by directional environmental drivers (A→B), or the addition or loss of an important species (A→C). II. Internal restructuring due to new biotic and abiotic interactions further alters community composition via changes in species abundances or species losses, and through changes in biogeochemical interactions.

Source: Adapted from Seastadt, Hobbs and Suding (2008).

Barrier Reef is probably no more than 6000–10 000 years old in its current form (although it is built on earlier reefs in many places) (Chappell *et al.* 1983; Veron 2008).

Since the European settlement of Australia, other novel communities have been produced, primarily as the result of invasions of new species (Hobbs *et al.* 2006), changes in fire regimes, and landscape transformations (Lindenmayer and Fischer 2006). Most of these recently formed novel communities are 'weedier', more open, with fewer interdependent relationships among species and less stability than those adapted to slow changes in climate, vegetation cover and fire regimes (Hobbs *et al.* 2006; Lindenmayer and Fischer 2006). Given the increased frequency and rate at which novel ecosystems are arising globally (e.g. Estades and Temple 1999; Gascon *et al.* 1999; Hobbs *et al.* 2006; Wethered and Lawes 2005; Williams and Jackson 2007), new methods and ways of thinking will be required to address biodiversity vulnerability in the face of climate change (Box 3.10). These issues are addressed in Chapters 6 and 7.

3.5 AUSTRALIA'S BIODIVERSITY UNDER PRESSURE

At the beginning of the 21st century, Australia's biodiversity remains under considerable pressure. Our continent's rate of extinctions is high in comparison with most other parts of the world, and many more species are on trajectories towards extinction. The flow-on effects of these changes in species diversity to the structure and functioning of ecosystems is equally serious. The proximate drivers of these changes – primarily the

Figure 3.14 Australia's 15 terrestrial 'biodiversity hotspots' as identified by a joint Commonwealth–state government study. Source: Department of Environment, Water, Heritage and the Arts.

transformation and modification of landscapes, and the introduction of exotic species – have the potential to continue well into this century. The ultimate and global drivers of these changes are the same activities that are driving human-induced climate change, and are likely to intensify over the coming decades at least.

In Chapter 2, it was stressed that biodiversity is not evenly spread, either worldwide or in Australia. In landmark studies, Norman Myers and colleagues (Myers 1988; Mittermeier *et al.* 1999; Myers *et al.* 2000) identified 25 'biodiversity hotspots' comprising only 1.4% of the land surface of the Earth, where as many as 44% of all species of vascular plants and 35% of all species in four vertebrate groups are confined and considered threatened with destruction (http://www.biodiversity-hotspots.org). The number of hotspots was later expanded to 34 (Mittermeier *et al.* 2005). One of the original 25 hotspots occurs within Australia – southwest Australia.

A study of Australian terrestrial biodiversity, using similar methodology to Myers *et al.* (2000) identified 15 'biodiversity hotspots' (i.e. areas with high species diversity that are also under threat; Fig. 3.14) (http://www.environment.gov.au/biodiversity/hotspots/index.html). Five of these lie within the global biodiversity hotspot of southwest Australia (Box 3.11). Concentrating limited conservation resources in these hotspots will ensure maximum cost-benefit.

Chapters 4 and 5 assess the vulnerability of biodiversity to the growing challenge of climate change. It is imperative that climate change is not viewed in isolation from the current status and trends in our biodiversity, and reasons behind these trends, which have been described in depth in this chapter. Although the future may appear bleak for Australia's biodiversity, the threat of climate change may provide an opportunity for new resolve and new approaches to secure our biotic heritage for the future. Chapters 6 and 7 explore some innovative policies, strategies and tools that Australian society can usefully employ to turn around the current downward trend in the status of our biodiversity. These come together in the final chapter, a summary of key messages and proposed policy directions.

Box 3.11 South-western Australia: a global biodiversity hotspot under stress. (Andrew A Burbidge)

Biodiversity

The south-west of Western Australia, a flat, stable, highly weathered low plateau dominated by old landscapes with nutrient-deficient soils, is the only designated global biodiversity hotspot in Australia (Mittermeier *et al.* 2005). Of the 25 such areas worldwide, few are in developed countries. There are more than 7400 named plant taxa and an estimated 6500 species of vascular flora in the south-west, of which >50% occur nowhere else (Hopper and Gioia 2004). Local species diversity and endemism is also high. For example, in Fitzgerald River National Park (329 000 ha) there are 1883 plant taxa of which 72 occur only in the park, while Tutanning Nature Reserve (2000 ha) has 850 species. Approximately 20% of plant species are listed as threatened, rare or poorly known; most threatened endemic taxa are woody species. Most plant species show very high levels of genetic diversity and population differentiation (Byrne 2007; Coates 2000) due to local persistence through geological time and natural fragmentation as a result of environmental heterogeneity. Many taxa have very small geographic ranges. About 100 species of vertebrates are endemic, including the honey possum *Tarsipes rostratus*, quokka *Setonix brachyurus*, red-capped parrot *Purpureicephalus spurius*, western swamp tortoise *Pseudemydura umbrina* and sunset frog *Spicospina flammocaerulea*. Some species that have become extinct elsewhere in Australia, such as the numbat *Myrmecobius fasciatus*, persist in the south-west. Many invertebrates occur nowhere else. Sixty-three wetlands of national significance are located in the south-west region. In 2007, 82 threatened ecological communities, 351 threatened plant taxa (111 of which are critically endangered) and 69 threatened non-marine animal taxa (16 mammals, 19 birds, 11 reptiles, 3 frogs, 3 fish and 17 invertebrates) were in the south-west.

Near the south coast, many species and some ecosystems are restricted to hilltops. In the Stirling Range National Park, 20 threatened plant taxa, two threatened animal species and one threatened ecological community are found only at high elevations. There are numerous Gondwanan relictual species in the south-west corner of the south-west, which indicate the region's role in harbouring organisms having an ancient genetic lineage. Almost all are dependent on mesic conditions. Many species of aquatic invertebrates are restricted to threatened ephemeral freshwater and brackish water ecosystems.

Because it has a Mediterranean climate, fire is a major component of south-west ecosystems.

Land use and threats

The south-west of Western Australia is the source of significant economic wealth, being Australia's largest cereal cropping area and having major mineral deposits including bauxite, gold and manganese. Other land uses include timber harvesting and tourism. The state's capital, Perth, lies within the south-west region, and urban expansion and tourism are having a negative impact on coastal and near-coastal lands. Clearing for agricultural and urban use has been extensive. In the Avon Wheatbelt 2 Interim Biogeographic Regionalisation for Australia (IBRA) subregion, 93% of land has been cleared. Small remnants make conservation management difficult and limit species movement. Less than 2% of this region is in protected areas. Salinisation is a major threat to both agricultural production and biodiversity conservation. Catchment management, revegetation and other new land use practices are increasingly important. In some areas it may become uneconomic to continue traditional farming practices, such as cereal cropping, due to reduced winter rainfall and increased costs. Further development of alternative crops, including those based on indigenous biodiversity, is needed. Some marginal land may be abandoned. Many plants and ecosystems are threatened by the introduced water mould *Phytophthora cinnamomi*. Many animals are threatened by introduced predators. Environmental weeds are an increasing threat to ecosystems and species. The opportunity to expand the protected area system from publicly owned land is limited; expansion by protecting private land is a developing concept.

Numerous translocations of both plants and animals have been carried out in the south-west and, building on this experience, translocations to adapt to climate change will become important. The development of vegetation corridors linking the relatively well-vegetated higher rainfall south-west corner and south coast with inland semi-arid and arid lands has been proposed, and is underway (e.g. Gondwana Link). Monitoring to evaluate the success of this strategy is vital.

The south-west has suffered a 10–20% decrease in winter rainfall over the past 30 years together with temperature increases, especially in winter and autumn (CSIRO and BOM 2007). This has resulted in sharp declines in both the quantity and quality of water flows (Fig. 4.13). The decline in precipitation was not gradual, but occurred quite abruptly in the mid-1970s. Groundwater-fed karst invertebrate communities, for example in the Yanchep Caves and on the Leeuwin-Naturaliste Ridge, are already under extreme threat from lowering groundwater.

Western swamp tortoise *Pseudemydura umbrina*, one of more than 100 species of vertebrates that are endemic to megadiverse south-west Australia.
Source: © Bert & Babs Wells/DEC.

4 The rate and magnitude of climate change

This chapter summarises climatic trends observed over the past few decades, both globally and in Australia, and describes the rate and magnitude of potential change over the next century.

4.1 THE NATURE OF CONTEMPORARY CLIMATE CHANGE

Climate change is altering the fundamental abiotic environment in which biological species and communities exist. Understanding the nature of climate change, from the human-driven changes that are observable now to the long-term patterns of variability within which contemporary ecosystems have developed, is essential for assessing the vulnerability of Australia's biodiversity to the rapidly changing environment of the 21st century.

The science of climate change has progressed significantly over the past two decades. The most recent report of the Intergovernmental Panel on Climate Change (IPCC) in 2007 concluded that the warming of the climate system over the past century is unequivocal. Figure 4.1 shows the changes in the mean surface temperature over the past 150 years. Global average temperatures have increased 0.74°C (1906–2007). Twelve of the 13 years in the 1995–2005 period rank among the 13 warmest years in the instrumental record since 1850. Warming has occurred across the globe but has been greatest in the northern high latitudes (IPCC 2007a).

The IPCC (2007a) concluded that it is *very likely* that anthropogenic (human-induced) greenhouse gas increases have caused most of the observed increase in globally averaged temperatures since the mid-20th century. The most important of the anthropogenic greenhouse gases in terms of its effect on climate is carbon dioxide (CO_2). The longest continuously monitored CO_2 site in a non-industrial area, on top of Mauna Loa, Hawaii, has documented increases from below 315 ppm to above 380 ppm in just over the past 50 years (Fig. 4.2). The current concentrations of CO_2 in the atmosphere far exceed pre-industrial values, rising from 280 ppm in 1750 to over 385 ppm by 2008, with 70% of the increase occurring since 1970 (IPCC 2007a).

Global average sea levels are rising at a rate consistent with the warming trends, increasing 1.8 mm/year in the period 1960–1983 and 3.1 mm/year since that time. Thermal expansion, melting glaciers and polar ice sheets have all contributed to the rise. It is unclear whether the apparent acceleration in rate since 1983 is due to decadal variability or to a longer-term trend (IPCC 2007a). The measurements of sea level over the past decade lie above the envelope of uncertainty described in the IPCC scenarios (Fig. 4.3). However, the model-based projections to date do not take into account the dynamic changes in the large polar ice sheets in Greenland and Antarctica (Oppenheimer *et al.* 2007).

The rate of change in the climate system over the past two decades is striking. In fact, CO_2 concentrations are rising faster than all previous scientific projections, including those published by the IPCC in

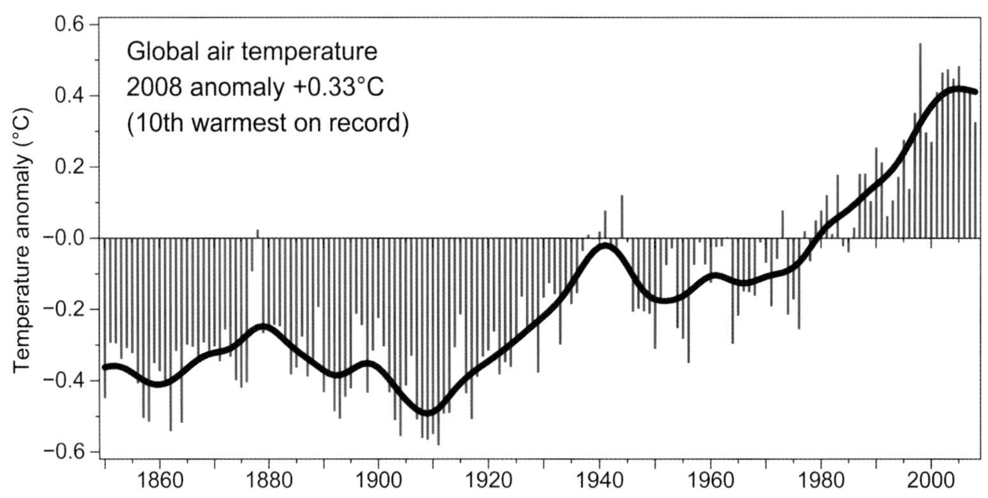

Figure 4.1 Global air temperature anomaly. The vertical bars are the annual mean surface temperature anomaly relative to an average temperature for the reference period of 1961–1990. The black line is a 20-year Gaussian (binomial) filter. Source: Climatic Research Unit (CRU) (2009).

2007 (IPCC 2007a), having almost tripled since the 1990s from 1.1% per annum to 3.1% in the 2000s (Fig. 4.4). During this period, global mean temperature (Fig. 4.5) was tracking within the envelope of IPCC projections, and sea level (Fig. 4.6) was tracking at the upper bounds of the IPCC projections. These observations confirm that the mean projected temperature increases published in the IPCC assessment (IPCC 2007a) are likely to be realised (Canadell *et al.* 2007; Raupach *et al.* 2007), and that the IPCC's projected rise in sea level, assuming that the relationship between increases in CO_2 concentration and sea-level rise observed over the past century continues, is likely to be an underestimate (Rahmstorf 2007).

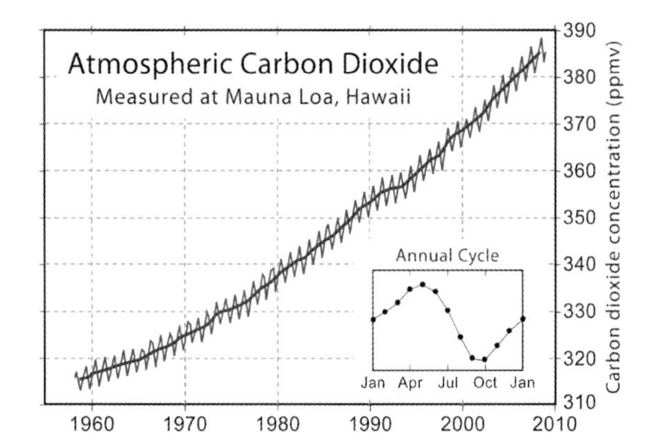

Figure 4.2 Atmospheric CO_2 concentration from the late 1950s to 2008, measured on top of Mauna Loa, Hawaii. Source: Robert A Rohde (Global Warming Art) using NOAA/ESRL data (http://www.esrl.noaa.gov/gmd/ccgg/trends/).

4.2 A LONG-TERM PERSPECTIVE ON CLIMATE CHANGE

4.2.1 Temperature

The mean temperature and CO_2 concentrations of the Earth have been positively correlated over geological time. During the past 2.5 million years, the rate of change during successive cooling (glacial) periods has been slow compared with the rate of subsequent warming. For example, during the most recent ice age, both temperature and CO_2 concentration declined steadily for more than 100 000 years but were restored to interglacial values within the relatively short succeeding time span of 5000–10 000 years (Lorius *et al.* 1990) (Fig. 4.7).

Today, increases in both atmospheric CO_2 concentration and global mean temperature are occurring at rates unprecedented in over 10 000 years at least (IPCC 2007a). Furthermore, the present value of atmospheric CO_2 concentration is higher than at any time in the past 650 000 years, as measured by polar ice cores. Indeed, the current atmospheric CO_2 concentrations are far above the approximately 300 ppm found during the warm period just prior to the most recent glacial period (IPCC 2007a).

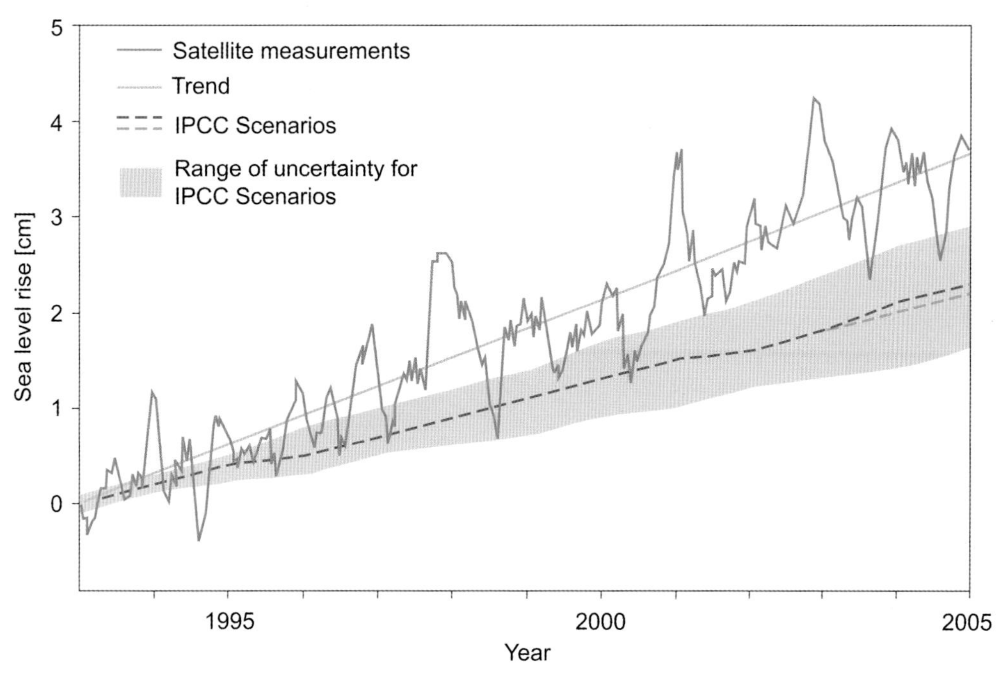

Figure 4.3 Actual vs projected sea-level rise for the recent 10-year period 1995–2005. Actual mean sea level from satellite measurements (blue) with trend line (red) are well above the IPCC average for all scenarios (black) as well as the entire range of various IPCC scenarios (brown shadow). Source: Derived from Cazenave and Nerem (2004).

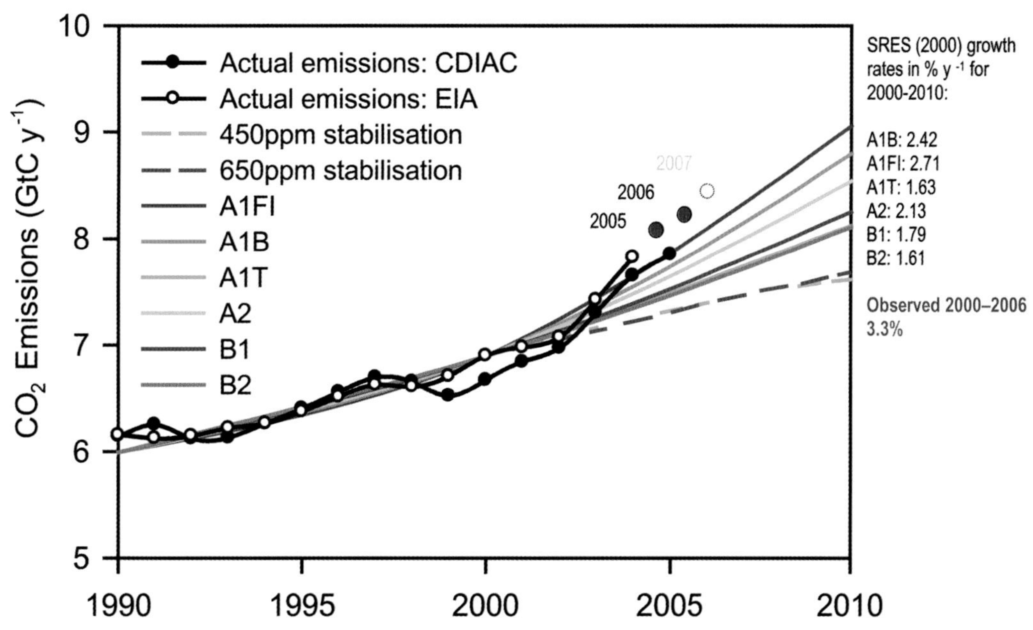

Figure 4.4 Observations of anthropogenic CO_2 emissions from 1990 to 2007. The envelope of IPCC projections is shown for comparison. Source: Raupach *et al.* 2007 with additional data points from Canadell *et al.* (2007) and Global Carbon Project annual carbon budgets; Copyright 2007 National Academy of Sciences, U.S.A.

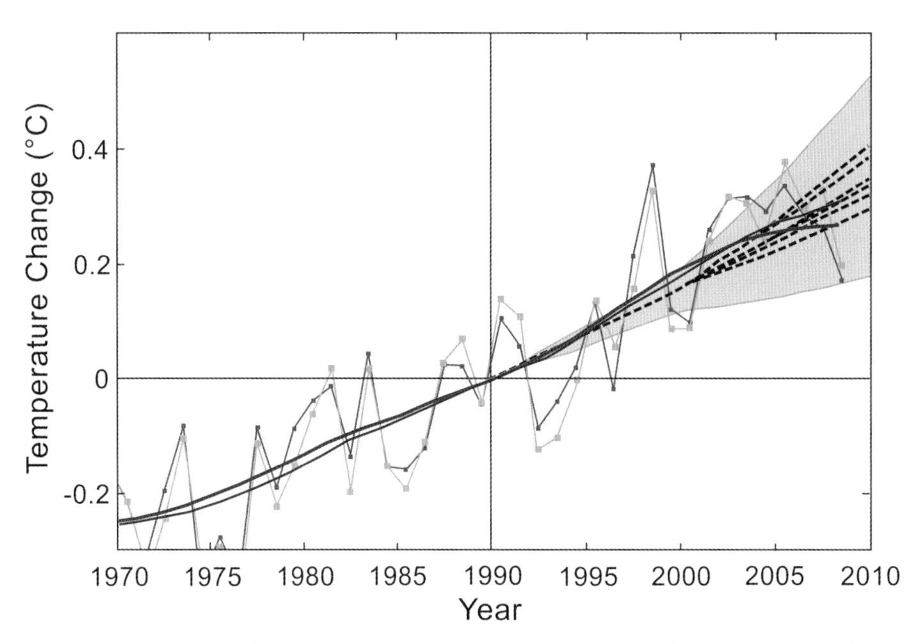

Figure 4.5 Changes in global mean surface temperature (smoothed over 11 years) relative to 1990. The blue line represents data from Hadley Center (UK Meteorological Office); the red line is GISS (NASA Goddard Institute for Space Studies, USA) data. The broken lines are projections from the IPCC Third Assessment Report, with the shading indicating the uncertainties around the projections. Source: Rahmstorf *et al.* (2007) with data for 2007 and 2008 from S Rahmstorf; Reprinted with permission from AAAS.

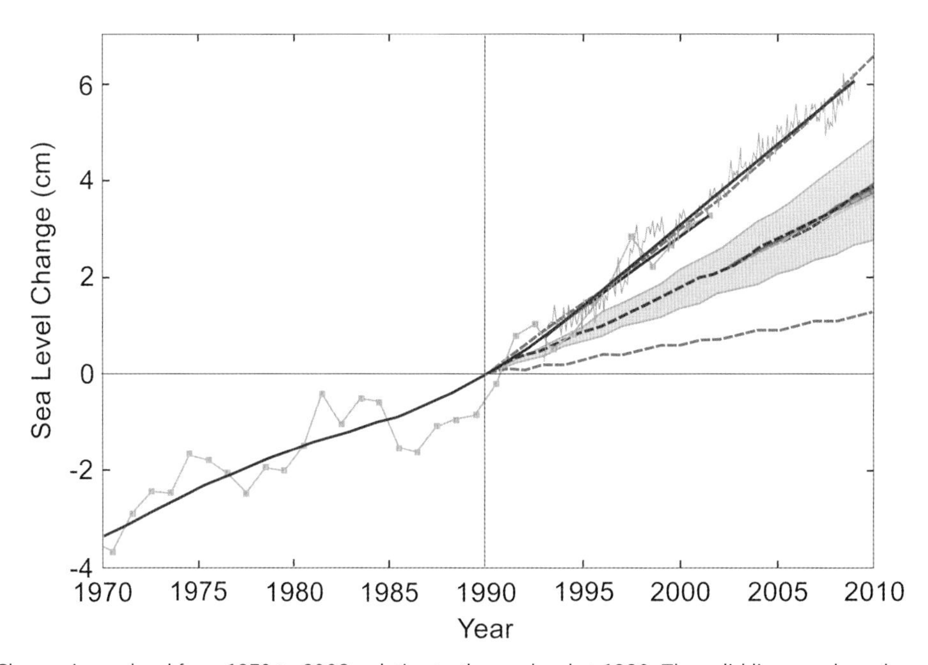

Figure 4.6 Change in sea level from 1970 to 2008, relative to the sea level at 1990. The solid lines are based on observations, smoothed to remove the effects of interannual variability. The envelope of IPCC projections is shown for comparison (broken lines with grey shading showing the uncertainty levels). Source: Rahmstorf *et al.* (2007), based on data from Cazenave and Narem (2004); Cazenave (2006) and A Cazenave for 2006–2008 data; Reprinted with permission from AAAS.

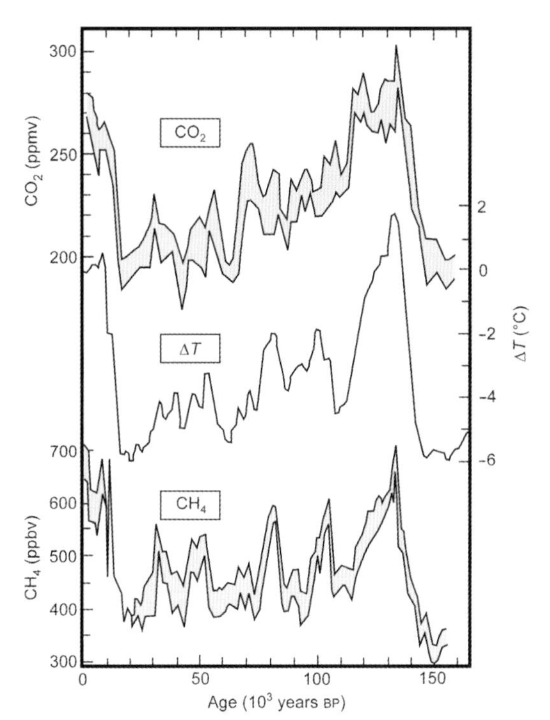

Figure 4.7 Change in global mean temperature and atmospheric concentrations of the greenhouse gases carbon dioxide (CO_2) and methane (CH_4) from 150 000 years b.p. to approximately 1800 AD. Data from Lorius *et al.* (1990) as modified by Cox and Moore (1993). Note that today the concentration of atmospheric CO_2 is over 380 ppm, off the scale.

Global mean temperature is also beginning to rise above the patterns of natural variability. Figure 4.8 shows a near-2000 year reconstruction of northern hemisphere surface temperature with the instrumental record of the past 150 years superimposed (red line). This puts the observed 0.7°C increase in global mean temperature over pre-industrial values into a longer time perspective. Unfortunately, an equivalent record does not yet exist for the southern hemisphere or for Australia. The figure shows that anthropogenic climate change is moving the Earth system out of the envelope of natural variability that the world's ecosystems have experienced over the past two millennia at least, and probably much longer (Alverson *et al.* 2003; Steffen *et al.* 2004).

4.2.2 Sea level

Sea level has changed significantly during geological time (Fig. 4.9). During the Last Glacial Maximum, with global mean temperature <10°C, the sea level was >100 m lower than today. During the Eocene epoch (53 million to 37 million years b.p.), with global mean temperature >18°C, sea level was >50 m higher than today. These extreme values, with intermediate values, form almost a straight line, suggesting the IPCC sea-level rise projection of up to about 1 m by 2100 may be

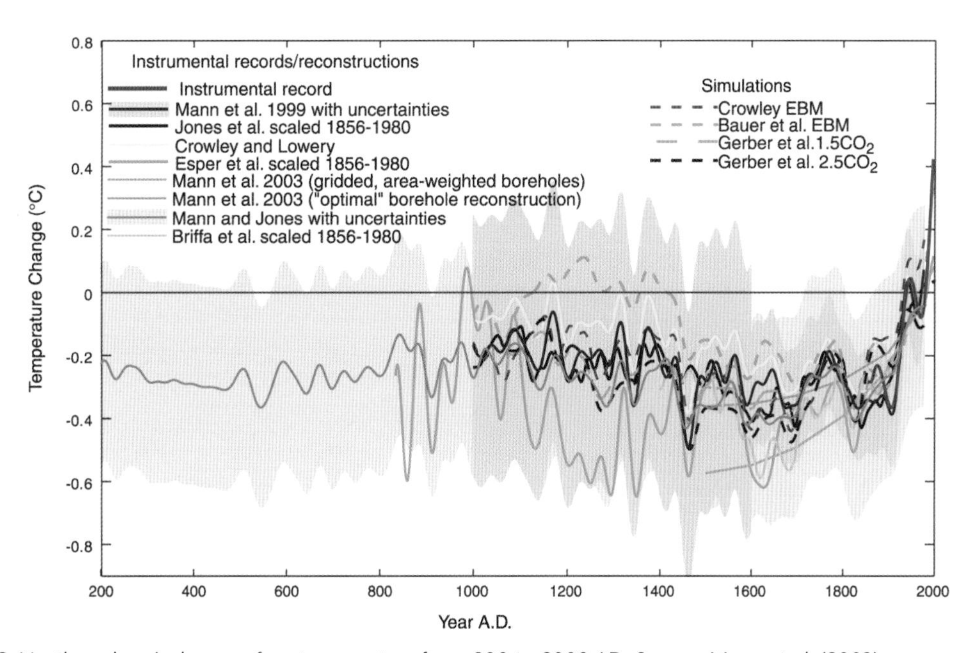

Figure 4.8 Northern hemisphere surface temperature from 200 to 2000 AD. Source: Mann *et al.* (2003).

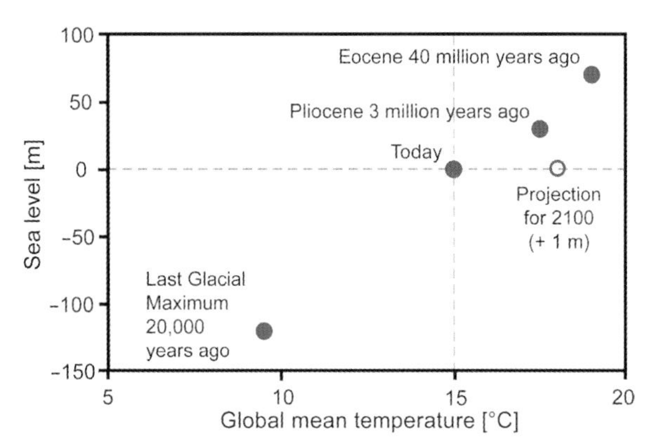

Figure 4.9 Actual vs projected sea level relative to global mean temperature over geological time. Source: German Advisory Council on Global Change (WBGU) (2006), after Archer (2006).

far too low in light of its own projected temperature change. However, the values of sea-level rise in the past are equilibrium values. Sea-level rise lags rapid climate change by centuries or millennia; once contemporary climate change and sea-level rise have eventually equilibrated, the point may lie much closer to the linear relationship.

4.3 PROJECTIONS FOR FUTURE GLOBAL CLIMATE CHANGE

It is very likely that the climate system will continue to warm through the rest of the 21st century and beyond, with associated changes to wind patterns, precipitation, sea level, extreme events, and many other aspects of weather and climate (IPCC 2007a). Figures 4.10 and 4.11 show the range of model projections of global mean temperature through the rest of the century. Three features are particularly important:

- The momentum in the climate system means that the Earth is committed to a further warming of at least 0.4°C regardless of human actions.
- There is a high probability that the Earth will warm beyond the 2°C level (compared with pre-industrial levels), which is sometimes considered to be the threshold of 'dangerous climate change' (van Vliet and Leemans 2006).
- Significantly higher temperature rises cannot be ruled out, and will become more probable if deep cuts in global emissions of greenhouse gases cannot be achieved in the next decade or two. The current global emissions trajectory is tracking at or near the upper limit of the IPCC suite of projections (Fig. 4.4; Raupach *et al.* 2007), increasing the risk that we will exceed a 2°C rise in global mean temperature during this century.

From the perspective of impacts on biodiversity, a crucial point is that the *rate* of projected climate change is almost certainly unprecedented for the last several million years (Steffen *et al.* 2004). For example, the transition from the Last Glacial Maximum to the present (Holocene) climate took at least 5000 years (Petit *et al.* 1999), compared with a projected change of similar magnitude this century. Within the perspective of the past 1000 to 2000 years, projected climate change will push the environment well beyond the range of natural variability to which ecosystems are adapted (Fig. 4.12). Selective pressure on many organisms, particularly long-lived organisms, will be extreme – especially as the incidence and severity of extreme events will increase even more rapidly than climatic means. Such dramatic rates of change have prompted some assessments to suggest that the Earth will experience a massive wave of extinctions this century, with rates of species loss about 1000 times background levels (MA 2005).

4.4 CLIMATE CHANGE IN AUSTRALIA

The observations and projections of temperature increase over Australia are similar to those of the Earth as a whole. That is, the rate of warming from the time of the Last Glacial Maximum (20 000 years b.p.) was several orders of magnitude slower than the rate of increase in temperature experienced over the past half century and also far slower than the expected rate of change from today to 2100.

4.4.1 Recent climatic trends in Australia

Australian average temperatures on land have increased 0.9°C since 1950, although with significant regional variations (CSIRO and BOM 2007) (Fig. 4.13). Minimum temperatures have been increasing faster than maximum temperatures (Nicholls 2006). The rate of

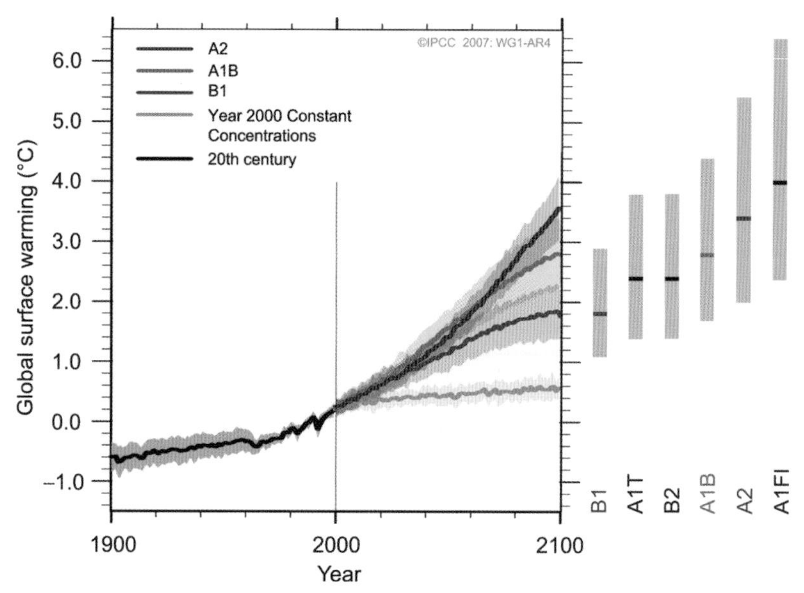

Figure 4.10 Projections of global mean temperature to 2100 (relative to 1980–1999). The projections are shown as continuations of the 20th century model simulations of IPCC (2007a) for various emission scenarios (see Box 3.1). The black line refers to actual measurements during the 20th century, and the gold line refers to results where greenhouse gas concentrations were held constant at 2000 values. The shaded areas around projection lines are plus/minus one standard deviation range of individual model annual averages. The likely range of warming by the year 2100 for various IPCC modelled scenarios plus results from independent models and observational constraints, with the mean warming value (solid line within each bar), are shown by the grey bars to the right of the graph (IPCC 2007a). Source: Modified from figure (IPCC 2007a, p. 762); http://www.ipcc.ch/pdf/assessment-report/ar4/wg1/ar4-wg1-chapter10.pdf.

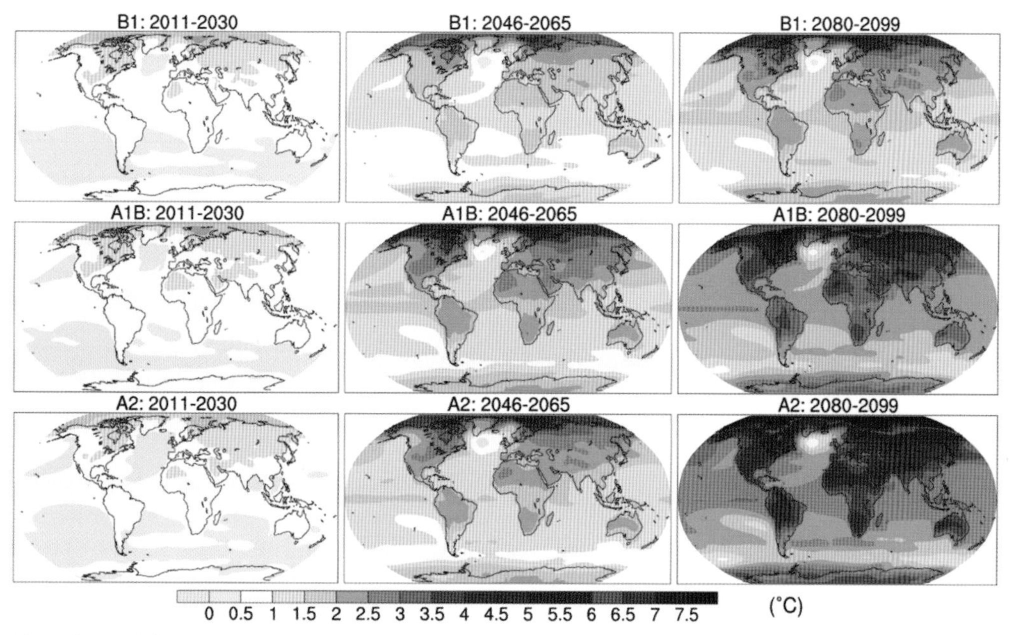

Figure 4.11 Multi-model mean of annual mean surface warming. Surface warming (surface air temperature change, °C) is given for the scenarios B1 (top), A1B (middle) and A2 (bottom), and three time periods 2011–2030 (left), 2046–2065 (middle) and 2080–2099 (right). Anomalies are relative to the average of the period 1980–1999. Source: IPCC (2007a), Figure 10.8.

Box 4.1 The IPCC projections of global climate change. (Lesley Hughes)

The IPCC (2007a) projections for global climate change were derived from a compilation of 23 global models, from 12 countries, used for simulations of 20th and 21st century climate (available at http://www.pcmdi.linl.gov/ipcc/info_for_analysts.php). The simulations were grouped into four major Emissions Scenarios in a future world, with variations on each (IPCC 2000). The major emission scenarios are:

- A1. Very rapid economic growth, with new and more efficient technologies and a global population that peaks in the mid-21st century and declines thereafter. There is convergence among regions, capacity building and increased cultural exchanges, resulting in reduced differences in income within regions. The A1 scenario is divided into three sub-groups based on alternative directions of technological change in energy systems: A1FI – fossil intensive; A1T – non-fossil energy sources; and A1B – a balance across all energy sources.
- A2. Economic development, primarily regional, with per capita economic growth and technological change more fragmented and slower than A1. The world is more heterogeneous economically, culturally and socially. Human population numbers rise continuously.
- B1. Rapid change in economic structures towards a service and information economy, with reductions in material intensity, and the introduction of clean and efficient technologies for energy production. Global population as in A1, which peaks in the mid-21st century and declines thereafter. The emphasis is on global solutions to environmental sustainability, and economic and social issues – including equity among peoples – but not additional climate initiatives.
- B2. Economic development is mainly local, less rapid and more diverse technologically than in A1 or A2. The emphasis is on local solutions to environmental sustainability, economic and social issues, with the result being an intermediate level of economic growth globally. Human population numbers rise continuously, but at a lower rate than A2.

Source: CSIRO and BOM (2007); IPCC (2007a).

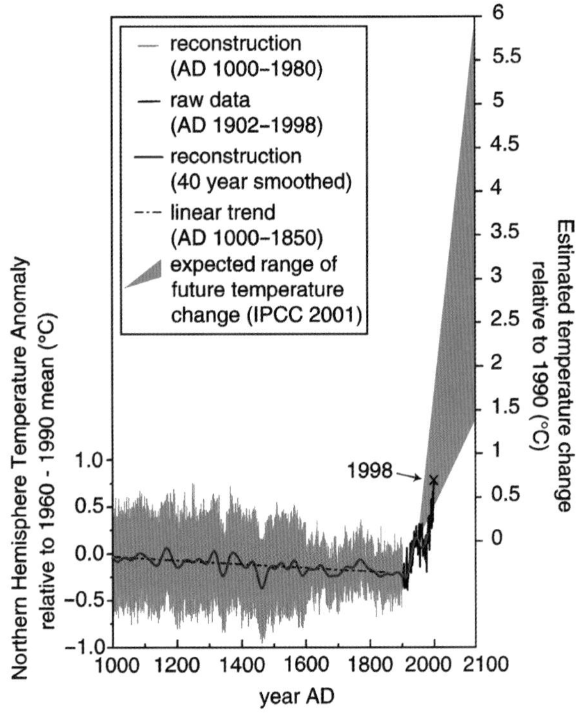

Figure 4.12 The range of future projections of temperature change compared to the envelope of natural variability for the past millennium. Source: Derived from Steffen *et al.* (2004). With kind permission of Springer Science+Business Media.

warming is accelerating, with twice the warming experienced since 1950 as in the first half of last century (CSIRO and BOM 2007). The warmest year on record in Australia was 2005, with an average temperature over 1.0°C above the long-term mean; 2007 was the sixth warmest year on record, and the warmest ever in southern Australia. Concomitant changes in intensity, distribution and seasonality of rainfall; snow cover and precipitation runoff; increasing acidity of oceans; and changes in extreme events such as floods, droughts and fire have also been documented.

The frequency of extreme hot and cold temperatures has also been changing. There has been an increase in hot days (over 35°C) since the late 1950s, as well as an increase in hot nights (>20°C), and a decrease in cold days (<15°C) and cold nights (<5°C) (Alexander *et al.* 2007; Chambers and Griffiths 2008; CSIRO and BOM 2007). There is some evidence that trends in the most extreme events of both temperature and rainfall are changing more rapidly in relation to

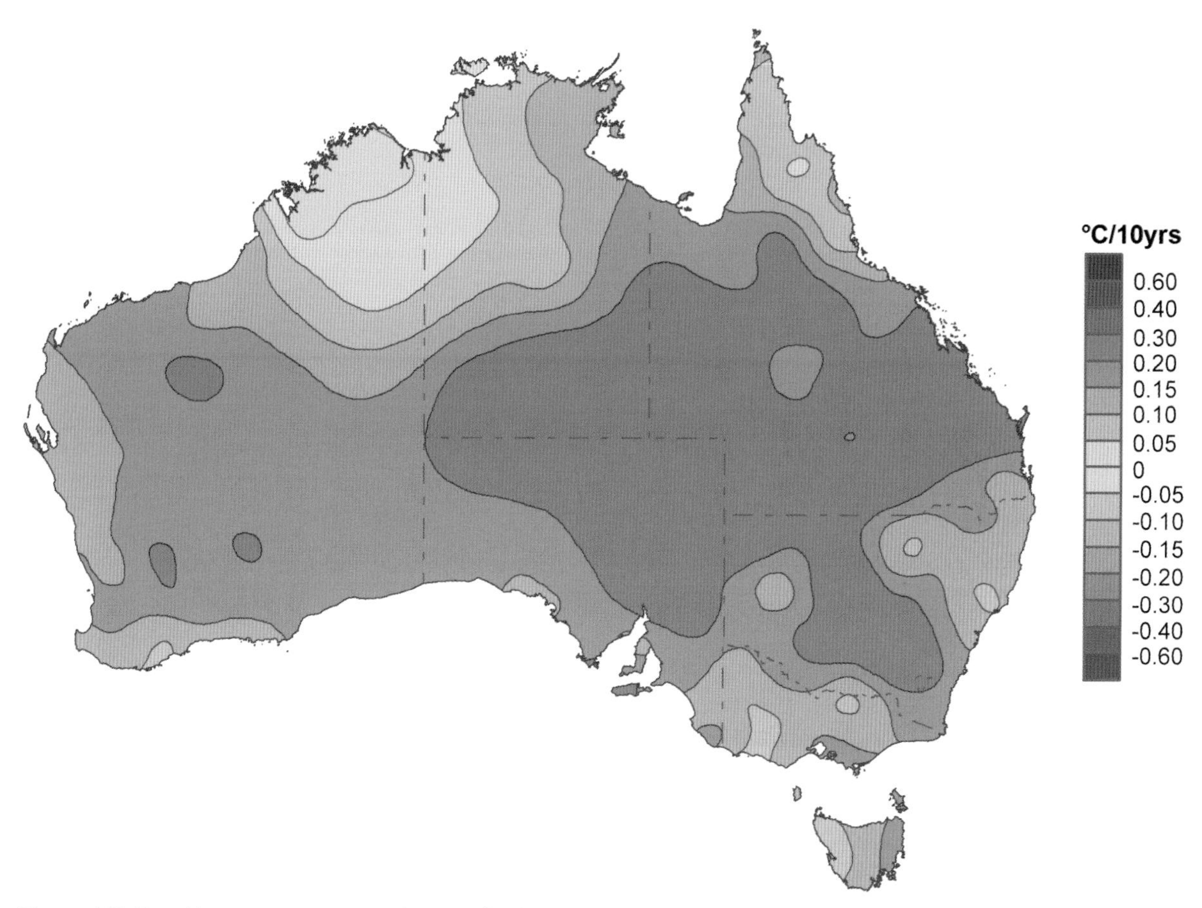

Figure 4.13 Trend in mean temperature in Australia, 1950–2007. Source: Australian Bureau of Meteorology.

corresponding mean trends than are the trends for more moderate extremes (Alexander *et al.* 2007).

Significant changes in regional rainfall patterns have occurred over the past century, especially when comparing the period 1910–1950 and the period since (Fig. 4.14). Since 1950, rainfall has increased over the north-west, and has declined in the south-west and along the eastern seaboard (Nicholls 2006). From 1950 to 2005, extreme daily rainfall intensity and frequency increased in central Australia and the north-west (mainly during the summer), and over the eastern tablelands of New South Wales. However, it has decreased substantially in south-eastern and south-western Australia, and along the central east coast. The rainfall decrease in south-west Western Australia is probably due to a combination of increased greenhouse gas emissions, natural climate variability and land use change (McAlpine *et al.* 2007; Nicholls 2006; Pitman *et al.* 2004). Pan evaporation has declined by about 3% since the mid-1970s but this may be due largely to changes in instrument exposure (Nicholls 2006).

Droughts in Australia have become more severe, with higher temperatures and increased evaporation (Nicholls 2004). The most recent drought has placed considerable stress on water resources in many regions. Declines in winter rainfall by 10–20% in the south-west have resulted in a 50% reduction in flows to the Perth water supply (Pittock 2003) (Fig. 4.15).

The effects of reduced rainfall and hotter droughts on fire frequency, duration and seasonality have not yet been detected throughout most of the agricultural and large settlement areas of Australia although hotter and drier years are generally associated with increased fire risk (Hennessy *et al.* 2005; Lucas *et al.* 2007). In the north where frequent fire is both natural and a management tool, changes in fire frequency, fire season and fuel loads as a result of climate change are currently under assessment (Box 5.11).

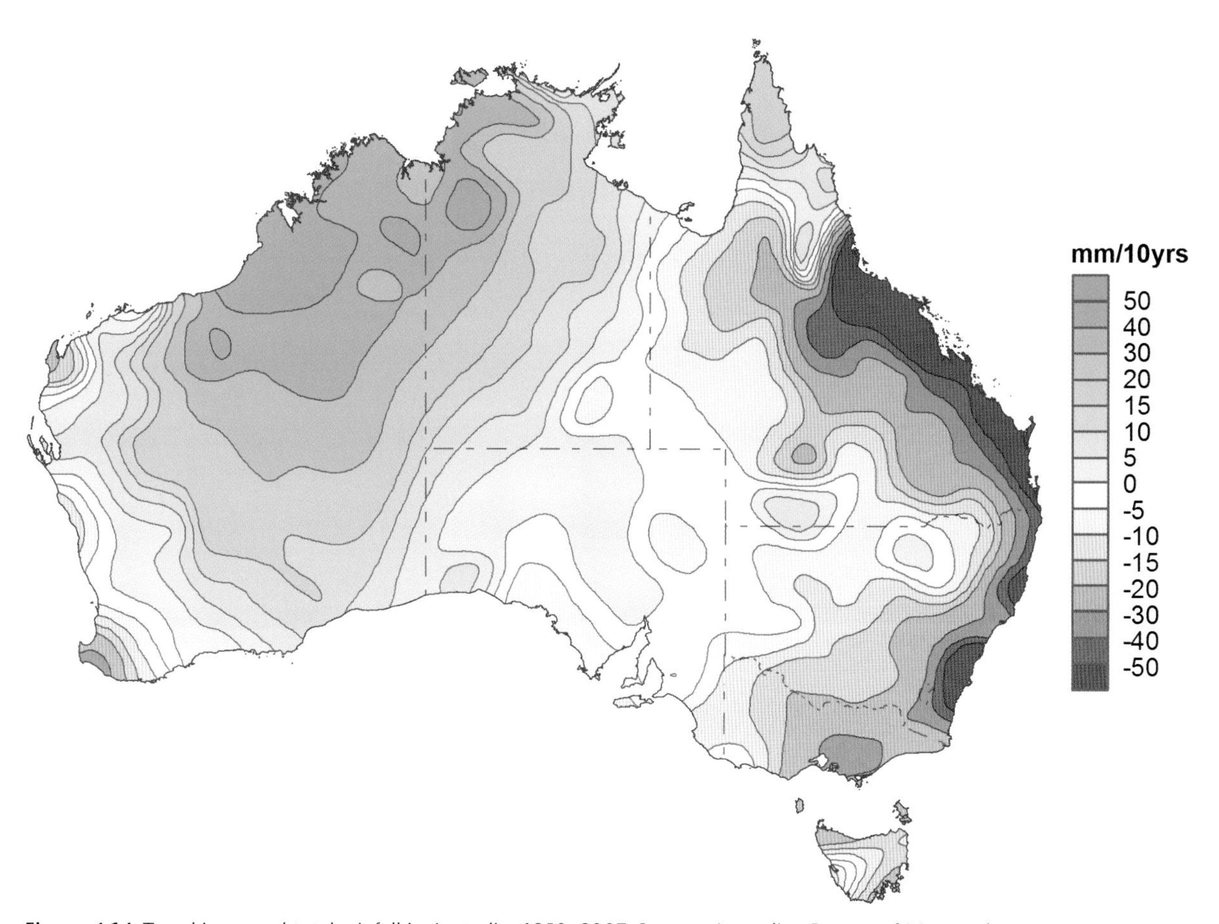

Figure 4.14 Trend in annual total rainfall in Australia, 1950–2007. Source: Australian Bureau of Meteorology.

Mean snow cover on the Australian Alps has declined significantly between the periods 1960–1974 and 1975–1989. Maximum winter snow depth has declined by about 40% since the 1960s (Green and Pickering 2002; Nicholls 2005; Osborne *et al.* 1998).

Substantial warming has also occurred in the three oceans surrounding Australia, particularly off the south-eastern coast and in the Indian Ocean (CSIRO and BOM 2007; IPCC 2007a), and the East Australian Current (EAC) has penetrated further to the south (Ridgway 2007). The region has become both warmer and saltier with mean trends of 2.28°C/century in temperature and 0.34 psu/century in salinity over the 1944–2002 period, which corresponds to a poleward advance of the EAC of 350 km (Ridgway 2007).

The relative frequency of the El Niño component of the ENSO phenomenon, a strong driver of Australian climate, has increased in recent years although it is not clear to what extent this is a natural or anthropogenically driven change (Power *et al.* 2006). The relationship between the Southern Oscillation Index (SOI) – a measure of ENSO activity – with rainfall and temperature has changed since the early 1970s. Temperature since that time has been higher for any given value of SOI (Nicholls *et al.* 1996), but there has not been a corresponding increase in rainfall. In fact, during the most recent 2007–2008 La Niña event, there was very low rainfall in southern Australia compared with other La Niña events.

Sea levels at sites monitored around the Australian coast from 1920 to 2000 rose by about 1.2 mm/year (Church *et al.* 2006), with large regional variation due in part to different rates and directions of land movement. Sea levels across northern Australia are

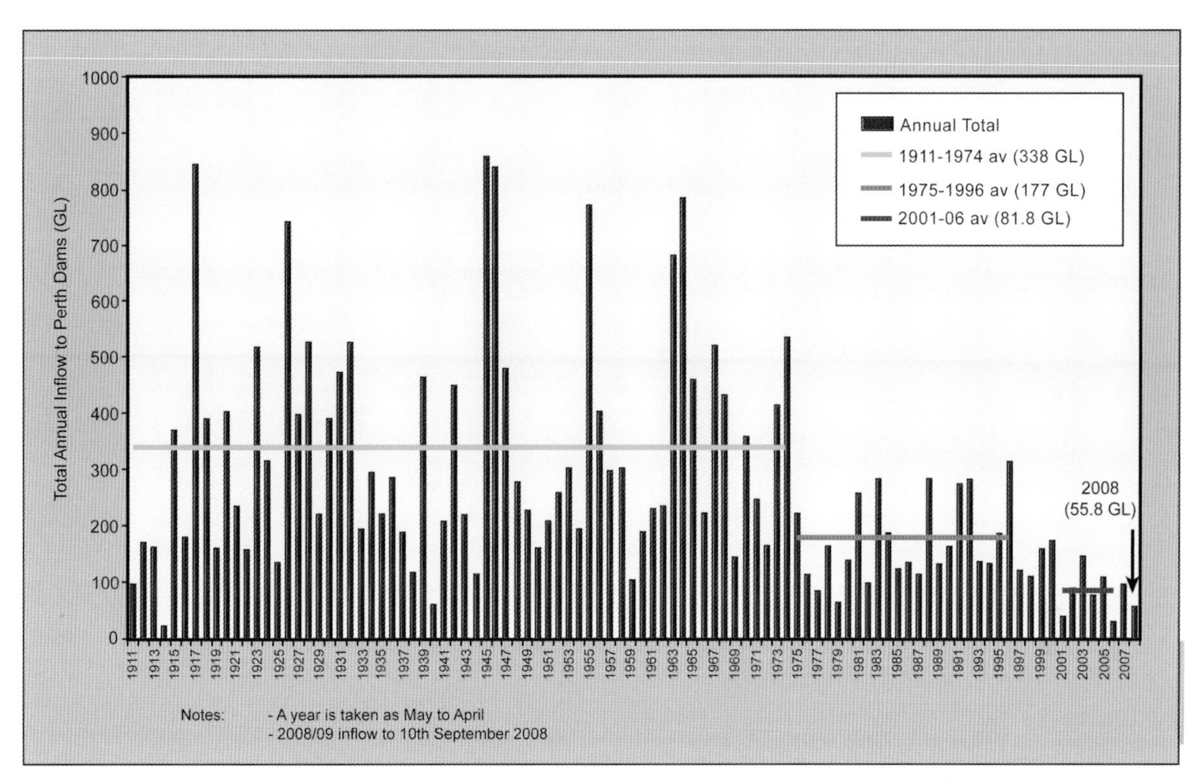

Figure 4.15 Trends in total annual stream flow into Perth dams 1911–2008. Source: Western Australian Water Corporation.

currently rising at 7.8–8.3 mm/year (National Tidal Centre 2006), about four times the global average. Ocean currents around the continent are also changing, with the EAC increasing in strength by about 20% since the late 1970s (Cai 2006).

The acidity of marine waters around Australia is increasing (lower pH). There is a north–south gradient in the rate of change in acidity, with southern seas becoming more acidic faster than northern seas because cold water absorbs more CO_2 (Hobday *et al.* 2007).

While there have been increases in the frequency of intense cyclones in both the Pacific and Atlantic oceans (Easterling *et al.* 2000), the total number of tropical cyclones along the east coast of Australia has declined since the 1970s (Kuleshov 2003). However, there has been an increase in the number of very intense systems (Kuleshov 2003).

4.4.2 Australia's future climate

The projections for Australia's future climate summarised below (Table 4.1) have been drawn principally from the IPCC Fourth Assessment Report (IPCC 2007a, b), and from joint research by CSIRO and the Bureau of Meteorology (CSIRO and BOM 2007; Suppiah *et al.* 2007). Detailed breakdowns of projected changes in temperature and rainfall for each of the 85 bioregions of Australia are presented in section 6 of Dunlop and Brown (2008). General Circulation Model (GCM) outputs are based on projected increases in atmospheric CO_2 to 540–970 ppm by 2100 (IPCC 2007a), depending on factors such as human population growth and adoption of different energy technologies. About half the uncertainty in the rate of future global warming is due to the uncertainty about future greenhouse gas emissions (Box 4.2).

Table 4.1 Projected climate changes and associated extreme events relevant to Australia. Source: IPCC (2007 a, b); Suppiah *et al.* (2007); CSIRO and BOM (2007) and some references contained therein.

Variable	Climate changes projected for Australia
Temperature	Maximum and minimum temperatures will increase in all regions and seasons. The best estimate of annual warming over Australia by 2030 relative to 1990 is about 1°C for the mid-range emission scenarios. Warming will be relatively less in coastal areas compared to the inland (by up to 2°C). Spring and summer temperatures are projected to increase more than those in autumn and winter. The range of uncertainty due to differences between the models is about 0.6–1.5°C for most of Australia, with the probability of the warming exceeding 1°C by 2030; being 10–20% for coastal areas and more than 50% for the inland. Later in the century, the warming projection is more dependent on the emissions scenario used. By 2070, the projections are for +1.8°C (range 1.0–2.5°C) for the low emissions case and around 3.4°C (2.2–5.0°C) for the high emissions case.
Precipitation	There remains considerable uncertainty about the patterns of future rainfall. Climate models show decreases and increases in rainfall depending on the location and on the model used. Decreases in rainfall are considered likely in southern areas in the annual average and in winter, in southern and eastern areas in spring, and along the west coast in autumn. For 2030, best estimates of annual rainfall change indicate little change in the north and decreases of 2–5% elsewhere (CSIRO and BOM 2007). Rainfall seasonality may also change, affecting moisture availability, patterns of runoff and moisture stress. There is a projected decrease in rainfall in central Australia of 10–20% and increased probability of extreme rainfall deficits (Taminiau and Haarsma 2007).
Solar radiation	Climate models indicate little change, although a tendency for increases in southern areas of Australia is evident, particularly in winter and spring.
Relative humidity	Small decreases are simulated over most of Australia.
Potential evapo-transpiration	There are likely to be increases in potential evaporation by up to 8% per degree of warming (CSIRO 2001). Taken together with rainfall changes, this will affect runoff, river flows and soil moisture in many regions, leading to substantial reduction in soil moisture and in water flows in many freshwater systems (Arnell 1999; Chiew and McMahon 2002).
Wind	Climate models indicate increases in most coastal areas except for a band around latitude 30°S in winter and 40°S in summer where there are decreases projected.
Snow and frost	Decreases in snow cover, average season lengths and peak snow depths are expected, and a tendency for maximum snow depth to occur earlier in the season (Hennessy *et al.* 2003, 2008). Frost occurrence is also expected to decrease.
Storms	Increases in storm intensity (5–10%) and/or frequency are projected (Pittock 2003; Walsh *et al.* 2004). The proportion of rainfall falling as intense events may increase (20–30%), leading to an increased probability of local flooding. Tropical cyclones may be initiated further to the south (Leslie and Karoly 2007).
Fire	Higher temperatures, regional reductions in rainfall, decreased relative humidity and higher fuel availability due to fertilisation of plant growth by increasing atmospheric CO_2 is likely to increase the intensity and frequency of wildfires (Beer and Williams 1995; Cary 2002; Pitman *et al.* 2007; Williams *et al.* 2001). Increases in the frequency of extreme fire weather days of 4–25% by 2020 and 15–70% by 2050 have been projected (Hennessy *et al.* 2005; Lucas *et al.* 2007). Changes in the seasonality of fire may also occur.
Sea-level rise	A global average rise of 18–59 cm by 2100 is projected (IPCC 2007a), with a possible additional contribution from ice sheets of 10–20 cm. However, further ice sheet contributions may push sea-level rise towards or above the upper limit of projections.
Ocean temperature	Higher sea surface temperatures are projected. The 2030 best estimate of temperature increase is 0.6–0.9°C in the southern Tasman Sea and off the north-western coast, and 0.3–0.6°C elsewhere (Church *et al.* 2006).
Ocean acidity	Increases in ocean acidity due to increased oceanic absorption of carbon dioxide will occur, with the largest increases in the high- to mid-latitudes.
East Australian Current (EAC)	The strength of the EAC may continue to increase, resulting in warmer waters extending further southward with possible impacts on severe storms and marine ecosystems.
El Niño–Southern Oscillation	The El Niño phase of the ENSO may tend to become drier and the La Niña phase wetter; that is, ENSO may intensify in magnitude.

Box 4.2 Assumptions and uncertainties in projections of climate change (Lesley Hughes)

There are four main areas of uncertainty in climate change projections: (i) the projected rate of increase in greenhouse gases (emission scenarios; see Box 4.1); (ii) the relationship between the rate of greenhouse gas emissions and their atmospheric concentrations; (iii) the magnitude of the global warming for a given change in concentrations; and (iv) identifying how global climate change plays out regionally.

The emissions scenarios depend on assumptions about future demographic changes, economic development and technological improvements. These uncertainties become greater further into the future, but the emission scenarios are fairly similar up to 2030.

Greenhouse gas concentrations in the atmosphere depend not only on the emissions, but also on the rates at which the gases are removed from the air by various processes. Most gases are removed by chemical reactions or ultraviolet radiation, but carbon dioxide is removed by absorption into the ocean and terrestrial biosphere, e.g. forests. Higher carbon dioxide concentrations and larger changes in climate tend to reduce the absorptive efficiency of these sinks, resulting in a positive feedback that has been observed in recent years (Canadell *et al.* 2007).

Increasing concentrations of greenhouse gases and changes in aerosol emissions affect the thermal radiation balance of the Earth and the average surface temperature. The radiative forcing due to greenhouse gases is well understood. The contribution from aerosols (microscopic particles in the air) is relatively poorly understood, but the net effect is a cooling (IPCC 2007a). Climate feedbacks are also important, such as: ice melt, which will lead to more absorption of solar radiation and greater warming; the ability of warmer air to hold more moisture (water vapour being a greenhouse gas); release of methane from melting permafrost in tundra regions; and changes in cloud properties (the largest source of uncertainty in the climate response). Global warming projections take account of these uncertainties. The estimated global warming in 2030 for the A1B emission scenario is 0.54–1.44°C, but as little as 0.45–1.20°C for B1 emissions and as much as 0.60–1.60°C for A1T emissions (CSIRO and BOM 2007).

Source: Hennessy *et al.* (2008, p. 21).

Figure 4.16 Intense storm events such as this storm over Healesville, Victoria, are likely to become more common. Source: AUSCAPE. Photo by Davo Blair.

5 Responses of Australia's biodiversity to climate change

This chapter outlines the ways in which climate change has already affected Australia's biodiversity and will potentially affect it in the future. The threats to biodiversity from climate changes are threats arising from changes in the basic physical and chemical environment underpinning all life, especially CO_2 concentrations, temperature, precipitation and acidity. Species will be affected individualistically by these changes, leading to flow-on effects on the structure and composition of present-day communities, and then potentially to changes in ecosystem functioning. Species, communities and ecosystems will also be affected indirectly, as climate changes affect important processes such as fire and disease. The chapter also outlines key information gaps and research questions for the future. While the chapter has attempted to present a broad overview across all sectors, there is a predominance of examples from terrestrial systems. This reflects firstly the paucity of knowledge about climate change and freshwater systems, and secondly the acknowledgment that climate change impacts on marine systems have been reviewed by Hobday *et al.* (2007).

5.1 A NEW AND DIFFERENT STRESSOR

A basic property of life is its ability to adapt to environmental changes through physiological, behavioural and genetic changes. At times, however, environmental change is of such magnitude or occurs at such a fast rate that biological and ecological systems become overwhelmed and cannot cope. In these cases, species can suffer shrinking geographic ranges, reduced population sizes and extinctions, leading to changes in structure and composition of communities and ecosystems. These events have occurred throughout Australia's geological history and, most recently, in response to human settlement, habitat loss and introduction of exotic species, as described in Chapter 3.

Australia is currently experiencing environmental change consistent with the global pattern of increasing atmospheric concentration of CO_2, higher temperatures, altered patterns of precipitation and sea-level rise (Chapter 4). This chapter describes observations of how the Australian biota is already responding and outlines the potential changes that may occur in the future.

Scientists have derived basic principles of biological and ecological responses to environmental change from case studies of species' life histories, physiologies and behaviour, and from studies of ecological systems and palaeoecology, as described in Chapter 3, Table 3.1. These principles are directly applicable to our assessment of biodiversity vulnerability and responses to climate change. Regarding climate change, however, there are two major differences in applying these principles, as this new challenge may require adaptations of the biota that are both qualitatively and quantitatively different from those that have been recorded to date:

- Firstly, the threats to biodiversity of climate changes are threats arising from changes in the basic physical and chemical environment underpinning all life – especially CO_2 concentrations, temperature, precipitation and acidity – unlike other threats such as land clearing and introduced species.
- Secondly, the rate of current warming and other associated changes in climate is unprecedented since the last massive extinction event 60 million years ago (with the possible exception of a very rapid cooling and subsequent rapid warming of more than 5°C in northern Europe at the start and finish of the Younger Dryas Period (from 12 000–11 500 years b.p.), caused by the disruption and subsequent restoration of the North Atlantic thermohaline circulation during deglaciation and freshwater runoff from North America and Greenland (Alley 2000; Severinghaus 1998).

Identifying and predicting the effects of such rapid climate change on biodiversity is complicated by the fact that many of Australia's species and ecosystems are already under great pressure from other drivers of change (Chapter 3). Indeed, it is difficult to attribute many of the recently observed changes in species characteristics and ecosystems to a specific cause, such as climate change. Nevertheless, scientists and managers have already documented changes in species, communities and ecosystems that carry a 'climate signal', being consistent with recorded changes in temperature, precipitation, CO_2 concentrations and/or sea level.

Predicting the future effects of climate change on Australia's biodiversity is complicated for a number of reasons, including, but not limited to, the following:

- Climate change will interact with other drivers that are currently affecting biodiversity.
- Responses to physical and chemical changes will be individualistic – they will occur at the level of the individual and be reflected in population dynamics of individual species. The component species or functional groups within an ecosystem will therefore not respond as a single unit, and interactions among species will have the potential to modify outcomes, sometimes in unpredictable ways.
- Even with application of general principles (Table 3.1), some properties of biological and ecological systems are inherently difficult to track. For example: (i) a change in the average value of a continuous environmental variable (such as temperature) may not be as important biologically as a change in variability or extremes of that variable; and (ii) responses of biological systems may be non-linear, with thresholds or 'tipping points' not yet identified.
- Basic knowledge about limiting factors, genetics, dispersal rates and interactions among species that make up Australian communities and ecosystems is generally lacking.
- Management actions taken to adapt and/or mitigate the impacts of climate change on human systems could have further adverse impacts on biodiversity.

5.2 BIODIVERSITY AND CLIMATE CHANGE: IMPACTS AND RESPONSES

5.2.1 Underlying physiological processes

Changes in the physical environment affect physiological processes in plants and animals such as respiration, photosynthesis, metabolic rate and water use efficiency. Individuals may also respond to environmental change by altering their behaviour or the timing of life cycle events (phenology) such as flowering, dispersal, migration and reproduction (Table 5.1). All organisms are able to cope with some degree of variability in their environment, and to maintain homeostasis and reproduction within the bounds of that variability. Beyond some physiological threshold, however, responses change quite dramatically and death may result (Fig. 5.1).

The future response of plants to rising CO_2 is particularly important – especially coupled with warming and/or altered rainfall patterns – as any differences among plant species could have large secondary

Table 5.1 Direct impacts of environmental changes on individuals, based on Table 3.1, and Principles 1 and 3.

Environmental change	Responses by individual organisms
Temperature	Metabolic and developmental rates in animals, and photosynthesis and respiration in plants, increase with increasing temperature until some upper limit. Increasing temperatures will interact with water stress for both plants and animals, and will affect the timing of important life cycle events such as reproduction and diapause (a quiescent period during a life cycle). Advances in spring events and delays in autumn events are probable for many species, and will result in a lengthening of the vegetative growing seasons in many regions. Animals also respond to temperature by altering their behaviour, for example, by seeking shade, altering the time of day they are most active or changing the position they occupy in the water column. In many reptiles, temperature during development affects sex ratios. Fundamental geographic ranges of many species are thought to be determined mainly by temperature extremes (e.g. hottest day in summer, coldest day or frost incidence in winter).
CO_2	Plants increase photosynthetic rate as the concentration of CO_2 increases in the atmosphere or in water (in the case of algae), until the CO_2 concentration or another factor (such as light, water or nutrients) becomes limiting – this process is known as the 'CO_2 fertilisation' effect (Box 5.1). Increasing CO_2 also reduces stomatal conductance, thereby increasing water use efficiency, particularly in C3 plants. CO_2-driven changes in productivity are usually accompanied by changes in plant chemical composition (such as increasing ratios of carbon to nitrogen, and altering the concentrations of secondary metabolites such as phenolics and tannins), as well as changes in plant structure and the allocation of biomass to various plant parts. Impacts of increasing CO_2 will vary considerably among different plant functional types and different vegetation types, and will depend on temperature and the availability of water and soil nutrients. As CO_2 is gradually absorbed by oceans and fresh water, the water becomes more acidic (lower pH), which increases the solubility of calcium carbonate, the principal component of the skeletal material in aquatic organisms.
Water	Water supply is critical for all organisms, and water – together with temperature – ultimately sets the fundamental distributional limit for all species. In plants, stomatal conductance declines as atmospheric CO_2 increases, resulting in lower transpiration rates. In regions where precipitation declines, increasing CO_2 may therefore mitigate water stress to some extent.
Extreme events	Extreme weather events such as floods, droughts, storms and fire can affect population dynamics, species boundaries, morphology, reproduction, behaviour, community structure and composition, and ecosystem processes. Changes in the frequency, intensity and seasonality of extreme events may have larger impacts on many species and communities than the directional shifts in temperature and changes in rainfall patterns.

impacts on plant community structure, animals that use plants as habitat or food, and even nutrient cycles in ecosystems (Box 5.1).

5.2.2 Will species stay or move?

Organisms experiencing environmental change may tolerate the change in situ (i.e. individuals acclimatise or populations adapt at the site) and/or disperse away from their current location (Box 5.2). The rate of environmental change is crucial in determining which response (that is, in situ adjustment vs dispersal) is more likely and/or more successful for any species within a particular time frame. All in situ and dispersal responses have the potential to change the population size (abundance) and/or distributional range of a species either positively or negatively.

In situ responses

Four types of responses allow an individual or species to meet the challenge of environmental change without dispersing to a new site. These responses are not mutually exclusive.

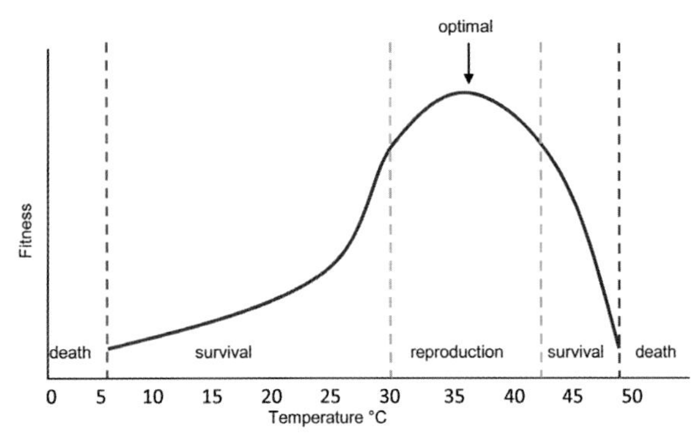

Figure 5.1 Response of an individual organism to a hypothetical environmental factor (e.g. temperature). Note the optimum where individuals thrive and reproduce. At suboptimal regions, individuals may survive but not be able to reproduce.

- *Acclimation* is the gradual habituation of an organism to a slowly changing environmental condition (such as temperature) by simple physiological or morphological means (e.g. less dense fur coat).
- *Behavioural change* occurs when some organisms respond to environmental changes simply by altering their use of microhabitat (e.g. by using deeper burrows, or moving a nest site from a north-facing to a south-facing aspect).
- *Phenotypic plasticity* is the range of variability shown by an organism's phenotype in response to environmental changes. These responses may be developmental (reproductive size of a fish, number of stomates on a leaf) or phenological, that is the timing of life history events (e.g. flowering, reproduction, survival, dispersal, migration, metamorphosis).
- *Genetic adaptation* applies to populations, and involves natural selection among individuals by means of differential survival and/or reproduction such that the relative frequencies of particu-

Box 5.1 How will Australian plant species and vegetation respond to rising atmospheric concentrations of CO_2? (Mark Hovenden)

Under current conditions, the rate of photosynthesis in plants is limited by the availability of atmospheric CO_2, which means that increasing the concentration of CO_2 in the atmosphere increases the rate of photosynthesis. Increasing CO_2 therefore has the potential to increase plant growth rate, in turn affecting ecosystem productivity and the function of trophic webs.

However, plant growth is not simply limited by CO_2, as nutrients, water and light are all limiting in different environments. Australian terrestrial ecosystems have chronically low nutrient availability and thus the increasing CO_2 is unlikely to increase growth in many instances, simply because the supply of essential nutrients prevents this. However, increasing plant carbon assimilation can lead to a raft of other changes, such as a decrease in leaf protein concentration and an increase in the production of carbon-based toxins. Both of these factors reduce leaf nutritional quality, which has a strong, direct impact on herbivorous animals – both vertebrates and invertebrates. Alterations of growth or survival of herbivores will in turn affect predators and pathogens, so the changes observed at the leaf level could cascade throughout the ecosystem. Changes in leaf chemical composition also affect leaf litter decay and therefore nutrient cycling. Thus, paradoxically, it is possible that the rising CO_2 could reduce ecosystem productivity by further reducing the already low availability of essential plant nutrients.

Increasing CO_2 also reduces the water loss from plants by reducing stomatal conductance; increasing CO_2 may therefore mitigate some of the impacts of reduced soil moisture in some areas. Plant species that are currently constrained by the availability of water may therefore be able to extend their ranges into neighbouring dry areas. This could lead to changes of community structure and function at vegetation boundaries – such as at the boundaries between deserts and grasslands, grasslands and woodlands, and woodlands and forests. Increasing CO_2 will also increase the growth of woody species more than that of grasses, therefore allowing shrubs and trees to invade grasslands and grassy woodlands – a process known as vegetation thickening, which is already being observed.

Box 5.2 Did Australian species stay or move when climate changed in the past? (Margaret Byrne)

Understanding ancestral patterns of diversity is one of the keys to predicting responses of species to climate change (Hewitt and Nicholls 2005). This is particularly important in biomes where persistence has been a key feature of biotic responses to historical climate change because it indicates adaptation and/or phenotypic plasticity as the main mode of response to changing conditions rather than migration tracking climatic niches.

Southern Australia is an ancient landscape with generally low geological relief that was not glaciated in the recent geological past but did experience significant climatic oscillations from warm, wet conditions in interglacials to cool, dry environments during glacial maxima (Chapter 1). So did species stay or move during climate change in the past? Contraction/expansion (move) and long-term isolation (stay) leave different genetic signatures in species, and phylogeography is a discipline that uses these genetic signatures to identify influences of past processes on species distributions.

In the Wet Tropics there is clear genetic evidence for contraction and expansion of species through time, and the location of refugia at high elevations (Schneider and Moritz 1999). Similar contractions and expansions at the Last Glacial Maximum are evident in eucalypts in glaciated areas of Tasmania (McKinnon *et al.* 2004), and in log-dwelling invertebrates (saproxylates) in wet forest at Tallaganda (Sunnucks *et al.* 2006). Thus, in areas of high topographical relief, moving is the most likely response of species to climate change.

But the majority of Australian landscapes are relatively flat and species would have had to move large distances to keep pace with changing climate in the past. Phylogeographic patterns in many of the biota of southern Australia reveal that some species did contract to and expand from major refugia in the early part of the Pleistocene epoch (1.8–0.7 million years b.p.) when significant aridity developed (Byrne *et al.* 2008). In the past 700 000 years, however, most species have 'stayed put'. That is, they have persisted through climatic changes in patchy, localised refugia rather than moving long distances. This emphasises the importance of maintaining a mosaic of habitats in heterogenous landscapes so that species can persist through changing conditions (Byrne 2007).

lar characteristics change within the population. Genetic change in response to climate change will be more likely in species with short generation times and large populations. To some extent, the previous three mechanisms will be mediated by genetic changes.

Dispersal responses: change in species distributions

Changes in species distributions can occur in two, non-mutually exclusive ways. In some species, particularly mobile species like birds, flying insects and pelagic aquatic organisms, changes in geographic range can occur when an environmental change (such as an increase in temperature) cues individuals to disperse to new, more suitable areas. Most geographic range changes, however, probably occur more gradually, as small numbers of individuals that naturally disperse during part of their life cycle are now able to survive in new areas because these sites have become more environmentally suitable. These sites then serve as foci for new establishing populations. Range contractions can also occur as local populations become extinct due to physiological or other stresses. Fragmentation of previous continuous populations may lead to genetic isolation and effects such as inbreeding.

Distributional shifts will be most obvious in the near term for those species whose limits occur at the edges of extreme environments such as coastal zones or snow lines. One question of considerable importance in assessing the vulnerability of Australian species to rapid climate change is whether they will be able to disperse fast enough to 'keep up' with shifting climate zones. Over the past 50 years, global average temperatures have increased about 0.13°C per decade. Based on the climatic gradients in continental Australia, Dunlop and Brown (2008) estimated that locations experiencing a given average temperature might be expected to shift at a rate of 3–17 km/

year across the landscape, though far less in mountainous areas.

Little is known about the potential dispersal rates for Australian species, although many plant species do not have long-distance dispersal mechanisms (Higgins *et al.* 2003; Hughes *et al.* 1994). Historical migration rates of plant species in the northern hemisphere after the last glacial period have been estimated from cores of fossil pollen. The estimated rates vary greatly between species but most are less than a few kilometres per year (range <0.1–20 km/year; e.g. Hughes *et al.* 1994). In most cases it is unclear whether these rates were maxima; that is, whether they were *limited* by the regional rate of warming or were a result of intrinsic biological limitations. If these rates can be translated to Australian plant species, they would appear to be inadequate for most species to maintain their ranges within their current climatic envelope over the next century. In addition, the high degree of fragmentation and degradation of suitable habitats in many regions (Chapter 3) means that new establishment sites where other important environmental factors such as soil type are suitable may not be accessible or exist at all, even for those species capable of relatively long distance dispersal (see also discussion of the role of connectivity in Box 3.3). Many freshwater organisms in lakes and rivers will face the additional challenge of moving between catchments or between areas of intermittent flow. Marine dispersal rates have the potential to be much greater; changes in distribution are the dominant response to climate change in the ocean (compared with changes in phenology on land) (Grantham *et al.* 2003).

Opportunities for terrestrial species to adapt to rapidly shifting climate zones will also be limited by the topography of the continent (Chapter 2). Australia's average elevation is only 440 m (Augee and Fox 2000), with about 13% of the country over 500 m and 0.01% over 2000 m (the tallest peak, Mt Kosciuszko, is only 2228 m above sea level). This means that relatively few terrestrial species will be able to keep pace with climate zones by moving small distances to higher elevations. Instead, most will be faced with large overland shifts to remain within their current climatic environments.

Responses of communities

Ecological communities are assemblages of interacting species that occur together in space and time (Table 3.1, Principles 4, 5 and 6). Any direct physiological or behavioural change that renders a species relatively advantaged or disadvantaged will have flow-on effects to other species with which it interacts (predators, competitors, mutualists, etc.). In this way, ecological communities will be affected by environmental change (Fig. 5.2). As the relative abundance of species at a local site changes due to individualistic responses of species to climatic change, changes in community structure (i.e. the relative importance of different species or functional groups) will occur. As climate change causes local extinctions of some species and/or the establishment of new species from other areas, changes in community composition will also become evident.

Individualistic responses of species to climate change mean that many communities and ecosystems will change in unpredictable ways. Differential rates of dispersal as climate zones shift, for example, mean that communities will not 'move' as units across a landscape. This process has been demonstrated many times in the geological past. A well known example is the recolonisation of glaciated land in the northern hemisphere as glaciers retreated 9000–11 000 years ago. During this gradual retreat, dispersal rates of individual plant species varied such that some of today's plant communities were formed as recently as 2000 years b.p. (e.g. Delcourt and Delcourt 1991). An Australian example is the loss of richness in Tasmania's biota following the last glaciation, when Bass Strait reformed and prevented the recolonisation of the island by slowly dispersing species from the mainland. As noted in Chapter 3, some of Australia's most iconic ecosystems, such as the Great Barrier Reef and the coastal wetlands of Kakadu National Park, are less than 6000 years old (Chappell *et al.* 1983; Chappell and Woodroofe 1985; Woodroofe *et al.* 1987).

Addition or loss of a single species may affect a community to various degrees, depending on the biological and ecological characteristics of the species and its role in the larger assemblage (e.g. the species' trophic level, competitive ability, life form, the number of interacting species, phenology). In general, the loss

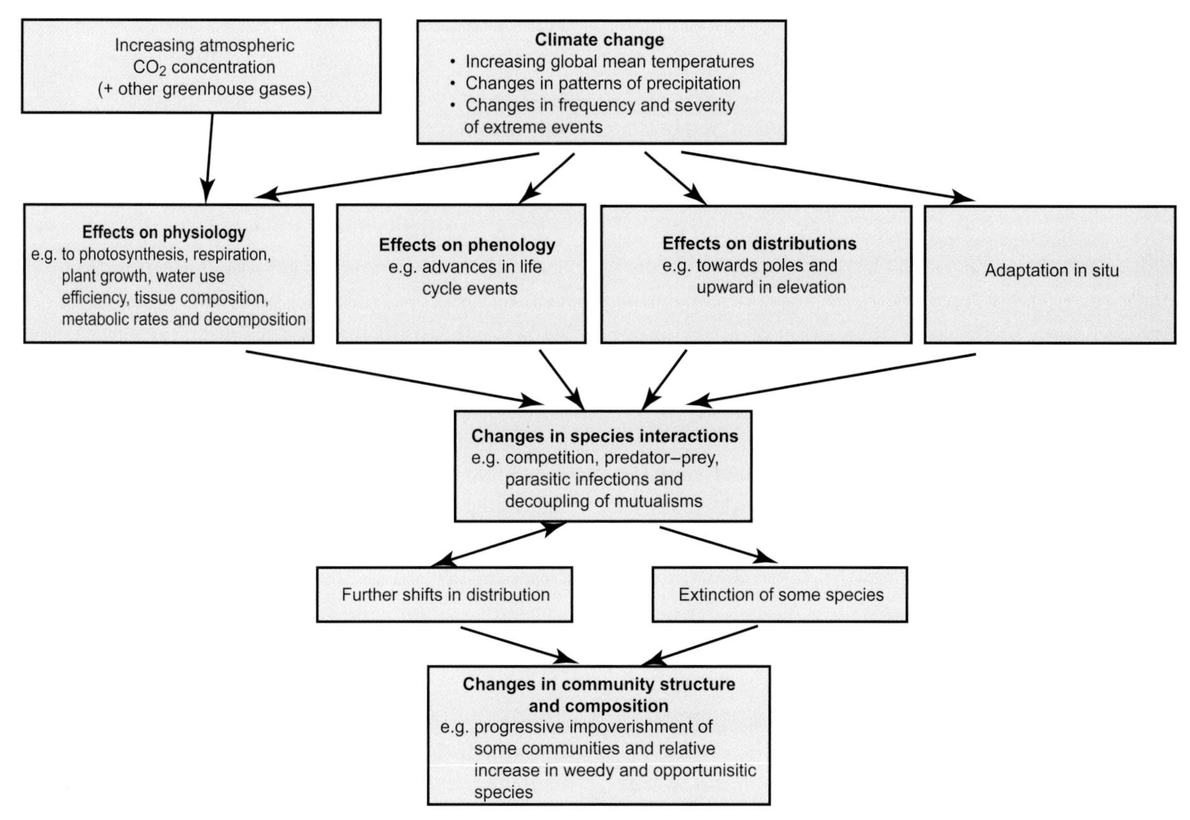

Figure 5.2 Example of the potential pathways of community change flowing on from individualistic responses to climate change. Increased CO_2 concentration will act on species directly (via physiology) and indirectly (via climate changes) (first tier). Individual species might respond in four ways (second tier), resulting in changes in species interactions (third tier). These changes might then lead either to extinctions or to further shifts in ranges (fourth tier), ultimately leading to changes in the structure and composition of communities. Source: Redrawn from Hughes (2000).

of a top carnivore, a structurally dominant plant or species at the base of the food chain in aquatic systems (Table 3.1, Principle 5) has significant consequences for the entire community and/or ecosystem.

As some species adapt to shifting climate zones by dispersing to new areas, local increases in the number of species (species richness or alpha diversity) may occur in the short to medium term (next few decades). This may be particularly evident at higher elevations as species move upwards over relatively short distances to adapt to higher temperatures. By its very success in extending its distribution into a new area, an 'invader' could cause declines in abundance of its competitors or prey populations. In only a few documented cases, mainly in successional landscapes, has the 'invader' filled a 'vacant niche' without negative effects on endemic populations (Werner 1976, 1977, 1979).

Community 'assembly rules' have been identified for only a handful of organisms, usually guilds of competing species (e.g. certain groups of birds or fish). Nevertheless, lessons about what functional groups are required to sustain a community may be gleaned from these studies, and could provide useful predictors to understand how communities may respond in rapidly changing environments.

Ecosystem responses

Ecosystems are made up of community assemblages interacting with each other and with abiotic components within a defined geographical region. The components of the biota are organised by trophic level related to their overall function in the system: primary producers, herbivores, carnivores and decomposers (Table 3.1, Principle 6). Tri-trophic interactions – such as those between herbivorous insects or grazing mammals, the plants on which they depend and their enemies (predators, parasites and pathogens) – are among the best known relationships within ecosystems. This is due in part to our vested interest in the middle level of the relationship (meat, fish), because grazing mammals account for a substantial portion of Australians' diets and economy, and herbivorous insects are major crop and forest pests – and thus competitors with humans for food and fibre.

Ecosystem processes include production, nutrient cycling, decomposition and energy transfers across trophic levels. These processes are highly dependent on abiotic factors such as temperature, water and nutrient availability, and/or substrate acidity. A change in one of these abiotic factors has the potential to change ecosystem functioning substantially. Any changes in species or community assemblages also have the potential to change ecosystem structure (trophic structure and/or relative abundance of community types) and/or ecosystem functioning substantially.

Ecosystem change has occurred naturally in geological time, with the repeated appearance of new species either via evolution or by dispersal to new areas, leading to novel species assemblages (Williams and Jackson 2007). With the current rapid rate of climate change, novel combinations of species will most certainly appear in the future, creating communities that have no present-day analogue (Hobbs *et al.* 2006; Lindenmayer *et al.* 2008a) (see also section 3.4, novel ecosystems).

As changes in Australian ecosystems occur, flow-on effects on ecosystem services to humans are potentially highly significant. These services include provisioning services such as food, fibre and water; regulating services such as pollination, pest control and water purification; and supporting services such as biogeochemical cycling and productivity (Dunlop and Brown 2008; see also Chapter 2). The degree of species redundancy or substitutability of species, without affecting ecosystem processes, is generally unknown (Fischer *et al.* 2007). There are few studies about how climate change will affect ecosystem services (but see Chapter 4; IPCC 2007b).

Biome responses

Biomes are 'climate-mediated, regional units of the Earth's biota'. They are the sum of plant and animal communities coexisting in the same region, or climatically equivalent geographic regions (Werner and Wigston 1994). In general, biomes are named by one or a few dominant life forms; examples include temperate sclerophyll forest, tropical rainforest, wooded savanna and coral reef. The life forms of the biota are the biological units used for the analyses, description and classification of biomes. Often the broad climatic regime is also included in the name of the biome.

Any particular biome includes all the constituent plant and animal community types (including all successional stages after disturbance), and the embedded geological features (e.g. watercourses), as well as landscapes influenced by human activity. At this level, it is useful to think of biodiversity as composed of a mosaic community or ecosystem (e.g. temperate forest) in which is embedded various other minor spatial elements (e.g. rock outcrop communities, swamp vegetation, floodplain ecosystem). Climate change may alter the matrix itself or, alternatively, only some of the elements or the ecotonal edges of the matrix and minor elements. For example, 'savanna' is the biome of Kakadu National Park, with component parts such as savanna grassland, savanna woodland, monsoon forest, rocky vegetation and floodplain ecosystems. Climate change may alter the matrix (e.g. produce a woodier grasslands and woodlands) or cause ecotonal edges of component parts to shift (e.g. monsoon forest increasing at the expense of grassland). Human-dominated or modified landscapes have become so pervasive since the industrial revolution (MA 2005) that they now also enter the matrix of elements that make up a biome. In fact, a new classification of biomes based on patterns of human settlement and use has been proposed (Ellis and Ramankutty 2008).

Earth's biomes have changed over millions of years as the climate changed over geological time scales, through changes internally among and within the component ecosystems, and by components moving individually to new sites and forming new biomes. A primary example was the shift in Australian biomes over the past 50 million years as the continent moved into the 30s latitudes; the dominant rainforest biome was replaced by arid and semi-arid biomes. A more recent example, on other continents, is the recolonisation of newly exposed areas as continental ice sheets melted after the most recent glacial period. The Earth's biomes have been relatively stable since that time. A very significant change in the climate is required to shift the distribution of a biome, or to create a totally new biome; nevertheless, it is an open question as to how they will change, or are beginning to change, with current and future rapid climate change.

5.3 ECOLOGICAL COMPLEXITY: MAKING PROJECTIONS DIFFICULT

As indicated above, even predicting change at the level of individual organisms and populations is difficult due to the lack of basic knowledge about the biology of Australian species and, in particular, about what currently limits their distributions. The following additional complex issues affect our ability to predict change at all levels of biological organisation.

5.3.1 Non-linearity, time lags, thresholds, feedbacks and rapid transformations

Biological systems, from individuals to ecosystems, generally respond to environmental changes in a non-linear fashion (Table 3.1, Principle 8). Time lags also play a large role, mainly due to development times and life spans of individual organisms. An example is the differential time lag among biota in response to the removal of feral water buffalo *Bubalus bubalis* in Kakadu National Park in the 1980s, which created large effects that differed among ecosystems. In some cases the systems recovered well, but in others permanent changes may have occurred (Box 3.9).

Physiological and life history thresholds are common at the individual and population levels of the biological hierarchy. Rapid transformations after a long period of no or little change are more the rule than the exception in populations, communities and ecosystems, perhaps due to complex internal interactions that are non-linear. These rapid transformations are related to 'tipping points' or 'critical thresholds' and occur when a relatively small perturbation qualitatively alters the state or development of a system (Lenton *et al.* 2008). Examples are numerous and found at all levels of the biological hierarchy (Box 5.3), although few have been described in Australia other than coral reef dynamics.

Human impacts on the environment also have the potential to modify feedbacks to the climate, and perhaps contribute to the approach to or crossing of thresholds. For example, modelling studies of the impact of 150 years of land clearing in eastern (McAlpine *et al.* 2007) and south-western Australia (Pitman *et al.* 2004) have indicated that replacing the native woody vegetation with crops and grazing has resulted in significant changes in regional climate, with a shift to warmer and drier conditions. Indeed, it has been suggested that the 2002–2003 El Niño drought in eastern Australia was up to 2°C hotter as a direct result of vegetation loss (McAlpine *et al.* 2007).

5.3.2 Averages vs extremes

Extreme climate and weather events are infrequent events at the high and low range of values of a climate or weather variable (Nicholls 2008). Trends in extremes are highly correlated with trends in means for both temperature and precipitation in Australia. However, the trend for daily rainfall extremes is often greater than the trend for the mean, indicating that the frequency of extreme rainfall events is changing faster than the mean (Alexander et. al. 2007). A small change in the average of a climate variable can cause a large change in the frequency of extreme events; for example, a change in the average temperature and the number of frosts (Nicholls 2008) (Fig. 5.3). Projections for the future suggest increases in the frequency and intensity of many extreme events such as drought, storm surges and fire (Table 4.1).

While aggregated community attributes like productivity are likely to change in response to changes in average values (means) of continuous environmental variables such as temperature or precipitation, many aspects of species responses will be affected primarily by changes in the extreme values of a climatic variable (Parmesan *et al.* 2000) (Fig. 5.3). The boundaries of a species distribution may be determined by the maximum summer temperature in a location, or by the minimum winter temperature. For example, the green ringtail possum *Pseudochirops archeri*, a species endemic to the Queensland Wet Tropics, cannot tolerate more

Box 5.3 Two examples of ecological 'surprise' following otherwise sensible management decisions. Both results are due to temperatures exceeding previously unknown thresholds, which then allowed two generations of insects to be produced in a single year. (Will Steffen)

Example 1. Wildfire management of coniferous forests.

The issue of wildfires in spruce forests in Alaska has been a long-standing one (pre-dating concern about climate change). As in many parts of the world, fire suppression has been practised in forests near human settlements to safeguard property and lives. Suppressing fires has led to stands of old growth/very mature forests that have not been subjected to a natural disturbance (fire) for unusually long periods of time.

About a decade ago, under the influence of a warming climate, a herbivore in the forests – the spruce bark beetle – passed a threshold in which it could complete two life cycles in one growing season rather than one. This resulted in a population explosion of beetles that quickly devastated the very old and vulnerable trees. The result was the death of most of the trees that had been protected from fire and, ironically, wildfires then quickly spread through the stands, threatening lives and property. A grassland has now replaced the forest. Frequent fires in this community are preventing the re-establishment of the spruce trees, and the conversion from forest to grassland appears permanent (Walker *et al.* 1999).

Example 2. Immunisation campaign against tick-borne encephalitis.

In the mid-1990s the county of Stockholm (Sweden) initiated an immunisation campaign to protect the population against tick-borne encephalitis. The immunisation campaign was highly successful in terms of the proportion of the population that was immunised, but yet the incidence of the disease did not decrease, as expected.

During the time that the immunisation campaign was being implemented, the climate in the Stockholm region warmed significantly and the tick that carried the disease now began to complete its life cycle in one rather than two years. Tick numbers exploded. In addition, the number of roe deer, the major host of the tick, had increased greatly due primarily to land cover changes around the region. The two processes – the immunisation campaign and the tick population explosion – offset each other. There was no change in the incidence of the disease (Lindgren 2000).

than 4–5 hours above about 30°C ambient temperature. Its geographic range is thus limited to cool montane rainforest uplands above 600 m where temperatures meet this requirement. This species will be extremely vulnerable to temperature rises in this region in the future. Another dramatic example of the impact of extreme temperatures is provided by observations of flying fox mortality in eastern Australia over the past two decades. Since 1994, over 30 000 flying foxes (mostly the grey-headed flying fox *Pteropus poliocephalus* and also the black flying fox *P. alecto*) are known to have died during 19 high temperature episodes (>40°C) (Welbergen *et al.* 2007). Extremely high temperatures also caused a significant number of deaths of Cape Barren geese *Cereopsis novaehollandiae* on islands in the Recherche Archipelago, Western Australia, in 1991 (Halse *et al.* 1995).

One important limitation in our ability to predict future distributions of species is that most current projections of future climatic conditions are couched in terms of average temperature and rainfall. There is thus a mismatch between the output of climate models, and the needs of researchers assessing impacts on particular systems and species. Furthermore, it is well known that while means and extremes of climatic factors such as temperature are correlated, the relationship is often highly non-linear (Chambers and Griffiths 2008). This problem is becoming increasing well recognised but is likely to remain a stumbling block in predicting impacts for some time to come.

5.3.3 Synergistic interactions and surprises

Drivers of change rarely act as single agents. A simple change in a single driver can initiate a whole suite of

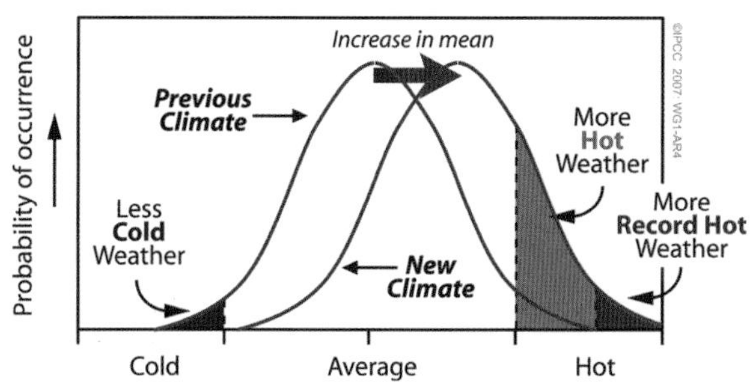

Figure 5.3 Relationship between means and extremes. Graph shows the relationship between a shifting mean and the proportion of extreme events when extreme events are defined as some fixed physiological or life history threshold. Here the environmental variable is temperature. When the mean becomes warmer (say, under climate change), the frequency of warm days past the threshold shows a disproportionate shift. Source: IPCC Fourth Assessment Report Technical Summary (Box TS.5, Fig. 1).

other changes, both positive and negative, with respect to an individual or a species, and produce a cascade of changes through an entire community and ecosystem (Table 3.1, Principle 7; and examples for ecosystems in section 3.4, community- and ecosystem-level changes). These indirect, cascading effects can be difficult to predict.

Potential large-scale 'surprises' in the Earth system, including climate, have been acknowledged for some time. They include abrupt changes (e.g. shutdown of the North Atlantic thermohaline circulation), gradual but irreversible changes (e.g. the melting of the Greenland ice sheet) and positive feedbacks such as the large discharge of methane into the atmosphere from warming permafrost. In reality, if we can anticipate such events, they are not really surprises, but nevertheless they are low-probability, high-impact events that are not expected or not well predicted in models of the Earth system. In addition, novel climates will almost surely emerge in the future; that is, combinations of temperatures, precipitation and seasonality that have no current analogues (Williams *et al.* 2007; Williams and Jackson 2007). Novel climates are particularly likely at regional scales where human modification of land cover also has an influence. Indeed, novel climates may have already emerged in some regions, such as south-west Western Australia.

Ecological surprises are also increasing, with unexpected outcomes becoming more common despite substantial knowledge of individual species biology and ecological interactions (Box 5.3). The surprises are not just due to lack of specific experience and knowledge about systems (which do contribute), but: because individuals and populations have physiological and life history thresholds; because communities exhibit keystone effects, ecological cascades and synergistic interactions; and because both deterministic and stochastic processes are involved in any environmental change. Some surprises will arise as climate change impacts on key species that structure habitats (like deep-rooted trees) or that are important for ecosystem processes (mycorrhizas or symbionts involved in nitrogen fixation). Such changes may not be as obviously visible as, for example, the local extinction of a charismatic endemic species. Other surprises arise when species encounter environmental conditions for which they have no previous experience; outcomes are virtually impossible to predict (Box 5.4).

5.4 IS BIODIVERSITY ALREADY RESPONDING TO CLIMATE CHANGE?

5.4.1 Observed changes in species and communities

On other continents, particularly in the northern hemisphere, the availability of long-term biological datasets has enabled the extensive documentation of recent climate and biological trends. The clearest evidence for such changes comes from observations of phenology (mostly advances in life cycle events) and geographic range shifts (mostly northwards and to higher elevations) (reviewed in Hughes 2000; IPCC 2007b; Parmesan 2006; Parmesan and Yohe 2003; Poloczanska *et al.* 2007, 2008; Root *et al.* 2003, 2005; Walther *et al.* 2002). Expansions at the colder edges of ranges appear to be occurring more rapidly than retractions at warmer edges (Parmesan 2006). It is not clear whether this is just a lag effect (colonisations occurring faster than local extinctions) or because minimum temperatures in many regions are increasing faster than maximum temperatures (IPCC 2007a), or simply because observations of colonisation are easier to observe than confident detection of local extinctions. There is also evidence that some organisms

Box 5.4 Climate, fire and the little penguin.
(Lynda Chambers, Leanne Renwick and Peter Dann)

In coastal regions, misty rain or fog following long spells of hot, dry and dusty weather can result in the ignition of powerpole cross arms, due to a build-up of salt and dust on the insulators. The red-hot salt crust can fall from the pole and ignite vegetation at its base. Corrosion of high-voltage power lines can also cause breakage and then fire when they fall to the ground. In recent years a number of such fires have occurred on Phillip Island, home to a large colony of little penguins *Eudyptula minor*.

These fires, as well as a lightning-initiated fire late in 2005 on Seal Island in Victoria, have caused the death or injury of many little penguins. In each case, the penguins did not avoid the fire, suggesting that their responses to fire are surprisingly inappropriate and maladaptive. In many cases, dead penguins were found either in their burrows (often collapsed) or within metres of burrows (see photo further on). Birds nesting under vegetation appeared to remain until they were severely burnt or killed. Penguins were also observed standing beside flames preening singed feathers, rather than moving away. Most live penguins suffered debilitating injuries including burns to their feet and legs, scorched feathers and blistered skin, swollen eyes, and many had difficulty breathing.

The synchronised breeding of seabirds such as penguins when large numbers are present in a colony makes them particularly vulnerable to such fires during their nesting seasons. This is particularly true for burrow-nesting species, such as the little penguin, which are disinclined to abandon eggs or chicks, or emerge from their burrows during daytime.

Increased occurrence of hot, dry and dusty weather is projected for the future and may result in increased fire-related risk of little penguin death and injury on Phillip Island if these periods are followed by rain or fog. As coastal development encroaches on little penguin colonies throughout south-eastern Australia, this risk is heightened. Risk reduction options include running the power underground, more regular pole inspections, improved insulator design and cleaning of the insulators. The risk can be reduced further by appropriate habitat management such as the planting of fire-retardant vegetation and quick response by agencies when a fire does occur.

Dead chick and eggs after fire.
Source: Photo by Peter Dann.

are responding genetically to the strong selective pressures imposed by climatic changes (Bradshaw and Holzapfel 2006). Most of the empirical evidence for rapid genetic adaptation comes from examples of increased frequencies of existing heat-tolerant genotypes, especially in populations in the centres of species ranges (Parmesan 2006). The precise role of genetic change vs manifestations of phenotypic plasticity in observations thus far, however, has been questioned (Gienapp *et al.* 2008).

To the extent that similar organisms respond to climate change in similar ways in Australia, we have confidence that many changes in species' life cycles and distribution recently observed in Australia are also due, at least in part, to climate change. It is important to acknowledge that many of these changes are likely to also have significant non-climatic components (Chapter 3). The precise role of different factors may continue to be almost impossible to quantify in most cases. Nevertheless, an increasing number of the

Figure 5.4 A wedge-tailed shearwater *Ardenna pacifica* at the entrance of its nesting burrow. Higher sea temperatures are leading to reduced breeding success in northern Australia. Source: Andrew A Burbidge.

Figure 5.5 Black flying foxes *Pteropus alecto* are extending their breeding range southwards. Source: Andrew A Burbidge.

changes in Australian biodiversity documented in recent times are consistent with having a climate change signal, mainly due to changes in temperature and rainfall.

Most of the recently observed changes in biodiversity have been at the species level, due partly to the visibility of larger mobile species such as birds, and partly to the nature of biological organisation itself. 'Fast' processes such as dispersal, migration and population growth in small organisms will be more obvious in many species. Greater time lags are predicted in responses of 'slow' processes such as vegetation change, reef building or reproduction in large organisms.

Table 5.2 provides some examples of recently observed changes in Australia that are consistent with having a climate change 'fingerprint'. This is not an exhaustive list, but highlights the wide range of species, taxa and communities that appear to be responding to recent changes in climate. What is most noteworthy about these observations, and those globally, is that in many cases significant impacts are apparently occurring with extremely modest increases in temperature compared with what is expected over the coming decades.

5.4.2 Observed changes in ecosystems

At the ecosystem level, observed changes in Australia cannot be definitively attributed to climate change to date, largely because of the many significant changes

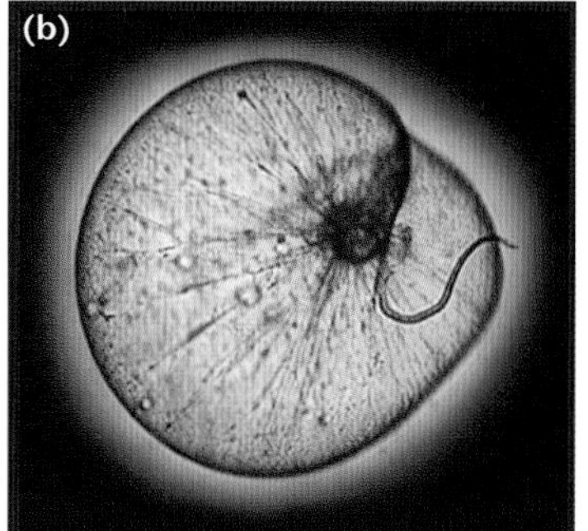

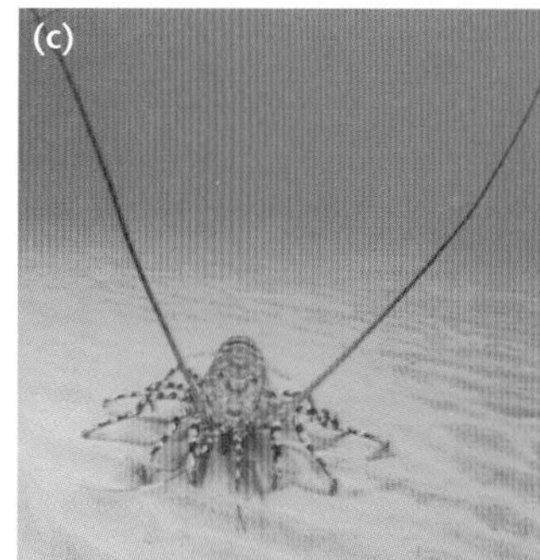

Figure 5.6 Examples of changes in south-eastern Australian marine ecosystems due to climate change. (a) Invasion of sea urchins native to NSW coast causing barrens (loss of kelp) off eastern Tasmania. Source: Nic Bax. (b) Changing composition of phytoplankton blooms off Tasmania – increased tropical species and occurrences of red tides. Source: Lisa Albinsson. (c) Rock lobster catch and distribution correlated with regional sea surface temperature changes around Tasman Sea. Source: Alistair Hobday.

Table 5.2 Examples of observed changes in Australian species and communities consistent with a climate change signal.

Type of change	Observations
Genetic constitution	Allele frequency shifts at the alcohol dehydrogenase locus in wild populations of the fruit fly *Drosophila melanogaster*, equivalent to a 4° shift in latitude (about 400 km) over 20 years (Umina *et al.* 2005); increases in warm-adapted forms of chromosome inversion arrangements in *Drosophila* spp. equivalent to a 5° shift in latitude over 20 years (Anderson *et al.* 2005); shifts in an inversion polymorphism and an allozyme polymorphism in *Drosophila* since the 1970s–1980s (Hoffmann and Weeks 2007).
Geographic ranges	Increased penetration of feral and native mammals to higher elevations in alpine and sub-alpine areas, and prolonged presence of macropods in winter (Green and Pickering 2002; Pickering *et al.* 2004).
	Range shifts and expansions of several bird species to higher elevations or higher latitudes (Chambers *et al.* 2005; Chambers 2007; Dunlop 2001; Garnett and Crowley 2000; Hughes 2003; McAllan *et al.* 2007; Olsen *et al.* 2003); shifts in the southerly boundaries of several species estimated to be approximately 100–150 km per decade with declines in some northerly parts of ranges, and southerly movement of the transition zone between tropical and temperate seabirds (Olsen 2007); establishment of new breeding colonies of several seabird species south of historical ranges off the Western Australian coast (Dunlop 2007).
	Black flying foxes *Pteropus alecto* have expanded their breeding ranges south by more than 750 km in the past 75 years, coinciding with a northerly range contraction of 250 km for the Grey-headed flying fox *P. poliocephalus* (Eby 2000; Welbergen *et al.* 2007; Fig. 5.5); the southerly range expansion of *P. alecto* may be associated with reduction in frosts (Tidemann 1999).
	Southern range extension of the barrens-forming sea urchin *Centrostephanus rodgersii* from the mainland to Tasmania, associated with increased southerly penetration of the East Australian Current (Ling *et al.* 2008, 2009; Fig 5.6a); coral reefs such as *Acropora* spp. establishing at higher latitudes at Rottnest Island, Western Australia, attributed to both warmer temperatures and lack of competition from macroalgae (Marsh 1993); changing composition of phytoplankton blooms off Tasmania (Blackburn 2005; Fig 5.6b); along the east coast of Tasmania, 17 of 32 (53%) intertidal species have shifted their distribution southward since the 1950s, an average increase of 145 km; on the west coast of Tasmania, 22% of the 32 intertidal species have shifted southward an average increase of 150 km in that time (Pitt 2008; Fig. 5.6c).
Life cycles	Earlier arrival and later departure times of migratory birds in Australian breeding and feeding grounds (Beaumont *et al.* 2006; Chambers 2005, 2008; Chambers *et al.* 2005); changes in number of breeding seasons per year (forest kingfishers *Todiramphus macleayii*) and altered patterns of over-wintering (white-throated nightjars *Eurostopodus mystacalis*, little bronze-cuckoos *Chrysococcyx minutillus*) in south-east Queensland (Chambers *et al.* 2005).
	Earlier mating and longer pairing of the large skink *Tiliqua rugosa* (Bull and Burzacott 2002).
Populations	Reduced reproduction in wedge-tailed shearwaters *Ardenna pacifica* on the Great Barrier Reef associated with higher sea temperatures (Smithers *et al.* 2003; Fig. 5.4); higher sea surface temperatures and recent El Niño events also associated with population declines in northern Australian birds (Devney and Congdon 2007).
	Population increases in black-browed albatrosses *Thalassarche melanophris*, King penguins *Aptenodytes patagonicus* and fur seals *Arctocephalus gazella* on Heard Island, associated with ocean warming but also with changed fishing practices (Weimerskirch *et al.* 2003; Woehler *et al.* 2002).
	Declines in mountain pygmy possum *Burramys parvus* and broad-toothed rat *Mastacomys fuscus* associated with declining snow cover, drought effects on food supply, loss of habitat after fire and increased predation (feral cats and foxes) (see Box 5.10).

Type of change	Observations
Populations	Otolith analysis indicates significant changes in growth rates of long-lived Pacific fish species; increasing growth rates in species that live in the top 250 m associated with surface warming; declining growth rates in deeper-dwelling species (>1000 m) associated with long-term cooling at these depths (Thresher *et al.* 2007).
	Widespread mortality of montane and central plateau eucalypts in Tasmania linked to a 30-year rainfall deficit (L Gilfedder, pers. comm.)
Ecotonal boundaries	Expansion of rainforest at the expense of eucalypt savanna woodland and grassland in the Northern Territory (Banfai *et al.* 2007; Bowman *et al.* 2001; Bowman and Dingle 2006; Fig. 5.7), Queensland and New South Wales (Harrington and Sanderson 1994) associated with increases in rainfall and changes in fire regimes.
	Encroachment by snow gums *Eucalyptus pauciflora* subsp. *niphophila* into sub-alpine grasslands at higher elevations (Wearne and Morgan 2001); woody shrub expansion into alpine heathlands and grasslands (R Williams, pers. obs.).
	Saltwater intrusion into freshwater wetlands since the 1940s in the Northern Territory, accelerating since the 1980s, and possibly associated with changes in sea level and rainfall, but also with impacts of introduced buffalo *Bubalus bubalis* (Bayliss *et al.* 1997; Eliot *et al.* 1999; Skeat *et al.* 1996; Winn *et al.* 2006; Woodroffe and Mulrennan 1993).
	Plant colonisation of areas exposed by glacial retreat, and changes in competitive relationships in plant communities on Heard Island (Bergstrom 2003); decline in area of sphagnum moss since 1992 on Macquarie Island associated with drying trend (Whinam and Copson 2006).
CO_2-related impacts	Woody 'thickening' of vegetation in some parts of northern Australia, consistent with a trend observed throughout the world's savannas and grasslands (Berry and Roderick 2002; Gifford and Howden 2001; Petty *et al.* 2007). In the Top End (northern Australia), this may be due to increases in precipitation but the patterns are not consistent, with fire and grazing regimes playing a large role (Sharp and Bowman 2004; Lehmann *et al.* 2008; Petty *et al.* 2007; Crowley *et al.* 2009).
Ecosystems	Eight mass bleaching events since 1979 on the Great Barrier Reef triggered by unusually high sea surface temperatures; no serious events known prior to 1979. Most widespread events occurred in the summers of 1998 and 2002, affecting up to 50% of reefs within the Great Barrier Reef Marine Park (Berkelmans and Oliver 1999; Done *et al.* 2003; Hoegh-Guldberg 1999); declines in live coral cover followed by declines in obligate coral-feeding fish (Pratchett *et al.* 2006).
Disturbance regimes	Fire regimes have changed over the past two decades throughout Australia, consistent with drier and hotter climate in the southern part of the country, with an increased incidence of days of extreme fire weather (Box 5.11).

to the structure and functioning of ecosystems caused by the range of proximate human drivers described in Chapter 3. However, there is one notable exception – coral reefs. Although reefs are subject to a number of direct human stressors, such as fishing and nutrient runoff from adjacent coastal regions, climate change has produced multiple, discernible impacts on the reefs, leading to overall decline in reef health and range (Box 5.5).

5.5 WHAT DOES THE FUTURE HOLD FOR AUSTRALIA'S BIODIVERSITY?

5.5.1 Which species will be 'winners' and 'losers'?

Responses of species to rapid climate change will be individualistic, with some species potentially advantaged ('winners') and others disadvantaged ('losers'). The vulnerability of an individual species (defined

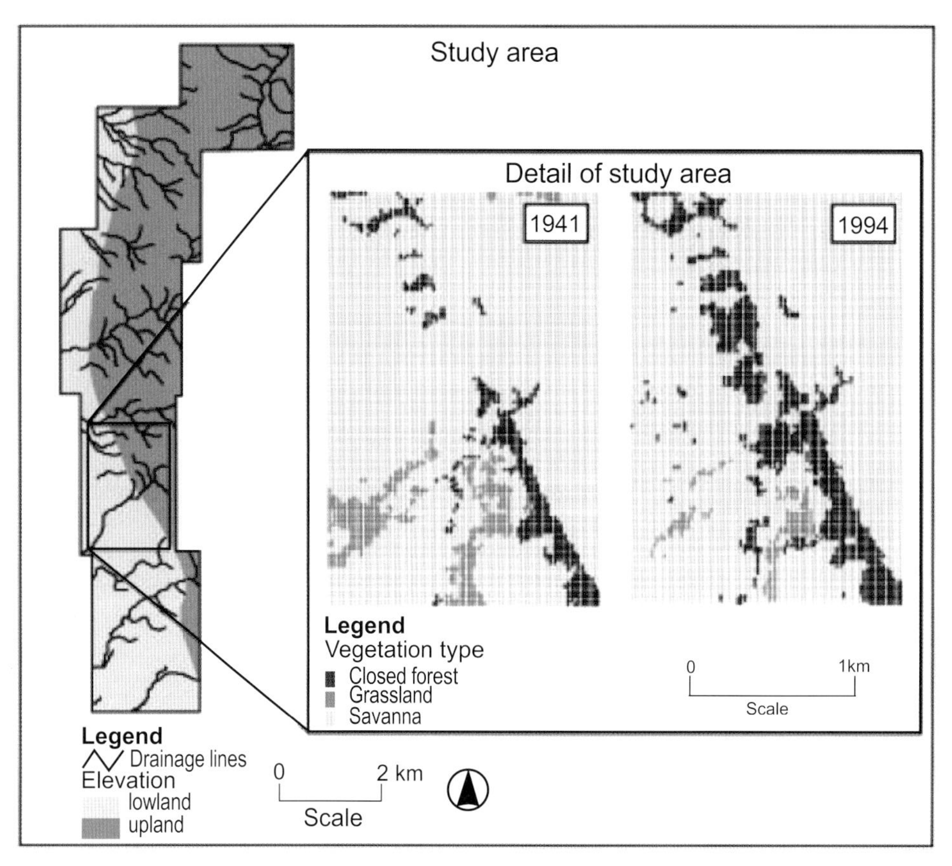

Figure 5.7 Introduced pasture grasses spreading into undisturbed northern savannas lead to fire regimes characterised by more frequent and intense fires (e.g. more, hotter late-season fires), which in turn can cause declines in tree cover, species diversity and lead to other broad ecosystem changes. Source: Bowman, Walsh and Milne (2001).

here as the susceptibility of a species to a negative impact, sensu Smit *et al.* 2000) will depend on a combination of several factors:

- *life history* traits and other traits of the species (McKinney 1997; Thuiller *et al.* 2005)
- degree of *exposure* to rapid climate change in the habitat and region where the species lives
- *capacity* of the species to adapt, either genetically or behaviourally (section 5.2).

Life history and other traits

The suite of traits that predispose a species to being vulnerable to disturbance and threats, such as land clearing or invasive species, overlap with the suite of traits that predispose species at risk from rapid climate change (Table 3.1, Principle 3; Fig. 5.8). Those species with traits that may advantage them in a rapidly changing climate may also be advantaged in disturbed areas, and may be successful invaders and colonists. Indeed, many species, both native and exotic, that are not currently considered to be invasive may expand their ranges and increase in abundance to such an extent that they have transforming, and sometimes negative, impacts on other species and ecosystems. Within wide-ranging species, individual populations may be affected differently, either because of ecotypic differences (i.e. population adaptations to local conditions), or regional variation in climate, or both (e.g. King *et al.* 1995).

Of particular concern is that many species of Australian flora and fauna are endemic, with restricted climatic and geographic ranges (Chapter 2). For example, over 50% of Australian eucalypt species – the tree

Box 5.5 Climate change and the Great Barrier Reef: impacts and adaptation. (Paul Marshall, GBRMPA)

The Great Barrier Reef World Heritage Area is one the most biologically diverse regions in the world. Climate change is now recognised as the greatest long-term threat to the Great Barrier Reef (GBR), with implications for nearly every part of the ecosystem. Coral reefs are among the most vulnerable of all ecosystems to climate change, due in large part to the high sensitivity of corals to small increases in water temperature. When sea temperatures exceed the long-term summer maximum by only 1–1.5°C for as little as six weeks, extensive coral bleaching occurs, leading to widespread coral mortality if temperatures do not return to normal levels. Unusually warm sea temperatures have now caused serious and lasting damage to 16% of the world's coral reefs. While the GBR has fared well by comparison, major bleaching events in 1998 and 2002 saw over 50% of reefs bleached and up to 5% considerably damaged each year (Great Barrier Reef Marine Park Authority (GBRMPA) 2007a).

However, the impacts of climate change extend beyond the immediate effects on corals. A vast array of organisms depend on corals for habitat and food, and many more will be affected directly or indirectly by shifts in environmental conditions brought about by climate change. In an effort to understand the full implications of climate change, and to build a knowledge base to inform adaptation strategies, the GBRMPA coordinated the publication *Climate change and the Great Barrier Reef: a vulnerability assessment*. This collaboration between over 80 leading climate and marine experts presents climate projections for the GBR region to 2100, assesses how these changes will affect the GBR and identifies strategies that can minimise climate change impacts. Climate projections for the GBR region show that sea and air temperatures will continue to increase, sea level is rising, the ocean is becoming more acidic, intense storms and rainfall will become more frequent, and ocean currents will change. These changes will have consequences for many reef species and habitats, as well as ecosystem processes, and the industries and communities that depend on the GBR.

The assessment extends our knowledge of the vulnerability of all parts of the GBR ecosystem, from marine microbes and plankton to fishes, seabirds and charismatic megafauna. The vulnerability of different species groups varies widely; for example, hard corals are extremely vulnerable to changes in sea temperature and ocean acidification, whereas many fleshy and turf macroalgae are likely to benefit from increased substrate and nutrient availability.

GBR habitats – coral reefs, pelagic environments, coasts and estuaries, and islands and cays – each of which encompass complex interdependencies and links between species and their environment are also vulnerable, with islands and coral reefs expected to undergo particularly substantial adjustments as the climate changes.

An overwhelming conclusion of the GBR vulnerability assessment is that key components of the GBR are highly vulnerable to climate change, and signs of this vulnerability are already evident. Some further degradation is inevitable as the climate continues to change, but the extent of the decline will depend on the rate and magnitude of climate change, and the resilience of the ecosystem.

Adaptive, resilience-based management offers the best hope of limiting the impacts of climate change on the GBR. Management and governance systems need to be flexible, so that managers can respond rapidly to opportunities to reduce local stresses, to protect sites with favourable characteristics or modify management practices as ecosystems change.

Understanding the social and economic implications of climate change for communities and industries that depend on the GBR, such as fisheries and tourism, is important to assist with adaptation and to ensure sustainable industries into the future. The focus must be on facilitating adaptation to bring about positive changes in the interactions between people and the GBR.

Management strategies

Recent management initiatives for the GBR have resulted in an increase in biodiversity protection, a multi-stakeholder agreement to address water quality and a multi-use marine protected area. Despite these landmark initiatives, there is an urgent need to identify ways to reduce impacts related to climate change. Strategies that enhance the resilience of the ecosystem, by reducing stresses from other human activities (e.g. from poor water quality or overfishing), are the most promising. Stressed ecosystems are less likely to recover from disturbances and more likely to collapse or flip to alternative states from which they might not recover. Similarly, industries and regional communities that adapt to changing conditions and continue to prosper have much greater potential for effective stewardship of the GBR into the future.

While climate change is certain to cause further degradation of the GBR, not all sites will be equally affected. Sites naturally resistant to climate-related stresses, and sites that could serve as climate change refugia, warrant consideration for special protection from other threats. Industries and regional communities vary in their dependency on ecosystem services provided by the GBR, and thus their sensitivity to changes in reef condition. An understanding of these dependencies, and of the resilience of social and economic systems, will help guide investment in adaptation initiatives at regional and national levels. Importantly for both the GBR and its stakeholders, the size of the GBR and the effective management regime in place render it more resilient than most coral reefs in the world.

The *Great Barrier Reef Climate Change Action Plan 2007–2011* (GBRMPA 2007) outlines a strategic response to the threat of climate change for the GBR. It identifies direct actions and partnerships that will increase the resilience of the GBR to climate change, and build the adaptive capacity of industries and communities that depend on a healthy reef. By facilitating adaptation, the Plan will help minimise impacts on GBR industries such as tourism, commercial fishing and recreational fishing that together contribute $6.9 billion to the national economy per annum, while also giving this international icon the best chance of coping with climate change.

Conversion of a healthy coral reef to an algae-dominated ecosystem via repeated bleaching events.

Source: Paul Marshall (GBRMPA).

Species least at risk	Species most at risk
• Physiological tolerance to broad range of factors such as temperatures, water availability and fire • High degree of phenotypic plasticity • High degree of genetic variability • Short generation times (rapid life cycles) and short time to sexual maturity • High fecundity • 'Generalist' requirements for food, nesting sites, etc. • Good dispersal capability • Broad geographic ranges	• Narrow range of physiological tolerance to factors such as temperature, water availability and fire • Low genetic variability • Long generation times and long time to sexual maturity • Specialised requirements for other species (e.g. for a disperser, prey species, pollinator or photosynthetic symbiont) or for a particular habitat that may itself be restricted (e.g. a particular soil type) • Poor dispersers • Narrow geographic ranges

Figure 5.8 Continuum of physiological and life history characteristics in response to disturbance. Source: Lesley Hughes.

genus that dominates all but the most arid landscapes – have distributions that span less than 3°C of mean annual temperature, with 25% spanning less than 1°C (Hughes *et al.* 1996). However, many of Australia's endemic species have adapted to a highly variable climate, especially in arid and semi-arid regions, and thus may have greater resilience to climate variability than their narrow ranges might otherwise suggest. In addition, their narrow ranges may be due to species–species interactions rather than to their fundamental environmental niche.

The ability to disperse, particularly across fragmented landscapes, will be a crucial factor determining a species' capacity to adapt to shifting climate zones. Plants with small, numerous seeds dispersed by wind will be better able to disperse to new sites, although they may be at a disadvantage during their establishment phase if they have to compete with existing vegetation. Plants with seeds dispersed by vectors that are themselves capable of long-distance movement, like birds or bats, may also be advantaged. Animals capable of flight will also disperse faster, and many observed range shifts attributed to climate change are in such species (e.g. birds, butterflies, flying foxes) (Table 5.2). In marine environments, the length of the larval development time is positively correlated with dispersal distance. There appears to be a bimodal distribution of dispersal, with some species tending to develop quickly (sometimes in just minutes to hours) and disperse less than 1 km during this time. Other species may spend weeks to months in the dispersive phase, and be able to disperse many kilometres (Shanks *et al.* 2003). Warming temperatures will tend to reduce development time and thus reduce dispersal distance (O'Connor *et al.* 2007; Munday *et al.* 2009). Species confined to freshwater lakes and waterways may only be able to disperse along temperature gradients if the geography permits, and may otherwise be trapped within increasingly hostile environmental conditions.

Species' generation times and longevity will also play a role. Those species with short generation times will have greater capacity for evolutionary adaptation in situ. Similarly, annual plants will have far more capability for rapid adjustment than long-lived trees that are only capable of reproduction at a mature age. However, long-lived species may also be more resilient to physiological stress in situ.

Species with specialist requirements for particular habitats, mutualists or hosts will be intrinsically more vulnerable to rapid change than those with more generalist habits. Generalist species may be buffered from change by, for example, switching to new hosts (in the case of herbivores on plants) or simply using other habitats.

Degree of exposure

The degree and rate of climate change in the future will vary from region to region. Relatively more warming is expected for the inland, compared with coastal

Box 5.6 Australia's islands and sea-level rise: Houtman Abrolhos example. (Andrew Burbidge)

Thousands of islands occur within Australia's jurisdiction, including oceanic islands that are volcanic in origin such as Christmas Island in the Indian Ocean, continental islands that were isolated from the mainland by rising sea levels during the Pleistocene epoch, and coral and sand cays that have accumulated on reefs. Many of these islands are extremely important for biodiversity conservation.

Low-lying islands are often important seabird nesting sites and sea turtle rookeries. As is the case elsewhere in the world, Australia's low-lying islands will be affected significantly by sea-level rise. Examples include the Houtman Abrolhos Islands off the west coast, the islands of the Capricorn group at the southern end of the Great Barrier Reef (including Heron, Lady Musgrave and North West islands), cays in the Coral Sea and islands in Torres Strait.

The Houtman Abrolhos comprises about 120 small, low-lying islands and has the greatest species diversity and largest concentration of breeding seabirds in the eastern Indian Ocean. It also provides a base for rock lobster fishers each year during the lucrative Abrolhos season and is a significant tourist destination. The mixture of species is unique, as the breeding islands are shared by subtropical (cool water) and tropical species, and both littoral and oceanic foragers. One listed threatened seabird, the lesser noddy *Anous tenuirostris melanops*, breeds only in three small areas of white mangrove *Avicennia marina* on three low Houtman Abrolhos islands.

The three largest islands in the Houtman Abrolhos are continental in origin and rise to a maximum of 14 m above sea level. They harbour a wide range of terrestrial animals, including Tammar wallabies *Macropus eugenii*, the threatened Abrolhos painted button-quail *Turnix varius scintillans*, the threatened Abrolhos dwarf bearded dragon *Pogona minor minima*, and other terrestrial birds and reptiles. However, even though sea-level rise may not have a major direct impact, these sandy islands may erode severely once sea levels rise due to exposure to increased wave energy.

Nesting colony of crested terns *Sterna bergii* on Pelsaert Island, Houtman Abrolhos. More than one million seabirds of 18 species nest on this low-lying island.
Source: Andrew Burbidge.

The vulnerable lesser noddy *Anous tenuirostris melanops* nests in only 4.1 ha of mangroves on three low-lying islands in the Houtman Abrolhos.
Source: Andrew Burbidge.

regions (Table 4.1). Australian arid–zone species, however, have evolved in a habitat with both great climatic extremes and high interannual variability, and may therefore be better equipped than mesic or coastal species to cope with rapid change. Many species, for example, have reproductive strategies that are highly opportunistic, being cued to occasional rainfall events. Australia also has a high proportion of species that are facultatively migratory, especially butterflies (Dingle *et al.* 2000) and birds (Chambers *et al.* 2005); these species are able to migrate relatively long distances to take advantage of favourable conditions. Warming is also expected to be relatively less at lower latitudes, so tropical species may be affected more slowly by warming trends than temperate species (CSIRO and BOM 2007; IPCC 2007a). These species, however, may have less intrinsic capacity to withstand even fairly modest changes, having evolved in a region with a more predictable environment (Deutsch *et al.* 2008). Predicted rainfall changes in northern regions are also relatively small compared with present patterns of temporal and spatial variability (Williams *et al.* 1995). On the other hand, these regions may well bear the impact of rising sea level sooner than many other parts of the continent; sea levels across northern Australia are currently rising at 7.8–8.3 mm/year (National Tidal Centre 2006), about four times the global average.

Within a habitat, some species may have more opportunity than others to take advantage of differences in microhabitats. Regions with high topographic relief, such as dissected plateaus with cool, moist gorges, may continue to provide refugia for some species as the regional climate warms. Species restricted to high elevations or high latitudes, to low-lying islands (Box 5.6), or to ephemeral habits such as intermittent streams and inland wetlands will be particularly at risk in the short term.

Capacity to adapt

As outlined in section 5.2, some species may adapt genetically or have sufficient phenotypic plasticity to tolerate new conditions in situ. Others may be able to cope, at least in the short to medium term, by altering their use of microhabitats or by shifting their geographic range. For mobile species physically capable of travelling some distance to more suitable areas, their capacity to do so will depend on the connectivity of suitable habitat and/or the 'permeability' of the landscape matrix between suitable habitats. Some species, although capable of shifting their range, will be prevented from doing so by physical barriers such as coasts or extensively cleared agricultural land. Clearly, species that are currently restricted to either Tasmania, to southern parts of the continent, to isolated lakes and waterways, or to mountain tops will simply have nowhere to go. Marine species confined to the continental shelf (waters <200 m) will also run out of habitat in the south of Australia, because warming will continue to occur from the north along both east and west coasts.

5.5.2 Predicted general trends

Species

From an assessment of the part played by species life histories, degree of exposure and capacity to adapt, we can make broad predictions as to what general trends to expect in species' responses to climate change over the next few decades:

- local extinctions of populations may occur at the lower elevations and northern or hottest edges of species' ranges as individuals become progressively more stressed by increasing temperatures, and/or physiological thresholds are exceeded on hot days
- colonisation of new sites at higher elevations and at the southern or cooler edges of species' ranges as establishment success increases due to reductions in frosts or limiting cold days
- continued observations of range expansions of the more mobile species to the south or at higher elevations
- progressive decoupling of present-day interactions between species as species respond at different rates to climatic and atmospheric changes
- spread of ecological generalists (both native and exotic) at the expense of native specialists via competition, predation and/or disease
- possible global extinctions of narrow-ranged endemics, especially species with low dispersal ability, and species already confined to the highest elevations of montane areas, southern coastal areas or low-lying coastal habitats
- geographical ranges of widespread species will become fragmented with smaller, patchily distributed populations.

Species interactions

Climate change is expected to affect many interactions between species. As any one species becomes advantaged or disadvantaged, all species with which it interacts (e.g. pollinators, competitors, predators) may be affected indirectly. Indeed, such indirect biotic effects may have greater impacts on many species than the direct impacts of changes in temperature and rainfall. Progressive decoupling of present-day interactions between species will result in changes in trophic interactions, food web structure and ecosystem processes (Table 3.1, Principles 4, 5 and 6).

Potential changes in species interactions include the following:

- changes in the synchrony of life cycles because one partner in an interaction is cued differently from another by warming temperatures. This type of change may be particularly important for species whose successful reproduction relies on seasonal food supplies – such as fledgling birds, which rely on a spring flush of emerging insects (Visser and Both 2005), and plants that rely on insect pollinators. In some interactions, photoperiod may be a proximate cue for the behaviour of one partner, while temperature may trigger behaviour in the other
- changes in plant–herbivore interactions mediated by changes in the chemical composition of plants as atmospheric CO_2 increases (Box 5.1). Plants grown at elevated CO_2 are generally less nutritious due to increased C:N ratios and reductions in digestibility (Kanowski 2001). The performance of insects and other herbivores feeding on plants grown at elevated CO_2 is generally reduced. Despite increased consumption rates, these insects often develop more slowly, have reduced body size and fecundity as adults, and are more vulnerable to parasitism and predation (e.g. Johns and Hughes 2002). All herbivores may potentially be affected, but impacts may be particularly severe for arboreal vertebrates (folivores) that may not be able to increase their consumption sufficiently to compensate for reductions in nitrogen content and digestibility (e.g. Kanowski 2001)
- changes in competitive interactions between plant species with different growth forms, different carbon-fixing pathways (C3 vs C4) or nitrogen-fixing capabilities (legumes vs non-legumes). For example, there is evidence from rainforests in the Amazon that vines and scramblers may be relatively more advantaged than the woody species on which they grow (e.g. Phillips *et al.* 2002), potentially leading to changes in the structure and functioning of many plant communities
- differential dispersal rates under the influence of shifting climate zones may result in curtailed geographic ranges of species that currently interact. For example, flying herbivorous insects may shift their ranges faster than the host plants on which they currently feed, potentially resulting in colonisation of new hosts
- changes in the virulence of parasitic infections and diseases, and in the susceptibility of hosts, may have large impacts. In particular, vector-borne diseases may spread upwards in elevation and southwards as mosquitoes and other vectors respond to warming. The potential impact of climate change on the incidence and distribution of the chytrid fungus may have significant impacts for the already threatened Australian amphibian fauna (Pounds *et al.* 2006)
- differences in the growth rates of coral species will affect their response to rising sea levels, with faster-growing species expected to take advantage of new colonisation sites, and cascading impacts expected for the structure and composition of reef communities
- changes in interactions between humans and other species; for example, if human-adaptive responses result in further modification of natural landscapes.

Vulnerability of different taxa

The general species-level predictions outlined above can be extended to further predictions about the vulnerability of particular taxa (Table 5.3). Of the taxa listed in the table, only the birds of Australia have received any systematic attention to identify particular species at risk (Olsen 2007).

Communities and ecosystems

The structure and functioning of Australian ecosystems will undoubtedly change as biotic and abiotic components of the system undergo change. Table 5.4 summarises projected responses of key Australian ecosystems to increased atmospheric CO_2 concentrations and anthropogenic climate change.

5.5.3 Biodiversity hotspots

Within the general classification of vulnerable ecosystems highlighted in Table 5.4 there are several specific

Table 5.3 Examples of factors that will increase vulnerability to climate change of particular taxa.

Taxa	Potential vulnerability
Mammals	Narrow-ranged endemics (particularly in montane regions) susceptible to rapid climate change in situ (Williams *et al.* 2003); changes in competition between grazing macropods in tropical savannas mediated by changes in fire regimes and water availability (Ritchie *et al.* 2008); herbivores affected by decreasing nutritional quality of foliage as a result of CO_2 fertilisation (Kanowski 2001)
Birds	Changes in phenology of migration and egg laying (Chambers *et al.* 2005); increased competition of resident species with migratory species, as the latter species stay at breeding grounds for longer periods; breeding of waterbirds susceptible to reduction of freshwater flows into wetlands; top predators such as some sea birds vulnerable to changes in food supply as a result of ocean warming (Smithers *et al.* 2003); rising sea levels will affect birds that nest on sandy and muddy shores, salt marshes, inter-tidal zones, coastal wetlands and low-lying islands; saltwater intrusion into freshwater wetlands, especially in northern Australia, will affect breeding habitat (Williams *et al.* 1995) [general references: Bennett *et al.* 2007; Chambers 2007; Chambers *et al.* 2005; Garnett 2007; Olsen 2007]
Reptiles	Warming temperatures may alter sex ratios of species with environmental sex determination (ESD) such as crocodiles and turtles (some species likely to modify use of microhabitats to cope with warming in situ)
Amphibians	Frogs may be the most at-risk terrestrial taxa. Altered interactions between frogs and the pathogenic chytrid fungus *Batrachochytrium dendrobatidis* may occur, with changes in both host susceptibility and pathogen activity; higher temperatures shift the growth optimum of the fungus, encouraging outbreaks (Pounds *et al.* 2006; but see Laurance 2008); susceptibility to competition and predation by cane toads as their range expands with warming; threatened alpine species such as the southern corroboree frog *Pseudophryne corroboree* at risk from drying and fire impacts on bogs used as breeding sites
Fish	Freshwater species vulnerable to reduction in water flows and water quality; limited capacity for freshwater species to migrate to new waterways; all species susceptible to flow-on effects of warming on the phytoplankton base of food webs
Invertebrates	Expected to be more responsive than vertebrates due to short generation times, high reproduction rates and sensitivity to climatic variables. Flying insects such as butterflies may be able to adapt by shifting ranges, as long as they are not limited by host plant distributions; non-flying species with narrow ranges are susceptible to rapid change in situ (e.g. Wilson *et al.* 2005 estimated that 25% of insect diversity in the Wet Tropics may be threatened this century); genetic changes already observed in some widespread species such as *Drosophila* spp. (Table 5.2); invertebrate herbivores also affected by reduced foliar quality under elevated CO_2
Plants	Longer-lived plants such as trees may be highly vulnerable if climate change 'moves' suitable establishment sites for seedlings beyond seed dispersal distance at a rate exceeding generation time. Narrow-ranged endemic plants requiring a very specific set of environmental characteristics (such as specific soil types) will have limited capacity to disperse to similar, rare sites. Elevated CO_2 will increase photosynthetic rates as long as other factors, such as water and nutrients, are not limiting (Box 5.1). There is potential for productivity to be boosted in some regions by a combination of increased CO_2 and longer growing seasons (e.g. Dunlop and Brown 2008). This effect, however, may not occur in regions where drying occurs. Increasing CO_2 will increase water use efficiency at an individual plant level, especially in C3 plants. But at an ecosystem level, total water use may not necessarily decrease, due to decreased total leaf area and increased evaporation from soil as a consequence of warmer temperatures. Competition between C3 and C4 plants may be affected by elevated CO_2 due to differential responses, but soil moisture may be a stronger influence than photosynthetic pathway (Box 5.1). Any changes in productivity and foliar nutrients will have flow-on effects to herbivores. Changes to fire regimes will have significant impacts on vegetation; increases in frequency and intensity of fires may disadvantage obligate seeders relative to vegetative resprouters (Box 5.11). Changes in the timing of plant phenology and insect life cycles will affect pollination and some forms of dispersal

Table 5.4 Projected impacts of CO_2 rise and climate change on key Australian ecosystems.

Key component of environmental change	Projected impacts
Coral reefs	
CO_2 increases leading to increased ocean acidity	Reduction in ability of calcifying organisms, such as corals, to build and maintain skeletons (Hoegh-Guldberg *et al.* 2007; Kleypas *et al.* 1999)
Sea surface temperature increases, leading to coral bleaching due to the death of symbiotic algae in coral polyps	If frequency of bleaching events exceeds recovery time, reefs will be maintained in an early successional state or be replaced by communities dominated by macroalgae (Hoegh-Guldberg 1999) Warming will increase susceptibility of corals to diseases such as White Syndrome, leading to outbreaks in locations of high coral density (Bruno *et al.* 2007) Potential for new reefs to develop at higher latitudes where suitable substrates are available (Greenstein and Pandolfi 2008) and until light becomes limiting (Hoegh-Guldberg 1999); potential decrease in beta diversity of coral communities as tropical-adapted taxa expand their range to the south (Greenstein and Pandolfi 2008), amplified by differential survival of different taxa (Done 1999)
Increases in cyclone and storm surge	Increased physical damage to reef structure
Rising sea levels	Fast-growing corals are advantaged over slow-growing species, leading to changes in structure and composition of reef communities (Buddemeier and Smith 1988)
Oceanic systems (including planktonic systems, fisheries, sea mounts and offshore islands)	
Ocean warming	Many marine organisms are highly sensitive to small changes in average temperature (1–2°C), leading to effects on growth rates, survival, dispersal, reproduction and susceptibility to disease; in particular, increasing temperatures reduce larval development time, potentially reducing dispersal distances during this phase (O'Connor *et al.* 2007); warm-water assemblages may replace cool-water communities, with coastal species endemic to the south-east most at risk; little is known about potential impacts on fish populations due in part to limited baseline information on fish stocks (Hobday *et al.* 2007)
Changed circulation patterns, including increase in temperature stratification and decrease in mixing depth, and strengthening of East Australian Current	Distribution and productivity of marine ecosystems is heavily influenced by the timing and location of ocean currents; currents transfer the reproductive phase of many organisms, thus playing a key role in dispersal and maintenance of populations; currents also play a role in nutrient transport by bringing cooler, nutrient-rich waters to the surface through upwelling, thus increasing productivity; climate change may suppress upwelling in some areas and increase it in others, leading to shifts in location and extent of productivity zones (Hobday *et al.* 2006; Poloczanska *et al.* 2007)
Changes in ocean chemistry	Increasing CO_2 in the atmosphere is leading to increased ocean acidity and a concomitant decrease in the availability of carbonate ions, the building blocks of calcium carbonate skeletons; many planktonic species and other species will be affected, as well as corals (see above); increased dissolved CO_2 may increase productivity (Hobday *et al.* 2007)

Key component of environmental change	Projected impacts
Alterations in cloud cover and ozone levels, which alter solar radiation reaching ocean surface	Potential negative impacts on phytoplankton production (Poloczanska *et al.* 2007)
Changes in timing of major climatic phenomena such as El Niño events	Changes in seasonal cycles of plankton abundance, with potential for mismatch between phytoplankton blooms and zooplankton growth, leading to cascading effects to the rest of the marine food chain (Hays *et al.* 2005)
Sub-Antarctic islands	
Extension of ice-free season for lakes and pools on Macquarie, Heard and MacDonald islands	Increased length of the stratified period with deeper mixing (Rouse *et al.* 1997).
Warming associated with glacial retreat on Heard Island	Exposure of previously ice-covered ground, and colonisation by plants and invertebrates
Estuaries and coastal fringe (including benthic, mangrove, saltmarsh, rocky shore, and seagrass communities)	
Sea-level rise	Landward movement of some species (especially mangroves) as inundation provides suitable habitat, which may be at the expense of other communities such as saltmarsh and freshwater wetlands (Bayliss *et al.* 1997; Eliot *et al.* 1999; Semeniuk 1994; Woodroffe and Mulrennan 1993); changes to upstream freshwater habitats will have flow-on effects to species such as wetland birds Rocky shore and saltmarsh species in areas of low topographic relief will be vulnerable to complete loss of habitat, especially when bounded by cliff lines or coastal development
Increased storm surges	Physical damage (erosion, slumping, rockfalls) to coastal zone, including beaches and rocky shores; changes to timing and magnitude of wrack (decaying plant material) washing up on estuarine and ocean shores
Increases in water temperature	Impacts on phytoplankton production will affect secondary production in benthic communities (Hobday *et al.* 2007)
Human adaptive responses to sea-level rise and beach erosion, including artificial beach nourishment	Changes in composition of soft-sediment communities with cascading impacts to higher trophic levels (Peterson and Bishop 2005)
Changes in upstream river flows	Alteration in quality and quantity of detritus flowing into estuaries and near-shore communities, leading to disruption in detritus-based food webs; patterns of primary production in estuaries may shift from light-dependent nutrient-intolerant seagrasses to fast-growing macroalgae (Bishop and Kelaher 2007); distribution and transmissability of many marine diseases closely linked to water temperature and salinity (e.g. susceptibility of Sydney rock oyster *Saccostrea glomerata* to QX disease is linked to periods of increased rainfall)

Key component of environmental change	Projected impacts
Savannas and grasslands	
Elevated CO_2	Shifts in competitive relationships between woody and grass species due to differential responses to CO_2 fertilisation (Ghannoum *et al.* 2001; Howden *et al.* 1999)
Increased rainfall in north and north-west regions	Increased plant growth will lead to higher fuel loads, in turn leading to fires that are more intense, frequent, occur over larger areas (compared with small-scale mosaic burning promoted by traditional Aboriginal practices) and occur later in the dry season; synergistic impacts of introduced grasses such as gamba grass *Andropogon gayanus*, which are intensifying fire regimes in general and leading to hotter late-season fires in particular (Fig. 5.8); change to ecotonal boundaries between savanna woodlands, grasslands and monsoonal rainforest patches; increased rainfall may increase overall productivity and shift productivity belts southwards, but predicted changes are small relative to the present degree of temporal and spatial variability (Williams *et al.* 1995); southerly range shifts of vegetation may be limited at about 16–18° due to significant changes in soil type (Williams *et al.* 1995). Overall, changes in rainfall seasonality are likely to be more important than changes in amount (Williams *et al.* 1995).
Tropical rainforests	
Warming and changes in rainfall patterns	Increased probability of fires penetrating into rainforest vegetation, resulting in shift from fire-sensitive vegetation to communities dominated by fire-tolerant species; cool-adapted species forced to higher elevations, altering competitive interactions; potential for reduced population sizes as area of potential habitat decreases (Hilbert *et al.* 2007a; Meynecke 2004; Shoo *et al.* 2005a, b; Williams *et al.* 2003); changes in distribution of different rainforest types (Hilbert *et al.* 2001); rise in basal altitude of orographic layer will reduce occult precipitation (cloud stripping) in cloud forests, exacerbating the effects of long-term drought
Change in length of dry season	Altered patterns of flowering, fruiting and leaf flush will affect resources for animals
Increased intensity of storms/ tropical cyclones	Increased physical disturbance to forests, thus altering gap dynamics and rates of succession; shallow-rooted tall rainforest trees are particularly susceptible to uprooting, breakage and defoliation (Corlett and Primack 2005; Gleason *et al.* 2008)
Rising atmospheric CO_2	Differential response of different growth forms to enhanced CO_2 may alter structure of vegetation; growth of vines could potentially be enhanced proportionally more than trees, leading to increased tree mortality (Phillips *et al.* 2002, 2004); enhanced CO_2 will increase overall productivity but decrease the digestibility and nutritional value of foliage for herbivores (Kanowski 2001); reduced stomatal conductance under elevated CO_2 will reduce rates of transpiration and alter catchment water balance – this may in turn affect runoff to estuaries and coastal ecosystems, including coral reefs (Gedney *et al.* 2006)

Key component of environmental change	Projected impacts
Temperate forests	
Potential increases in frequency and intensity of fires	Changes in structure and species composition of communities with obligate seeders may be disadvantaged compared with vegetative resprouters
Warming and changes in rainfall patterns	Potential increases in productivity in areas where rainfall is not limiting (Kirschbaum 1999b; Lucas and Kirschbaum 1999); reduced forest cover associated with soil drying projected for some Australian forests (Notaro *et al.* 2007)
Increasing atmospheric CO_2	Overall increases in productivity and vegetation thickening
Inland waterways and wetlands	
Reductions in precipitation, increased frequency and intensity of drought	Reduced river flows and changes in seasonality of flows, exacerbated by competition with the needs of agriculture and urban settlements (Arnell 1999; Kothavala 1999; Schreider *et al.* 1997; Walsh *et al.* 2001); reductions in area available for waterbird breeding (Johnson 1998; Kingsford and Auld 2005; Kingsford and Norman 2002; Roshier *et al.* 2001); increased rainfall in the catchment of Lake Eyre could transform it from an ephemeral to a permanent wetland (Roshier *et al.* 2001); 55% of Australia's wetlands of international importance and 26% of wetlands of national importance are considered at high to very high risk in the near term (Jones *et al.* 2008); changes in species composition and community structure (Bunn and Arthington 2002) More intense rainfall events will increase flooding, affecting movements of nutrients, pollutants and sediments, riparian vegetation, and erosion Groundwater-dependent ecosystems, such as cave streams and mound springs, may be negatively affected; organic wetlands likely to suffer more frequent burning of peaty sediments (Horwitz and Sommer 2005)
Changes in water quality, including changes in nutrient flows, sediment, oxygen and CO_2 concentration	May affect eutrophication levels, incidence of blue-green algal outbreaks; loss of cool-adapted aquatic species and increase in populations of warm-adapted species; increased eutrophication at permanent waterholes will have negative consequences for dependent fauna, including grazing animals
Sea-level rise	Saltwater intrusion into low-lying floodplains, freshwater swamps and groundwater; replacement of existing riparian vegetation by mangroves
Warming of water column; increase in depth of seasonal thermoclines in still water	Changes in abundance of temperature-sensitive species such as algae and zooplankton; reduction in depth of lowest, oxygenated zones in some cases, leading to local extinction of some fish and other vertebrates

Key component of environmental change	Projected impacts
Arid and semi-arid regions	
Increasing CO_2 coupled with drying in some regions	Interaction between CO_2 and water supply is critical, as 90% of the variance in primary production can be accounted for by annual precipitation (Campbell *et al.* 1997) Altered competition between C3 and C4 grasses in sub-tropical regions; increased productivity
Shifts in seasonality or intensity of rainfall events	Any enhanced runoff redistribution will intensify vegetation patterning and erosion cell mosaic structure in degraded areas (Stafford Smith and Pickup 1990) Woody vegetation may be favoured, leading to encroachment of unpalatable woody shrubs ('woody weeds') in many areas (Gifford and Howden 2001). Changes in rainfall variability and amount will also have important impacts on fire frequency that greatly increases after wet periods (Griffin and Friedel 1985) Dryland salinity could be affected by changes in the timing and intensity of rainfall
Warming and drying, leading to increased frequency and intensity of fires	Reduction in patches of fire-sensitive mulga *Acacia aneura* in spinifex *Triodia* spp. grasslands, potentially leading to landscape-wide dominance of *Triodia* spp. (Bowman *et al.* 2008)
Alpine/montane areas	
Reduction in snow cover depth and duration	Potential loss of species that are dependent on adequate snow cover for hibernation and protection from predators; increased establishment of plant species at higher elevations as snow pack is reduced; potential displacement of high-elevation species by competition as lowland species shift upwards; potential extinctions of summit-restricted species; changes to composition of snowbank/snow patch communities (Venn and Morgan 2007); changes in hydrology will affect distribution and persistence of fens and bogs, and their dependent species; potential increases in fire frequency following prolonged dry periods in non fire-adapted vegetation types (Pickering *et al.* 2004); changes in timing of acclimation and deacclimation rates of snow gums *Eucalyptus pauciflora* subsp. *niphophila* to freezing temperatures (Woldendorp *et al.* 2008); increased predation pressure on small mammals at higher latitudes from foxes and cats as snow declines; early spring thaw may advance phenology of many species including emergence of insects, flowering and bird migration. Increased growth rates of extant shrubs may promote expansion of woody vegetation into areas currently dominated by herbaceous species (Williams and Costin 1994); reduction in extent of persistent summer snowdrifts may allow shrubs, grasses and sedges to expand at the expense of cushion plants and rushes (Edmonds *et al.* 2006) Potential increases in mammalian herbivory at higher elevations (macropods and feral horses), where herbivory has previously been dominated by invertebrates

Figure 5.8 Introduced pasture grasses that have spread into undisturbed northern savannas lead to intensified fire regimes (e.g. hotter late-season fires), which in turn can change the balance between trees and grasses, leading to broad community changes. Source: Photos by Michael Douglas (left) and Samantha Setterfield (upper and lower right), Charles Darwin University; obtained from Richard Williams.

locations where climate change may have a disproportionately large impact on all aspects of biodiversity, and especially on extinctions of narrow-ranged endemics (Allen Consulting Group 2005; Jones and Preston 2006; IPCC 2007b; Dunlop and Brown 2008). With the exception of south-west Western Australia, these areas are largely protected in reserves. Although they are not entirely free of the existing stressors described in Chapter 3, they are less affected than most areas. Thus, projections based primarily on climate change considerations alone may be more relevant.

These 'biodiversity hotspots' include:

- Great Barrier Reef (Box 5.5)
- North Queensland Wet Tropics (Box 5.7)
- South-west Western Australia (Box 5.8)
- Kakadu World Heritage Area (Box 5.9)
- Australian Alps (Box 5.10).

5.5.4 Interaction of climate change with fire regimes, introduced species and water resource development

Australia is one of the most fire-prone continents in the world, with all but the wettest areas subject to frequent burning (Bradstock *et al.* 2002). Fire regimes (incorporating intensity, frequency and seasonality of individual fires) are affected by physical factors such

Box 5.7 Climate change in the rainforests of the North Queensland Wet Tropics. (Steve Williams)

The Wet Tropics bioregion in north-eastern Queensland is a biodiversity hotspot of global significance, with unique regional biota, that was listed as a World Heritage Area in 1988. The region is dominated by mountain ranges varying from sea level to 1600 m. Altitude is the strongest environmental gradient affecting species composition and patterns of biodiversity in the region (Williams 1997; Williams *et al.* 1996; Williams and Hero 2001; Williams and Pearson 1997). Tropical mountain systems are expected to be extremely vulnerable to climatic change due to their fragmentary nature, relatively small area, high rates of endemism and specialisation, and the compression of climatic zones over the elevation gradient (see figure further on). It is these montane forests that contain the highest numbers of endemic species in the Wet Tropics and are the most threatened by climate change. Cloud forests and other highland rainforests types are predicted to become greatly reduced in area and more fragmented across the Wet Tropics, even under a conservative scenario of 1°C temperature increase and a small reduction in rainfall (Hilbert *et al.* 2001).

The biogeographic and evolutionary history of the region predisposes the biota to being vulnerable to climate change, as most regionally endemic species are adapted to a cool, wet and relatively aseasonal environment. The spatial pattern of long-term habitat stability in the cool, wet uplands of the region has been a major factor determining the current spatial patterns of distributions and species richness (Graham *et al.* 2006; Schneider and Williams 2005; van der Wal *et al.* 2008; Williams and Pearson 1997; Winter 1997). A significant body of ecological and molecular-phylogeographic research provides evidence that previous climatic changes during the Quaternary ice ages resulted in significant levels of localised species extinction followed by periods of recolonisation as the rainforest expanded again, with little evidence for rapid evolutionary adaptation (Graham *et al.* 2006; Moritz *et al.* 2000; Schneider and Williams 2005; Williams and Pearson 1997; Williams, Shoo *et al.* 2008). Within-year climatic stability is also an important factor; areas with lower seasonality (more stable rainfall pattern) support higher densities of birds (Williams and Middleton 2008). Therefore, increasing seasonality and dry season harshness under climate change scenarios could be predicted to cause declines in population density.

Bioclimatic modelling of the distributions of regionally endemic rainforests vertebrates in the Wet Tropics predicted that with a >2°C temperature increase, most species will undergo dramatic declines in distribution, with some completely losing their current climatic environment (Williams *et al.* 2003). However, these predictions are being re-evaluated, given the propensity of BIOCLIM predictions to be exaggerated when attempting to predict into novel climate space. Current predictive modelling by Williams and van der Wal at the Centre for Tropical Biodiversity and Climate Change (CTBCC), using MAXENT models based on both climate and vegetation, still predict dramatic declines in distribution with increasing dry season severity producing impacts comparable to increases in temperature. Invertebrate fauna are also likely to suffer major impacts as they too have strongly structured assemblages across altitude, with many endemic species restricted to high altitude (Monteith 1985, 1995; Monteith and Davies 1991; Yeates *et al.* 2002). Increasing CO_2 is anticipated to have direct impacts on the region's herbivores via decreasing leaf nutritional content and digestibility of food plants for rainforest herbivores (Kanowski 2001). This impact is likely to be further exacerbated by the predicted range shifts onto less productive granite soils at higher altitudes that currently support folivores at comparably lower densities (Kanowski 2001; Williams *et al.* 2003). However, it is vital to also consider spatial patterns of abundance and population size rather than just distribution area, as population sizes are predicted to decline faster than area, given the spatial configuration of abundance patterns for most of the regionally endemic vertebrates (Shoo *et al.* 2005a, b). These analyses predict that 74% of rainforest bird species will become threatened under IUCN criteria as a result of projected mid-range warming in the remainder of this century.

There is some potential for amelioration of impacts via buffering of climatic extremes in topographic and microhabitat refugia such as boulder

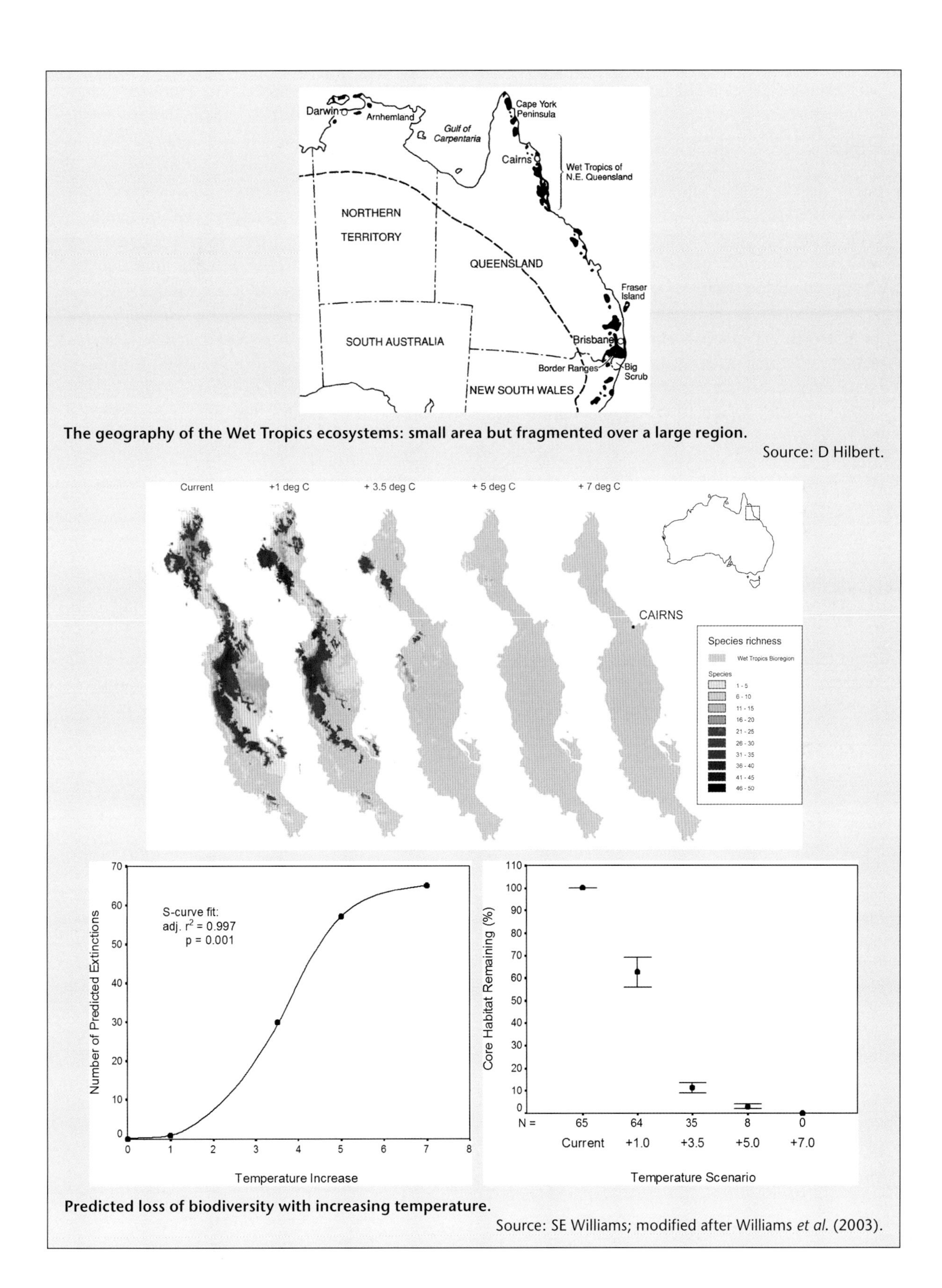

The geography of the Wet Tropics ecosystems: small area but fragmented over a large region.

Source: D Hilbert.

Predicted loss of biodiversity with increasing temperature.

Source: SE Williams; modified after Williams *et al.* (2003).

piles, dense vegetation and under logs. The degree of buffering and its potential to reduce impacts on the region's biota is the subject of ongoing research within the CTBCC. There may also be some potential for minor amelioration of impacts due to an increase in temperature, thus leading to an increase in net primary productivity that could reduce somewhat the predicted negative effects of climate change on upland rainforest birds.

The biodiversity contained within the Wet Tropics bioregion is highly vulnerable to global climate change. Prioritisation of adaptation actions should be based on informed assessment of relative vulnerability of species, habitats and ecosystem processes. We need to determine which species, habitats and ecosystems will be most vulnerable, exactly what aspects of their ecological and evolutionary biology determines their vulnerability, and what we can do to manage this vulnerability and minimise the realised impacts. Fire is a priority for further research, as increasing temperature, dry season severity and decreased levels of cloud stripping will undoubtedly increase the frequency and severity of rainforest fires, and potentially pose a major threat in the longer term. Management action can initially focus on any actions that maintain or increase general system resilience, such as weed and feral animal control, and on maintenance of habitat extent and integrity. Priority should be given to maintaining the upland, high biodiversity, refugia areas such as Bellenden Ker, Thornton Peak, and the Carbine and Windsor Tablelands.

Box 5.8 Vulnerability of biodiversity in Western Australia to climate change. (Colin Yates)

Mean annual temperatures throughout Western Australia have increased during the 20th century. Average annual rainfall has also changed, but the patterns vary across the state. Since the 1950s, there has been an increase in summer rainfall and extreme daily rainfall intensity and frequency in monsoonal and semi-arid northern and inland Western Australia. In contrast, in Mediterranean climate south-west Western Australia (SWWA), there has been a significant decline in autumn and early winter rainfall since the 1970s, and this has been attributed, at least in part, to global climate change (Bates *et al.* 2008; CSIRO and BOM 2007). Further increases in temperature and changes in rainfall similar to those already observed in SWWA are predicted (Bates *et al.* 2008; CSIRO and BOM 2007).

Observed impacts of climate change on Western Australian biodiversity are very limited and largely restricted to the conspicuous avifauna. All studies are correlative, and consequently it is not always easy to distinguish climatic effects from other human pressures. However, the limited data available indicate that some terrestrial species and ecosystems in Western Australia are being affected. Changes in ocean temperatures and climatic conditions along the Western Australian coast have contributed to the southerly extension in the breeding distribution of some tropical seabirds (Dunlop 2001). In SWWA the decline in rainfall is associated with changes to the seasonal movement of bird species, particularly waterbirds (Chambers 2008) and, together with groundwater extraction, has lowered water levels in cave streams, threatening aquatic communities (English *et al.* 2003).

Limited climate change impact models predict that the ranges of many species in SWWA, a global biodiversity hotspot (Box 3.11), will contract, but there are many uncertainties (Fitzpatrick *et al.* 2008; McKellar *et al.* 2007). Some species have persisted in situ through multiple sequences of climate change in the Pleistocene and their current distributions may not reflect their climatic tolerances (Yates *et al.* 2007). If contraction is the primary response of many species, then identifying refuges will be as important for some species as establishing ecological linkages (Fitzpatrick *et al.* 2008). Warmer and drier conditions in SWWA may have a large impact on the region's many narrowly distributed endemic species because, with

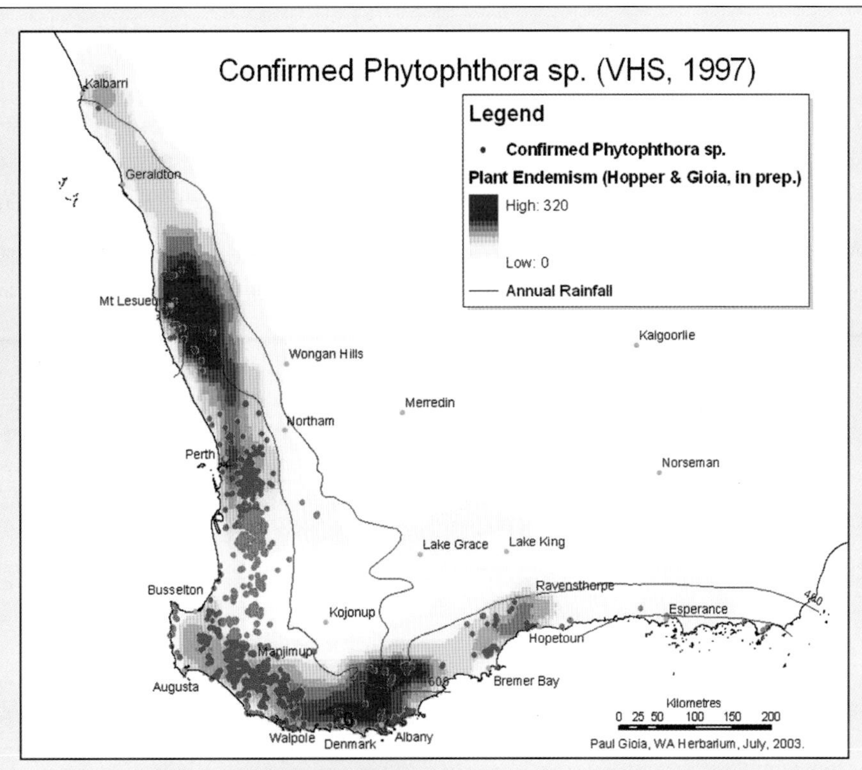

South-west Western Australia is already experiencing the impacts of multiple, interacting stressors, including climate change. Shown here is the outbreak of *Phytophthora* sp.

Source: Paul Gioia.

Species-rich kwongan (heath) in Fitzgerald River National Park, a hotspot within a biodiversity hotspot. More than 1800 vascular plant taxa occur within this national park.

Source: Andrew Burbidge.

the exception of the Stirling Ranges – which are just 1109 m above sea level – the region is of low relief offering limited scope for altitudinal migration into montane refuges. Many relictual mesothermic lineages are restricted to the coolest and wettest climate zones on the south coast and these climate zones may disappear (Hopper *et al.* 1996).

Extensive habitat fragmentation and limited dispersal ability will create substantial barriers to migration; and the presence of invasive weeds, introduced diseases such as *Phytophthora cinnamomi* (see figure) and introduced predators may further limit natural adaptation responses (Hobbs and Yates 2003; Shearer *et al.* 2004). Changes in fire regimes and lower rainfall may threaten particular species and functional types, especially non-sprouting serotinous plant species that are notably prevalent among narrow range endemics in the diverse kwongan plant communities (Cowling *et al.* 1994; Yates *et al.* 2003). Synergies among threats are likely to reinforce current declines in biodiversity and lead to tipping points much sooner than hitherto realised.

Box 5.9 Kakadu: a climate change hotspot. (Barry W Brook)

When ecologists, policy makers or the public think about the visceral impacts of climate change on Australia's natural systems, World Heritage-listed Kakadu National Park (KNP), located in the seasonal tropics of the Northern Territory, is high on the at-risk list. But looking deeper into the human-driven processes now threatening KNP, there is actually a synergy of interrelated problems requiring simultaneous management – a situation common to most biomes threatened with global warming (Brook *et al.* 2008). The big issues for KNP are changed fire regimes (impacting savanna and rainforest communities), rising sea levels (affecting the floodplain wetlands), and a suite of invasive weed and feral animal species, operating across all three major ecosystems. All three threats have a climate change component – although for fire and feral animals, not wholly.

The savannas, which by area make up the largest part of KNP, are at first glance apparently intact. There has been relatively little clearance of the woody component (dominated by *Eucalyptus tetrodonta* and *E. miniata*). Indeed, analysis of historical aerial photography has documented some vegetation thickening that may be linked to elevated atmospheric CO_2, which favours the growth of woody C3 species (Banfai and Bowman 2005), although other factors are probably also important (P Werner, pers. comm.). However, an emphasis by park managers on avoiding hot, late dry season fires has meant that a large proportion of KNP is burnt during the dry season, with a return time of one to five years (Williams *et al.* 1999). The impact of regular early-season burning on the park's biota and on the structuring of understorey vegetation is a topic of ongoing debate and research. Nevertheless, some long-term studies have explicitly linked high fire frequencies to species declines (Anderson *et al.* 2005; Lehmann *et al.* 2008; Pardon *et al.* 2003). Climate change, via increased temperatures or shifts in the timing and intensity of monsoonal rainfall, will likely enhance future fire risk (Parry *et al.* 2007).

The KNP wetlands support a rich and spectacular biota, including vast flocks of magpie geese *Anseranas semipalmata*, which congregate in millions to feed on *Eleocharis* chestnuts growing on the floodplains of the Alligator Rivers system. These wetlands formed around 6000 years ago after sea level stabilisation, following a post-glacial rise of 120 m (Woodroffe *et al.* 1987). Ironically, additional sea level change associated with anthropogenic global warming threatens their future viability (Brook and Whitehead 2006). About 20 cm sea-level rise occurred during the 20th century. At least double that amount – and potentially >1 m due to accelerated melting of polar icesheets – is predicted by 2100. Rising sea levels, in combination with intense tropical storm surges, increases the regularity and severity with which saline flows penetrate the low-lying freshwater wetlands. At the mouth of the Mary River to the west of KNP, extensive earthen barrages have already been built in an attempt to alleviate the damage caused by saltwater intrusion.

A complex network of low-lying natural drainage channels, enlarged or cross-connected by movement of feral animals such as Asian water buffalo *Bubalus bubalis* and pigs *Sus scrofa*, means that even a few tens of centimetres of additional sea-level rise may be sufficient to degrade or eliminate a large fraction of the floodplain communities. What remains will be isolated patches of freshwater wetlands within a mire of brackish swamps and saltwater mangroves. Beyond their impact in facilitating saline intrusion of the wetlands, feral ungulates help to spread weed species such as introduced pasture grasses and mimosa (Bradshaw *et al.* 2007; Skeat *et al.* 1996), which compete with native plants for space and nutrients. Climate change will also cause shifts in the relative ability of invasives to compete with indigenous species, especially if natives are also under stress from herbivore grazing, changing habitat quality and altered fire regimes (Brook 2008; Rossiter *et al.* 2003). KNP inevitably faces a tangled web of mutually amplifying processes associated with global change.

South Alligator wetlands in flood, Nourlangie Creek.

Source: Lochran Traill.

Box 5.10 The Snowy Mountains story. (Ken Green)

Australia's alpine environments make up only a little over 0.001% of its total continental area These environments consist of a group of 'island' ecosystems, centred on the peaks of the highest mountains in the relatively cool and wet south-eastern regions (Green 1998). Their isolation from other alpine environments sets them apart from mountainous ecosystems elsewhere in the world (Wardle 1988) while their lower temperature, higher precipitation and regimes of regular annual snow cover differentiate them from other Australian environments (Green and Osborne 1994; Williams and Costin 1994).

Alpine environments are extremely vulnerable to climate change. The Australian Alps have warmed at a rate of about 0.2°C per decade over the past 35 years (Hennessy *et al.* 2003). With a temperature lapse rate of 0.77°C per 100 m (Galloway 1988), this warming has caused the vegetation in the lowest 100 m of the alpine zone to lose its competitive advantage against the incursion of woodland trees. Projected temperature increases (Table 4.1) suggest that another 100 m will lose that advantage by 2020, as will the remainder with a rise of up to 2.9°C by 2050. The worst case sce-

The Southern Corroboree frog, a highly endangered species in the alpine zone.

Source: Hal Cogger.

nario is the complete loss of the alpine zone during the next century.

Snow cover in the Snowy Mountains has declined by about one third since 1954 (Nicholls 2005). The advance of the alpine treeline has not yet been observed, although trees are moving into frost hollows where they were absent historically (Wearne and Morgan 2001; see figure further on). Snow patches (sites where snow may persist throughout the summer) are declining in extent and duration, threatening the plant communities that depend upon this long-lasting snow (see figure further on). Many pest animals have been excluded in the past from alpine areas because of winter snow, but already as a response to declining snow cover a number of species such as feral horses have moved to higher altitudes (Green and Pickering 2002).

Weeds are becoming more common in alpine areas (McDougall *et al.* 2005), and subalpine native plants are already moving to higher altitudes. Climatic variability at the end of the winter snow season is also stressing animal communities. Population numbers of the endangered broad-toothed rat *Mastacomys fuscus* crashed in the winter of 1999; the earliest thaw on record to that date was followed by late snowfalls that inhibited small mammal movement and made them more susceptible to fox predation. Late-season instability in the weather also made the area inhospitable for migratory birds (Green 2006).

The spread of chytrid fungus, which may be exacerbated by warming temperatures (Table 5.3) is potentially the greatest current threat to alpine frog species, with southern corroboree frogs *Pseudophryne corroboree* close to extinction in the wild and alpine tree frogs *Litoria verreauxii alpina* no longer present in alpine areas (see photo below). Climate change impacts are not acting in synchrony for all species and a mismatch in responses to timing of

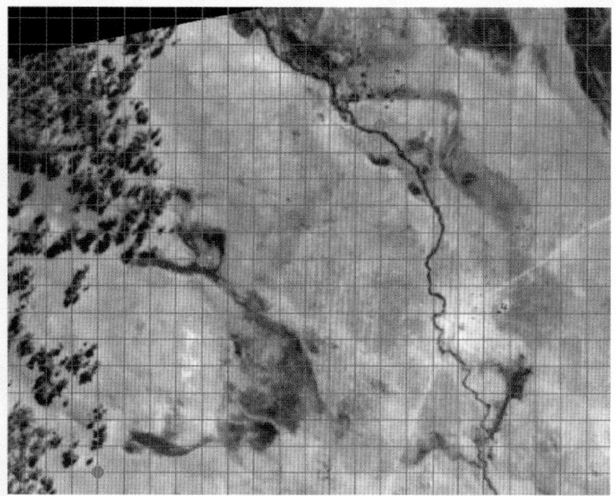

1944

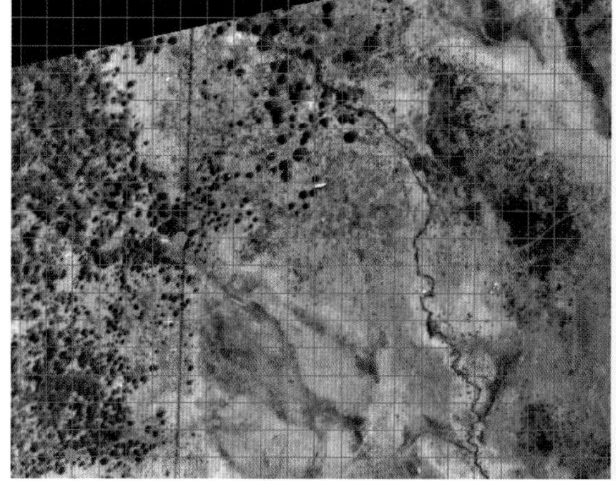

1998

Incursion of woody species into frost hollows. The example shown is from Currango Plain, Kosciuszko National Park.

Source: National Library of Australia; Keith McDougall.

events is having a number of effects. Bogong moths are a major source of food for many mammals (including the threatened mountain pygmy possum *Burramys parvus*), birds and invertebrates in the alpine zone. The arrival of the moths in the alpine zone has historically occurred at about the same time as the spring thaw. Over the past 30 years, however, the spring thaw has occurred progressively earlier, but the timing of moth arrival has not. In 2006, for example, the spring thaw occurred at the earliest time ever recorded, while the moths arrived later than previously recorded. This mismatch in timing, increasing with climate change, will also affect relationships between migratory birds, insects and plant flowering, with implications for nesting success in birds and pollination of flowers.

The frequency of alpine fire is likely to increase. In 2003 there was the rare incursion of fire into alpine areas on the mainland. There was a repeat episode in 2006 in alpine Victoria (Williams *et al.* 2006). While the primary response to fire by government agencies is to protect life and property, the cost of rehabilitation – and the downstream costs of fire in terms of erosion and loss of water potential – make the alpine zone an increasingly valuable property from which fire must be excluded.

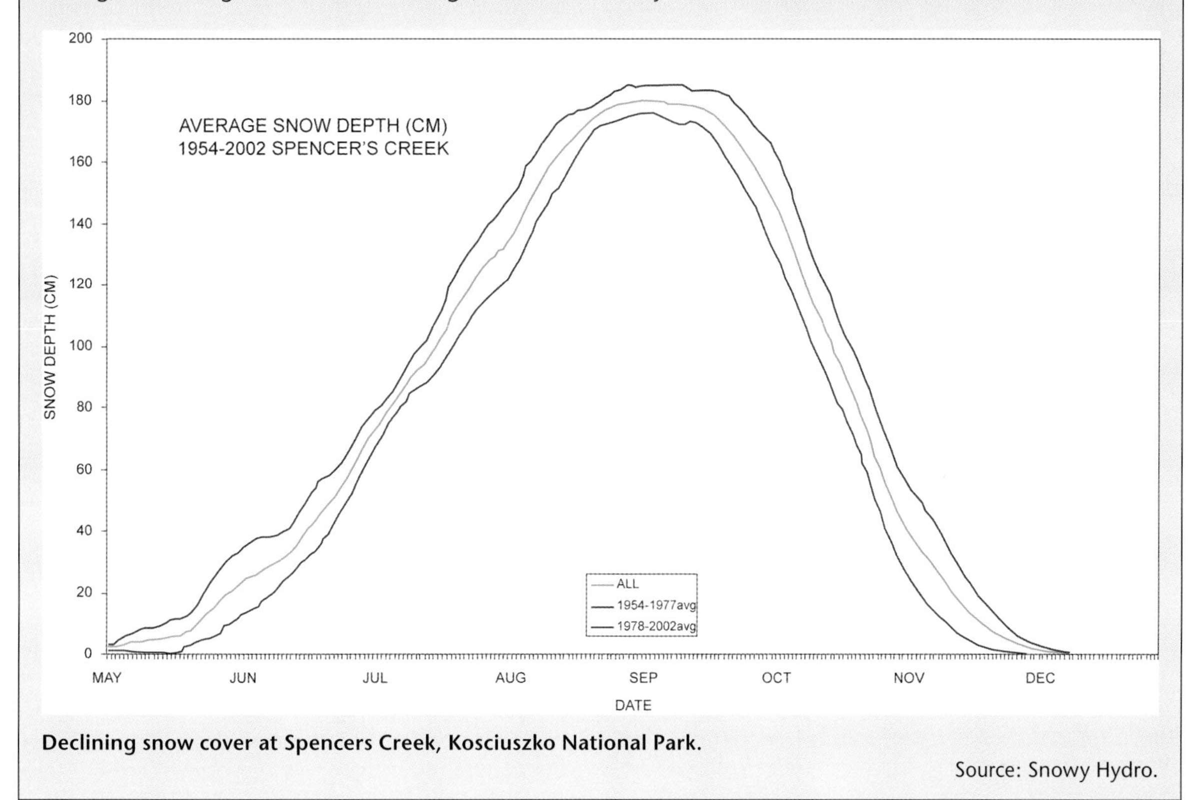

Declining snow cover at Spencers Creek, Kosciuszko National Park.

Source: Snowy Hydro.

as temperature, humidity, rainfall and wind speed, as well as biotic factors such as fuel load (Gill 1975). With warming temperatures, potential drying in some regions, and increased fuel loads possibly due to CO_2 fertilisation (as well as increased rainfall in the north), it is likely that fires will increase in both intensity and frequency in most regions (Box 5.11).

Many introduced species that are already considered pests will have an advantage under climate change because they possess life history and other characteristics that give them an 'edge' under any disturbance (Fig. 5.8). Vacant niches may also be created by declines in local populations as some species become stressed by rapid climate change, thus enhancing the colonisation of newly introduced species or expansion of 'sleeper' invasive species already present in low numbers (Box 5.12).

The development of water resource infrastructure and management has had significant impacts on

Box 5.11 Climate change and fire regimes. (Richard Williams and Ross Bradstock)

A preliminary assessment of climate change–fire regime–biodiversity interactions (Williams *et al.* 2009) has examined how climate change may affect fire weather and fuels, how these changes may alter fire regimes, and how this in turn might affect the dynamics and management of biodiversity. The assessment reviews the predicted impacts of climate change on fire weather and fuels; proposes a broad, national framework for addressing the problem; and reviews and interprets 'climate change–fire regime–biodiversity' interactions for four selected vegetation types/regions, for which at least some pertinent data and analyses are available (alpine ash forests, sclerophyllous vegetation of south-eastern and south-western Australia, and the tropical savannas).

Historical trend analysis of weather data over the period 1973–2007 suggests a change towards increased fire danger, with an increased incidence of days of extreme fire weather. Model simulations for future climate indicate continued warming and decreased humidity across the whole continent throughout the 21st century; reduced rainfall in south-eastern, south-western and central Australia (although no change or slightly wetter conditions in northern Australia); and some changes in wind speed. Further increases in the intensity and frequency of extreme fire weather events are therefore likely based on these projected changes to climate. However, translating these projected climate trends into changes in local and national patterns of fire regimes is complex, because of the uncertainties surrounding the impact of climate change on other critical drivers of fire events (such as fuels and ignitions) and the effects of increased CO_2 on vegetation.

A sensitivity analysis was undertaken at a national, broad-biome scale, and via select regional case studies in vegetation types in south-eastern, south-western and northern Australia, where the most detailed data were available to assess climate change–fire regime–biodiversity interactions. The analysis showed that the likely increases in fire danger indices in these regions of the country may result in shorter intervals between fires – particularly in southern Australia. The risk to biodiversity as a consequence of changing fire regimes is not equal across the country. In particular, the biodiversity values of sclerophyllous vegetation of south-eastern and south-western Australia appear to be at higher risk than those of the savanna woodlands of northern Australia. In all settings, pockets of vegetation

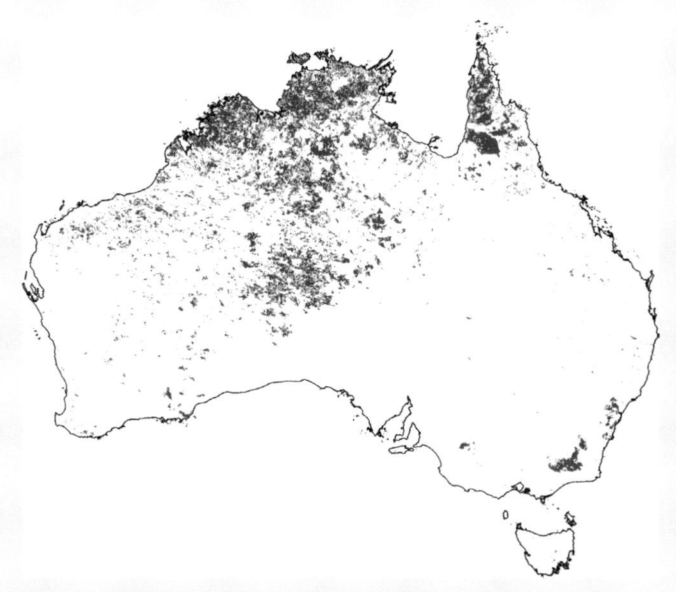

Fire incidence across Australia in 2002–2003 as observed by NOAA satellites.

Source: Satellite information provided courtesy of Satellite Remote Sensing Services, Western Australian Land Information Authority (Landgate), WA, 2009.

that are sensitive to higher-intensity fires, and occur within flammable matrices, will also be at risk.

Further research is needed to clarify risk and vulnerability of biodiversity to altered fire regimes at the landscape, regional and national scales. There is some evidence that risk to biodiversity values posed by some current or proposed land use changes (e.g. more frequent prescribed burning in sclerophyllous vegetation in south-eastern Australia, the spread of exotic pasture grasses in the savannas) is greater than the risk posed by altered fire regimes as a consequence of climate change. However, there will be complicated and poorly understood interactions among these variables and fire activity.

The primary implication for the management of protected areas is that fire management, an already complex issue, will become more complex in the coming decades. This will be due to both the potential trade-offs that will be required to manage biodiversity values in the face of (either perceived or actual) more frequent and/or intense fires; and the need to account for fire regime effects on additional values such as carbon abatement, smoke management and water yield, in the matrix of protected area management objectives.

In the face of such complexity and potentially competing demands on resources, it is vital that adaptive management systems for fire management be developed, enhanced, reviewed and evaluated in the coming years. An essential task for all regions and agencies will be to quantify risk to biodiversity values posed by changing fire regimes, using approaches outlined in the assessment of fire regimes and climate change. Key ingredients include developing comprehensive fire mapping programs to track fire activity across the landscape (see figure on previous page), and rigorous evaluation of the benefits and costs of all risk-mitigation decisions. In this process, identifying thresholds of potential concern will be as important as prescriptions and targets. Much further research is needed to quantify such risks, benefits, costs and returns on investment of fire management strategies within protected areas.

Box 5.12 Climate change and invasive species. (Tim Low)

Unwanted species invasions are expected to become a major consequence of climate change due to their response to disturbance regimes and their typical life history traits.

The responses of invasive plants to flood disturbance, for example, provide a preview of potential events in the future. Exceptional floods in 1974 and 1988 allowed a weed tree from Africa and Asia, athel pine *Tamarix aphylla*, to replace dislodged river red gums *Eucalyptus camaldulensis* along 600 km of the Finke River in central Australia (Griffin *et al.* 1989). Another dramatic flood in 2000 facilitated invasion of the Karinga Creek catchment (Brown and Grace 2006). Prior to 1974, Athel pine was hardly rated a weed at all but now rates as one of Australia's 20 worst weeds – one of the Weeds of National Significance (WONS). It drains waterholes, increases surface soil salinity and erosion and, unlike river red gums, provides no nectar or hollows for wildlife (ARMCANZ 2000).

Of Australia's 20 WONS, 11 are known to recruit following floods (Low 2008). Floods are expected to become more frequent and intense under climate change (Hennessy *et al.* 2007). Floods benefit environmental weeds partly by stripping away competing native vegetation. Cyclones, droughts and fires – which are also expected to increase in severity or frequency or both (Hennessy *et al.* 2007) – benefit weeds in the same way. Fire-promoting pasture grasses such as buffel grass *Cenchrus ciliaris* and gamba grass *Andropogon gayanus* create positive feedback loops in which they promote fire and are promoted by fire (D'Antonio and Vitousek 1992) – an extremely serious problem in northern and inland Australia.

Even without extreme weather events, climate change will provide many opportunities for weed establishment wherever native plants are killed by heat or moisture stress. Most environmental weeds are escaped garden plants, originating from many different climatic zones. Australia's pool of weeds is very large, with 2700 exotic species present (Humphries *et al.* 1991), a number that will keep growing as nurseries import more drought-hardy plants from abroad (Low 2008).

In addition to an increasing weed problem, introduced species replacements are likely in freshwater environments where rising temperatures will suit aquarium fish dumped in streams, and in coastal waters where marine-invasive species that arrive in ballast water represent a large pool of immigrant species (Hewitt *et al.* 2004). Many invasive animals such as cane toads *Bufo marinus* and mosquito fish *Gambusia holbrooki* do best in disturbed environments (Low 1999).

Native species can also increase in numbers and become 'invasive' in response to human impacts (Low 2002). Several noteworthy examples are attributed to recent climate change. In Tasmania, a sea urchin *Centrostephanus rodgersii* from the mainland is forming degraded areas where species-rich kelp beds once grew (Johnson *et al.* 2005). On the Great Barrier Reef, the coral disease White Syndrome causes mortality during times of high water temperatures (Bruno *et al.* 2007). In the Australian Alps, kookaburras *Dacelo novaeguineae* are hunting at higher altitudes than before, preying on alpine skinks that fail to recognise them as predators (Ken Green, pers. comm.). And with less snow, swamp wallabies *Wallabia bicolor* and red-necked wallabies *Macropus rufogriseus* are reaching higher altitudes, placing alpine herbfields at risk (Ken Green, pers. comm.). Newcomers can pose management problems because migration to new regions is considered a positive adaptive response to climate change, but in some cases they may reduce rather than enhance local biodiversity.

One native bird, the noisy miner *Manorina melanocephala*, could threaten the capacity of some eucalypt species to survive climate change. Noisy miners have multiplied dramatically in eastern Australia by exploiting fragmented landscapes (Dow 1977; Low 1994). These aggressive sedentary honeyeaters repel other birds in eucalypt woodland (Grey *et al.* 1998). Noisy miners, which have extremely small home ranges (Dow 1979), greatly reduce potential pollen flow by defending nectar supplies against migratory and nomadic honeyeaters.

Climate change will also have impacts on species via its effects on introduced and indigenous pathogens, parasites, predators and competitors. Concerns about this are growing, as reflected in recent reports about chytrid fungus *Batrachochytrium dendrobatidis* causing mass frog extinctions in Latin America (La Marca *et al.* 2005; Pounds *et al.* 2006), native beetles killing trees over vast regions in North America (Logan *et al.* 2003), predatory Humboldt squid *Dosidicus gigas* colonising the seas off California (Zeidberg and Robison 2007), the prospect of sharks and crabs invading Antarctic waters (Aronson *et al.* 2007), and marine species from the Atlantic Ocean flowing into the North Pacific Ocean via the North-west Passage (Reid *et al.* 2007). The biodiversity losses could be enormous, and could precede many of the direct impacts of warming and altered rainfall.

Australia's biodiversity by changing the amount and timing of flows through ecosystems. This is most pronounced in the Murray–Darling Basin, where the ecology of floodplain wetlands has been affected significantly by altered wetting and drying regimes. Climate change will exacerbate many of these management-driven impacts (Box 5.13). Although in many cases the projected impacts of climate change are smaller than those already created by water resource development, climate change could nevertheless lead to severe impacts if it drives ecological systems across critical thresholds.

5.6 PREDICTIONS ABOUT FUTURE IMPACTS: HOW CAN WE DO BETTER?

5.6.1 Coping with complexity

The previous sections have outlined the many complexities and uncertainties that face us in trying to better predict the responses of the Australian biota to climate change. One method that has been proposed for simplifying ecological complexity is a 'functional analysis' approach. This is commonly used in analyses of ecological communities facing disturbance (e.g. Noble and Slatyer 1980). It uses a functional

Box 5.13 Floodplain wetlands in the Murray–Darling Basin: ecological implications of hydrologic change. (Bill Young)

The Murray–Darling Basin has a diverse range of floodplain wetlands, many of which cover large areas and support high biodiversity. They range in character from large ephemeral lakes, billabongs, extensive floodplain forests and terminal wetlands supporting mosaics of different vegetation communities, to the naturally estuarine environments of the Coorong and Lower Lakes on the Murray River.

Water resource development over the past century has led to considerable changes to the hydrologic regimes of many of these environments, resulting in extensive ecological change. The most important hydrologic changes are: (i) changes in the frequency and magnitude of floods reaching floodplain ecosystems; (ii) changes in the volumes of the lower flows reaching terminal wetlands and lakes; and (iii) changes in the seasonal patterns of flow.

Coombool Swamp upstream of Renmark.
Source: CSIRO. Photo by Willem van Aken.

Changes in the period between ecologically beneficial floods can lead to significant ecological change because of the importance of the temporal pattern of wetting and drying for ecological processes in wetlands, and the importance of the temporal patterns of habitat connectivity across the riverine landscape. Altered wetting and drying regimes are likely to alter ecological processes and, ultimately, the composition of the biological community. Both the average (see chart further on) and maximum periods between floods are ecologically important. The volumes of lower flows reaching terminal wetlands and lakes are important; it is these flows that maintain the limited, but critically important, permanent aquatic environments that provide refugia for aquatic animals. The seasonal patterns of flow are important, as they provide cues for biological processes such as fish migration and spawning.

These changes vary considerably across the Murray–Darling Basin, from the Paroo River where the flow regime remains essentially unchanged by water resource development, to rivers such as the Murrumbidgee where river regulation and water abstraction – coupled with land management changes including levees and drainage systems – have led to major alterations to the hydrologic regimes of floodplain wetlands.

In addition to the impacts of water resource development, future climate change is likely to lead to significant changes to these aspects of floodplain wetland flow regimes. CSIRO has made detailed assessments of the changes to river flow regimes that could occur with climate change by 2030, assuming current water sharing arrangements. These hydrologic assessments provide an information base that can guide ecological interpretations and assessments.

Tragowel Swamp near Echuca.
Source: CSIRO. Photo by Willem van Aken.

Currently, there remains considerable uncertainty in the regional-scale hydrologic implications of global climate change – even over the time frame of the next few decades. However, very considerable additional changes in the hydrologic regimes of floodplain wetlands are possible and moderate additional changes are likely. The changes in flood regimes that would result from the median of the climate change range by 2030 would be less than the changes already imposed by water resource development. However, the lower flows that are

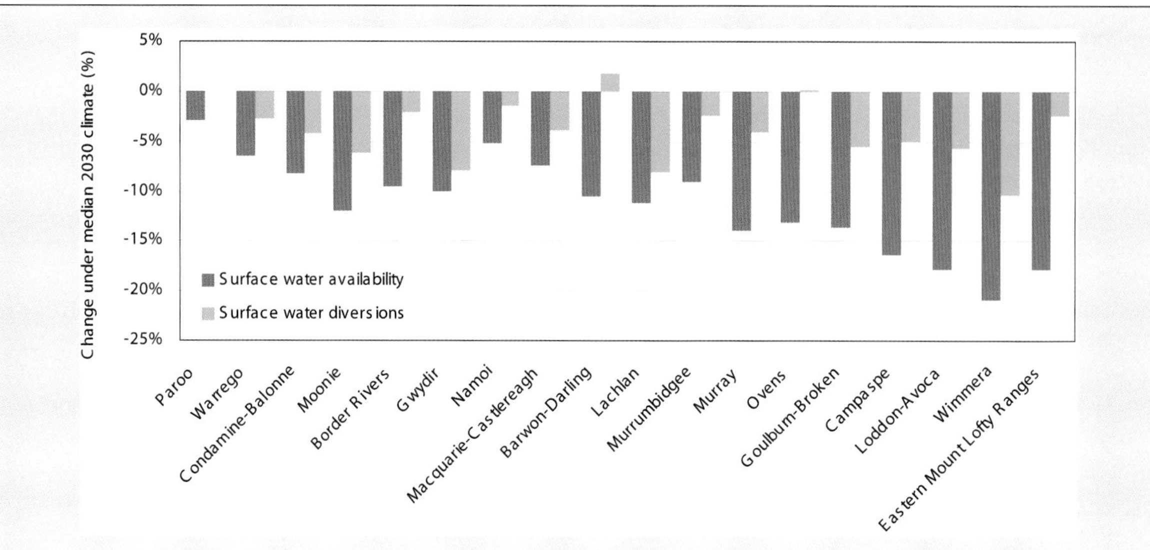

Percentage change in surface water availability and surface water diversions as a result of median climate change by 2030, assuming current water sharing arrangements.

important in maintaining ecological refugia within terminal wetland systems (such as the Macquarie Marshes and the Great Cumbung Swamp) would be greatly affected by even this median climate change scenario. This is because under current water sharing arrangements, any reduction in surface water availability caused by climate change would have a comparatively small impact on total consumptive water use and a disproportionately large impact on the water available to the environment.

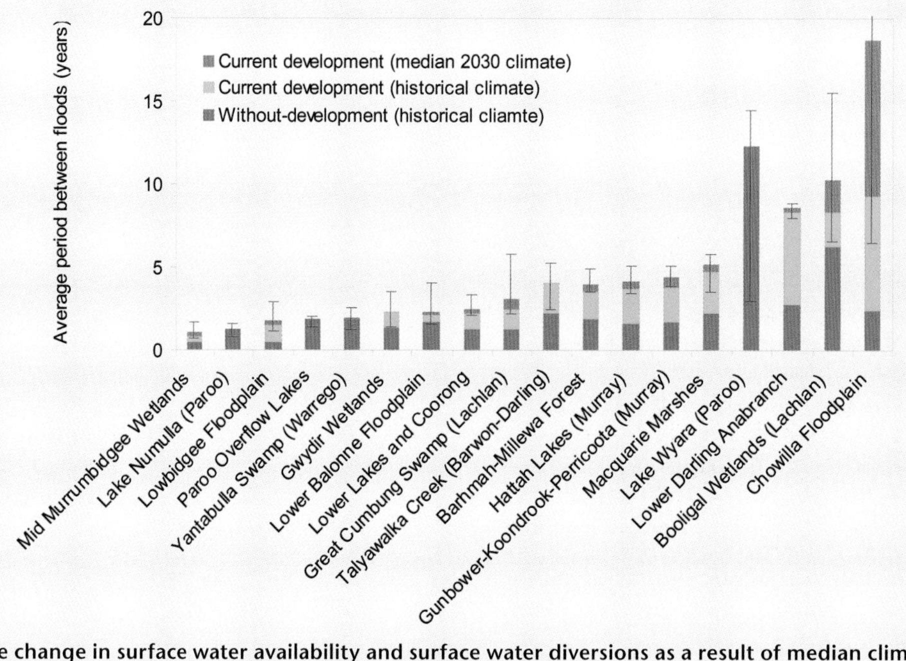

Percentage change in surface water availability and surface water diversions as a result of median climate change by 2030, assuming current water sharing arrangements.

In the Paroo and Ovens rivers where the level of use is low, water use would not be affected by a change in availability. In the Eastern Mount Lofty Ranges where the level of use is also low, the change in water availability would have only a minor impact on use. In the Barwon–Darling, where the level of use is moderate, increased temperatures would increase consumptive demand, thereby increasing water use. Across the other regions of the Murray–Darling Basin (excluding the Warrego River) where the level of use is moderately high to extremely high, the impact of reduced water availability varies according to the different water sharing arrangements currently in place. In all the high water use regions in the southern Murray–Darling Basin (Murrumbidgee, Murray and Goulburn–Broken), consumptive use is protected from much of the impact of reduced water availability (although the impacts in dry years are much greater).

Although in most cases the flood regime changes anticipated with median climate change by 2030 would be smaller than those already brought about by water resource development, when superimposed on the changes that have already occurred, the consequences could be very considerable and potentially catastrophic. This is because important ecological thresholds may be crossed and resulting changes may well be largely irreversible. For example, the current average periods between beneficial flooding events for many major wetland systems are already considerably affecting the reproductive opportunities for some waterbird species, particularly colonial breeding species such as ibis, herons and egrets. Any significant increase in these average periods may mean individuals of certain species do not get an opportunity to breed within their lifetime. The population and wider ecosystem consequences of such changes could clearly be catastrophic. In the case of Chowilla Floodplain, for example, median climate change would lead to large increases in the average period between floods, with major ecological consequences likely.

classification of individual species (or species groups) based on species characteristics, and the range of their possible physiological states or life history stages, and then analyses the potential impacts of the disturbance on those classifications and states. Landsberg and Stafford Smith (1992), for example, used such a functional analysis approach to make predictions about the potential outbreak of herbivorous insects under global atmospheric change. This result has provided guidance and a framework for setting priorities for research and management (Box 5.14).

5.6.2 Modelling

In Australia, most projections of impacts on species and communities have involved the use of bioclimatic models to predict changes in geographic range. These are mainly correlative models that have both strengths and weaknesses (Box 5.15).

Modelling of individual species distributions to date has generally indicated that as the climate continues to change rapidly, most species will suffer reductions to and/or fragmentation of their current geographic ranges. Bioclimatic modelling of weed species has also indicated that some may expand their ranges (e.g. prickly acacia *Acacia nilotica*, Kriticos *et al.* 2003a; and rubber vine *Cryptostegia grandiflora*, Kriticos *et al.* 2003b). However, modelling efforts need to be much better integrated with observations and experiments. Currently, we are a long way from routinely incorporating biological information such as life history and species interactions into these models, even for individual species.

Although many qualitative predictions have been made of the vulnerability of certain systems (e.g. Wet Tropics, alpine zones, coral reefs; Table 5.4), quantitative, community- and ecosystem-level projections for Australian systems are rare. One exception is the work by Hilbert *et al.* (2001), who used artificial neural network techniques to model potential changes in rainforest types in North Queensland under various combinations of temperature and rainfall; large changes in rainforest types were predicted with relatively small climatic changes. Modelling effort to project changes in vegetation types is gearing up but preliminary results may still be a year or more away. This is in contrast to research in the northern hemisphere, where modelling of changes in major vegetation types/biomes has been underway for at least 10–20 years.

Box 5.14 An example of the functional analysis approach; assessing the likelihood of insect outbreaks under climate change. (Patricia Werner)

Landsberg and Stafford Smith (1992) used a functional analysis approach to make predictions about tri-trophic structures, and in particular the potential for outbreak of herbivorous insects under global atmospheric change. They started with critical functional attributes of plants, insects and insect 'enemies', as well as possible states of each attribute (e.g. dormant, active). For each attribute/state, the direct effects of climate change were derived from published literature, with added notes as to which were temperature-sensitive, rainfall-sensitive, or CO_2-sensitive (Table 5.3).

From their analyses, they concluded that global changes that increase environmental stress on host plants are most likely to favour sap-feeding insects. They also concluded that elevated CO_2 was not likely to have a major influence on the probability of devastating insect outbreak, except possibly in systems where nitrogen-based defensive compounds are produced by plants (e.g. on all other continents). Thus, because Australian plants tend to use carbon-rich defensive compounds (Orians and Milewski 2007), increased outbreaks of herbivorous insects (above the already high levels) will not likely happen as a consequence of demonstrated effects of higher CO_2 on plants in Australia.

In sum, the analysis yielded critical insights that were not immediately obvious, simplified the complex interaction without compromising any scientific knowledge or principles, and provided guidance and a framework for setting priorities for research and management.

Dunlop and Brown (2008) provided a preliminary regional assessment of biodiversity impacts under climate change based on agro-climatic zones. This ecosystem-based approach assesses how seasonal patterns in plant growth, and therefore productivity, may vary and how these changes in growth may drive or contribute to a range of other changes affecting species and ecosystems. Ten agro-climatic zones were used (Hobbs and McIntyre 2005), and each of the 85 IBRA regions were placed in a zone based on seasonal growth and moisture indices. The CSIRO and BOM (2007) projections were then used to predict environmental changes. The zones identified as most likely to experience significant ecosystem-level changes were 'Temperate cool-season wet' (southern Victoria and north-eastern Tasmania) and 'Temperate subhumid' (north-eastern New South Wales). Both zones currently have limited growth in winter due to cool temperatures despite available moisture; warming is therefore expected to lead to marked changes in vegetation structure and composition. These zones are also likely to be affected by changes in fire regimes and are already subject to extensive habitat modification (Dunlop and Brown 2008).

Australian modelling of impacts such as fire, however, which will clearly be an important component of community response to climate change in many terrestrial systems, is receiving substantial research attention (Box 5.11).

5.6.3 Monitoring

On-site monitoring projects need to distinguish between monitoring to inform management (a form of hypothesis testing) and monitoring to simply watch species go extinct and ecosystems fail. A national framework to identify suitable species and sites, and to execute appropriate monitoring aimed at informing management and building on any long-term records where useful, would greatly assist both researchers and managers (Chapter 7).

It is also important to increase the role of the community in monitoring; a side-effect is increased awareness of the sensitivity of biological systems to relatively small amounts of climatic change. Phenological monitoring is particularly well suited to community involvement (Chapter 7).

5.6.4 Dealing with uncertainty: ecological resilience and transformation

It will be very important to focus on the biological and ecological qualities that give biodiversity (at all levels) increased resilience; that is, the capacity to experience shocks while retaining essentially the same functioning, structure, feedbacks and therefore identity. This can apply to an individual species, or to a community

Box 5.15 Where will species go? Bioclimatic models: a tool for predicting future changes in geographic ranges. (Lesley Hughes)

Bioclimatic models (also known as niche models and species distribution models) utilise occurrence records from museum and herbarium collections to predict changes in geographic ranges of species under scenarios of future climate. The models have been used in Australia and elsewhere to predict the impacts of climate change on the ranges of thousands of species, including vascular plants, butterflies, amphibians, reptiles, birds and mammals. They have also been used to assess the future efficacy of protected area networks (reviewed in Peterson *et al.* 2006). In general, most such modelling exercises have indicated that the distributions of many species will contract in the future (although some may expand) and/or become more fragmented. In some cases, the modelled range contractions are so severe that significant extinctions are projected (e.g. Thomas *et al.* 2004; Williams *et al.* 2003).

All bioclimatic models have inherent assumptions including:

- Species distributions are primarily determined by climate; that is, a species' realised niche is essentially the same as, or highly correlated with, its fundamental niche (see Chapter 3).
- Present-day relationships of species with climate and other factors will remain the same as the climate changes; that is, the species will move with shifting climate zones in a predictable way.
- There will be no lag effect between a particular climatic change and the response of a species range; that is, climate and species ranges are in equilibrium with each other.
- Species abundance, genetic variability, and the nature of demographic processes such as birth and death rates, are uniform throughout the geographic range.

Further limitations include the fact that the models at present do not incorporate species interactions, the potential for species to use microhabitat buffering or behaviour to mitigate the impacts of regional climatic changes, or to evolve in response to climate change as a selective pressure.

Despite these obvious imperfections, the dominant role played by climate in controlling biological activity, and species distributions in particular, means that 'potential niches' identified by bioclimatic envelope models are, at the very least, a good first guide to 'realised' niches in both current and future climates, and can provide a 'first cut' in understanding future changes.

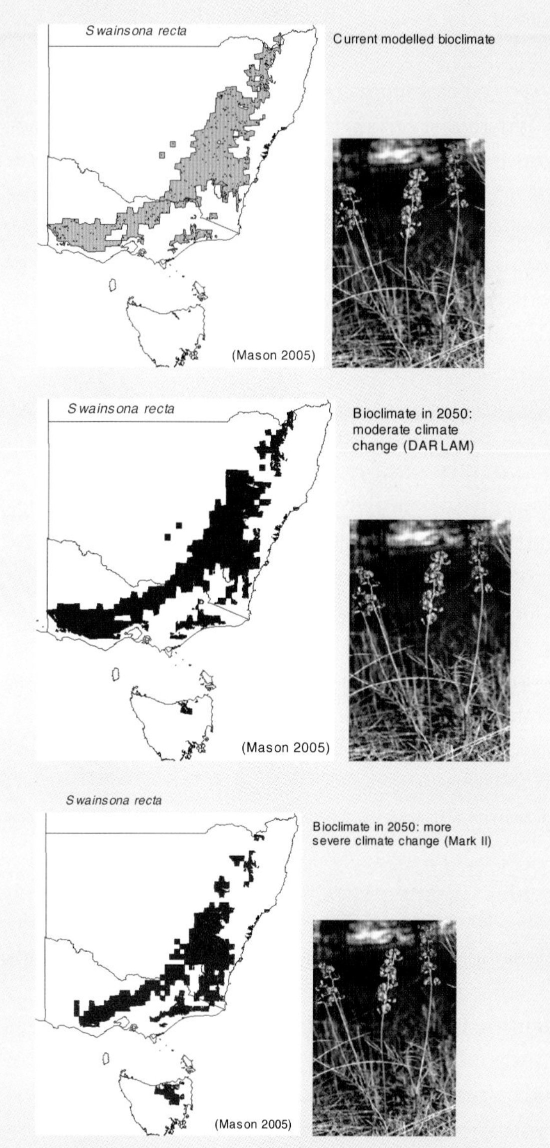

An example of the use of bioclimatic modelling to project the impacts of climate change on the flowering plant *Swainsona recta*.

Source: Mason (2005).

or ecosystem where the resilience refers to particular trophic structures or functioning. It also applies to maintenance of evolutionary processes through preservation of genetic diversity. Preservation of heterogeneity and diversity of environments is particularly important in providing avenues for resilience of biodiversity. In turn, society will be required to take appropriate action, in the face of uncertainty, that will increase the resilience of biodiversity. It must also be recognised that current conservation policies and practices are not adequate, given the continuing documented species losses and communities at risk (Chapter 6).

It is almost certain that large and rapid transformations in both climate and Australian biodiversity will take place. Ancillary questions as to how society plans for transformations and predicts them wherever possible, and the degree to which society can assist in transformations to new states beneficial to both natural biodiversity and human society, also need addressing with urgency (Chapter 7).

5.6.5 Ecological knowledge gaps and research questions

There are substantial gaps in our ecological knowledge and many research questions of direct relevance to the climate change challenge. Many relate to an improved understanding of how the ecological principles described in Table 3.1 are expressed in real-world situations under rapid environmental change. The gaps are grouped below by level in the biological hierarchy relevant to current and future impacts of climate change on the Australian biota (Tables 5.3 and 5.4). This is certainly not an exhaustive list, but serves to highlight the wide range of issues requiring more information.

The types of knowledge gaps and research questions described in Table 5.5 present a considerable challenge to the research community. Some questions have been tackled, and key knowledge gaps recently identified, for specific biomes and geographical areas of Australia (e.g. forests – Lindenmayer and Franklin 2002; arid zone – Stafford Smith and McAllister 2008; Great Barrier Reef – Johnson and Marshall 2007). To our knowledge, however, there is no general set of guidelines for identifying knowledge gaps in the face of climate change applicable to Australia as a whole, although a number of publications have identified many areas worthy of research (e.g. Hilbert *et al.* 2007b; Dunlop and Brown 2008). The principles set out in Table 3.1, and the analysis of observed and projected climate change impacts outlined in this chapter, provide a framework within which to build a coherent, integrated research program to eliminate these knowledge gaps.

5.7 THE CHALLENGE FOR POLICY AND MANAGEMENT

The Australian biota is beginning to experience anthropogenically driven climate change at an unprecedented rate. Changes in species distributions and abundances, and in community structure and composition, have already been detected that are consistent with recent changes in temperature, rainfall and sea level. Further climate changes will put tremendous pressure on our biodiversity, especially those organisms that cannot disperse rapidly, or cannot adapt in their current location to rapid change. The species-level changes already observed, and ever more rapid changes in climate through this century and beyond, will undoubtedly cascade through to affect entire communities and ecosystems so that novel ecosystems will be formed. The services provided by these ecosystems will also be affected. While some of these impacts can be anticipated, cascading effects will produce many unanticipated and surprising outcomes (Fig. 5.9).

The stress of rapid climate change is being imposed on top of existing stressors to biodiversity such as land use changes and species introductions. Chapter 3 described the current state of Australia's biodiversity, the stressors that have driven (and are continuing to drive) these changes, and the ultimate causes that lie behind our record of significant biodiversity loss over the past two centuries. We are thus facing the climate change challenge with a biotic heritage that is already impoverished in many ways and which continues to face most of the historic stressors that have operated over the past two centuries.

This situation calls for strategic planning in the face of uncertainty, and for innovative ways of thinking and doing. Fortunately, many policies and management strategies, implemented by a large community of dedicated practitioners, have been implemented over

Table 5.5 Examples of key knowledge gaps and research questions (see also Box 7.4).

Species
Knowledge gaps
Genetic structure and ecotypic variation within species/populations
Extent of phenotypic plasticity in populations, especially with regard to physiological temperature responses and behaviour
Key threshold values for keystone and habitat-structuring species in relation to the changing climate
Dispersal mechanisms and dispersal rates (capabilities) for most organisms
The role of biotic vs abiotic factors in controlling species distributions and population dynamics
Research questions
To what extent will species be able to cope with regional climatic changes by altering use of microhabitats in situ?
What is the level of ecotypic variation in key life history traits and physiological tolerances for Australian species in general? Are there discernible relationships between ecotypic variability and present-day geographic range size and, if so, are these relationships consistent among different functional groups?
What will be the role of genetic adaptation relative to phenotypic plasticity and behavioural responses in different functional groups or different taxa in adaptation to climate change?
How can we link bioclimatic models with demographic and population viability models to better predict future responses?
Communities
Knowledge gaps
The relationship between temperature, CO_2 and water availability, and key structural and functional (e.g. keystone) species; and the flow-on effects to species interactions such as competition
Community assembly rules that will be useful predictors for assessing responses to rapid climate change
Research questions
Are some taxa/functional groups more likely to be limited by biotic interactions than others? Can we begin to generalise as to how closely the fundamental and realised niches correspond in relation to other characteristics?
What are the physiological thresholds of temperature and water availability (and pH for marine organisms) of key species (e.g. keystone, foundational and habitat structuring species)?
Which key species and ecotones should be targeted for monitoring to best inform future management and provide information about the rate of important changes?
Ecosystems
Knowledge gaps
Ecological cascades and trophic structures within ecosystems
Ecosystem functioning rules: what basic components are required to ensure sustained ecosystem functioning?
Key physical elements in ecosystems (e.g. water holes)
The range of ecosystem services obtained from various types of ecosystem
Magnitude of interaction and synergies with existing stressors such as fire, clearing, grazing, invasive species, salinity, disease and water extraction
Research questions
What will be the role of fire management in mitigating threats including climate change?
To what extent can species be lost and/or substituted in an ecosystem without affecting ecosystem functioning?

Palaeoecology
Knowledge gaps
Biotic responses to past climate change, particularly periods of rapid or abrupt climate change
Location and role of refugia in the past, especially in areas of high topographic relief (e.g. the Wet Tropics)
Role of multiple localised refugia in terms of resilience
Research questions
How can we better identify past climatic refugia and use this information to target conservation planning?
Will climate change differentially affect known centres of endemism compared with other regions?

the past several decades to begin to address the array of non-climatic stressors that have historically diminished Australia's biodiversity. These are described in the next chapter, and provide a useful foundation for dealing with the additional stressor of climate change. However, much more is required. Chapter 7 provides an overview of the challenge that rapid climate change poses for our unique biodiversity, and points towards some innovative approaches that might meet this challenge and reverse the ongoing decline in the richness of Australia's biotic heritage.

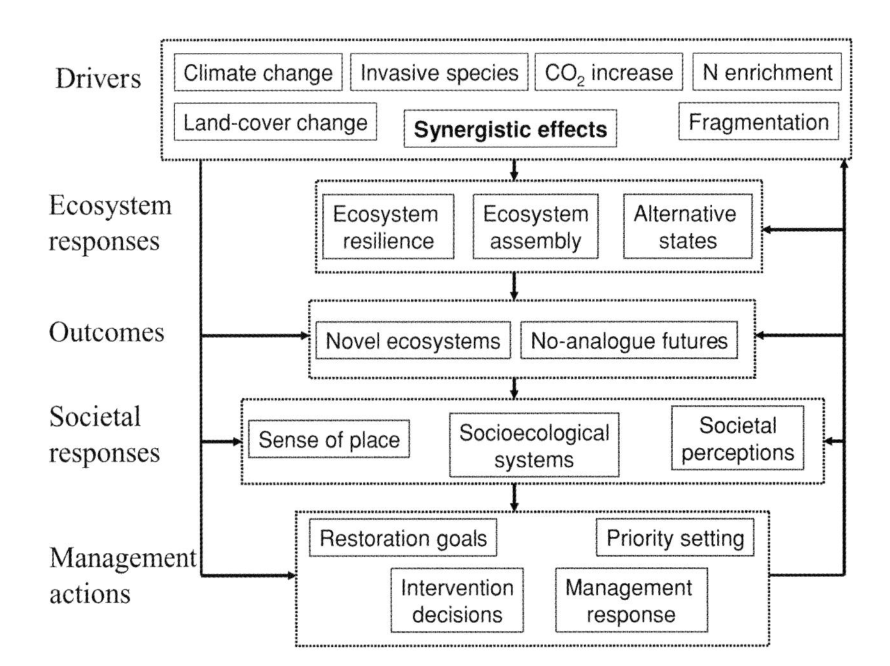

Figure 5.9 Diagrammatic representation of the challenge ahead. Source: Hobbs and Cramer (2008).

6 Current biodiversity management under a changing climate

This chapter provides an overview and discussion of existing biodiversity planning and management mechanisms, and considers their strengths and weaknesses under a changing climate. It firstly examines the principles that underpin current approaches to biodiversity conservation, the policies that are based on these principles and the governance structures that deliver them. The chapter then discusses management strategies and tools that are used to implement current policies. Limitations of the current policies and practices in the context of climate change are then outlined. The chapter concludes by exploring those aspects of current management practices that could provide a platform for adaptation to climate change.

6.1 READY FOR THE CLIMATE CHANGE CHALLENGE?

Chapters 4 and 5 have set out the nature and magnitude of the challenges to biodiversity conservation that a changing climate presents over the next few decades and beyond. Today the historical stressors on biodiversity, such as landscape fragmentation and invasive species, still dominate as drivers of biodiversity change that are observable now. However, the potential for accelerating change due to climate is large and growing. As global emissions of CO_2 continue to rise at or near the upper bounds of projections, the resulting rate and magnitude of climate change will place ever more severe pressure on biodiversity in Australia and worldwide. Is Australia ready to meet the climate challenge?

As detailed in Chapter 3, the conservation of Australia's biodiversity has been, and continues to be, significantly challenged by historical stressors. Climate change is an additional stressor. The national *Australia State of the Environment 2006* report (Beeton *et al.* 2006) identified land clearing, altered fire regimes, total grazing pressure, weeds and feral animals, and changes to the aquatic environment as major pressures on biodiversity. According to that report, the rate of land clearing has accelerated, with as much land cleared during the past 50 years as in the 150 years before 1945. Furthermore, only four other countries exceeded Australia's estimated rate of clearance of native vegetation in 1999. By any measure, the number of threatened and extinct plant species in Australia is high. Assessments of the state of Australia's biodiversity consistently point to the daunting battle that we face to halt the current rate of biodiversity loss. Perhaps this record is not surprising, as most of the underlying drivers of biodiversity change (Chapter 3) have not yet been adequately addressed by conservation managers.

Despite the poor historical outcomes, Australia has made some progress in recent years in dealing with current stressors that threaten biodiversity. Some of these tools and approaches, knowledge, and expertise – although developed before general awareness of climate change – are appropriate to the conservation of biodiversity under a changing climate.

This chapter examines:

- the principles that underpin current approaches to biodiversity conservation (section 6.2)
- the suite of policies that are based on these principles and the institutional structures that support them (section 6.3)
- the management strategies and tools deployed to implement the policies (section 6.4)
- the limitations of the existing principles-policies-institutions-tools framework for biodiversity conservation to meet the additional challenges that arise from climate change (section 6.5).

Where appropriate, the chapter describes how current approaches are being used or adapted now to deal with climate change. It also examines the opportunities that these existing approaches provide. Successfully meeting the challenge of climate change, however, will require a fundamental reappraisal of how we approach biodiversity conservation, which is the subject of Chapter 7.

6.2 CURRENT MANAGEMENT PRINCIPLES

Although there have been many definitions of biodiversity (for reviews see Bunnell 1998; Delong 1996), the term is now usually defined as the variety of all life forms: the different plants, animals and micro-organisms, their genes, and the communities and ecosystems of which they are part. Biodiversity is usually recognised at three levels: genetic diversity, species diversity and ecosystem diversity. Biodiversity conservation has primarily concentrated on species (e.g. recovery plans) and ecosystems (e.g. protected areas), although there is increasing recognition that maintaining genetic diversity is an essential component of conservation practice. However, although still insufficiently recognised, modern biodiversity conservation involves much more than protecting specific sets of entities (genes, species and ecosystems). It should also ensure the maintenance of ecological processes and the delivery of ecosystem services.

One example that goes beyond the maintenance of specific sets of entities is a view of forest biodiversity conservation with the goal of '... perpetuating ecosystem integrity while continuing to provide wood and non-wood values; where ecosystem integrity means the maintenance of forest structure, species composition, and the rate of ecological processes and functions within the bounds of normal disturbance regimes' (Lindenmayer *et al.* 2006, p. 434). These authors argued that management should aim primarily for the maintenance of the *processes* that lead to ecosystem integrity in order to underpin the biodiversity, rather than aiming at biodiversity directly.

Climate change – because it will lead to shifts in species ranges, and transformed and novel ecosystems – raises three related issues to be considered in a possible restating of the objectives of biodiversity conservation, at all levels of policy and strategy setting:

- Should the focus move more to maintaining the diversity of all biodiversity entities (genes, species, communities and ecosystems) rather than the specific current suite of species?
- In so doing, should the focus move further to (or at least explicitly encompass) maintaining the ecological *processes* that underpin the diversity of these entities, rather than the diversity of the entities themselves?
- Furthermore, should consideration be given to minimising losses of biodiversity or to maximising biodiversity at various scales, with regard to variety (diversity) rather than any particular set of entities?

Because of climate change, all levels of policy for biodiversity conservation – starting at the national level – should be clear about which qualities of biodiversity are being conserved.

Over time, biodiversity conservation has developed principles, approaches and tools in response to the historical drivers of change. In Australia, the definitive approach to biological conservation policy is found in the National Strategy for the Conservation of Australia's Biological Diversity (NSCABD) (Department of the Environment, Sport and Territories 1996) (Box 6.1). This document defines biological diversity as the variety of life forms, including genes, species and ecosystems, with the goal 'to protect biological diversity and maintain ecological processes and systems.' The NSCABD also contains basic principles that serve as a basis for the Strategy's objectives and actions (Box 6.1). The consultation paper for a revised national strategy, Australia's Biodiversity Conservation Strategy 2010–2020 (National Biodiversity Strategy Review Task Group 2009), now recommends the goal be articulated as 'to ensure our biodiversity is healthy, resilient to climate change and valued for its essential contribution to our existence'. It is up to individual state, territory and local governments to coordinate their planning, approaches and goals with the NSCABD (or its replacement strategy once this is in place), and this is being done to various degrees. For example, the Liverpool City Council Biodiversity Strategy (Liverpool City Council and Ecological Australia 2003) comes very close to the definitions, goals and objectives of the NSCABD, as does the report *National Approach to Addressing Marine Biodiversity Decline* (Marine Biodiversity Decline Working Group 2008). In Western Australia, the consultation paper for a new Biodiversity Conservation Act (Government of Western Australia 2002) defines biodiversity conservation as the 'combination of actions targeting the protection, restoration and sustainable use of our native plants, animals and other native organisms', with the Bill intending to provide their protection, restoration, and sustainable use. South Australia has focused on a 'no species loss' policy (Department of the Environment and Heritage 2007); biodiversity is considered 'a measure of South Australia's biological wealth', with the stated goal that 'all threatened species and ecological communities be improved and where possible restored.'

Historically, most approaches have focused on two levels of biodiversity, with overarching goals of:

- saving all species in situ as the highest priority
- protecting a diversity of communities or ecosystems and landscapes/seascapes, some of which are of particularly high value in their own right (e.g. the Great Barrier Reef).

Furthermore, virtually all conservation efforts to date have assumed an unchanging climate. That is, they have assumed no changes in the long-term climatic parameters such as mean temperature and rainfall. Some of the current management approaches will be directly relevant to climate change, while others may need to be modified or could even become counterproductive.

Principles and approaches to management for biodiversity conservation have tended to be based on one or more of the following underlying goals (Table 6.1).

Species should continue to live in their current locations

The corollary of this principle is that the replacement of particular species, such as during revegetation, should be done using the same provenance that has historically been found in the particular location. Again, this principle assumes a static climate. Its application to biodiversity conservation may become increasingly irrelevant as the climate shifts over the coming decades. In fact, it could easily become counterproductive if scarce resources are used in futile attempts to maintain current species distributions, or are used to reintroduce or replant species and/or genetic provenances in locations that are historically appropriate but are becoming increasingly marginal under a changing climate. This principle, which has been a conceptual pillar of biodiversity conservation for a long time, requires significant rethinking (Hobbs and Suding 2009; Lindenmayer and Hobbs 2007).

Offsets can help species to survive elsewhere

This principle, closely related to the previous one, applies when development or disturbance of some type in one region or location will affect a threatened species or community. In compensation, additional effort and resources are applied elsewhere, often to maintain or enhance particular habitats to ensure that the species or community is conserved in another location. Using offsets has become particularly important for marine species, in which new or enhanced marine protected areas are established to conserve a species under pressure from exploitation elsewhere. Offsets are also increasingly applied as conditions placed on land users where their activities

Box 6.1 The National Strategy for the Conservation of Australia's Biological Diversity. (Patricia Werner)

The National Strategy for the Conservation of Australia's Biological Diversity was released in 1996 (Department of the Environment, Sport and Territories 1996) with a review published in 2001 (ANZECC 2001). The goal of the Strategy is to 'protect biological diversity and maintain ecological processes and systems.' The glossary definition of biodiversity, however, is more limited: 'The variety of life forms: the different plants, animals, and micro-organisms, the genes they contain, and the ecosystems they form. It is usually considered at three levels: genetic diversity, species diversity, and ecosystem diversity.' From this, it would seem that the stated goal to maintain ecological processes would be done *in order to* preserve the entities of the biota, rather than to preserve the processes themselves. This may need a rethink in the face of climate change, where the processes themselves may, of necessity, become the primary object of biodiversity conservation, with secondary consideration for the entities (e.g. particular species) themselves.

With the addition of that caveat, the principles underlying the NSCABD's objectives are guides for implementation of plans and actions:

- Biological diversity is best conserved in situ.
- Although all levels of government have clear responsibility, the cooperation of conservation groups, resource users, Indigenous peoples, and the community in general is critical to the conservation of biological diversity.
- It is vital to anticipate, prevent and attack at source the causes of significant reduction or loss of biological diversity.
- Processes for and decisions about the allocation and use of Australia's resources should be efficient, equitable and transparent.
- Lack of full knowledge should not be an excuse for postponing action to conserve biological diversity.
- The conservation of Australia's biological diversity is affected by international activities and requires actions extending beyond Australia's national jurisdiction.
- Australians operating beyond our national jurisdiction should respect the principles of conservation and ecologically sustainable use of biological diversity, and act in accordance with any relevant national or international laws.
- Central to the conservation of Australia's biological diversity is the establishment of a comprehensive, representative and adequate system of ecologically viable protected areas integrated with the sympathetic management of all other areas, including agricultural and other resource production systems.
- The close, traditional association of Australia's Indigenous peoples with components of biological diversity should be recognised, as should the desire to equitably share benefits arising from the innovative use of traditional knowledge of biological diversity.

affect threatened species or ecosystems (Gibbons and Lindenmayer 2007); however, these often implicitly assume a static climate. An offset approach also raises issues of maintaining the genetic variability and integrity of a species across its entire range, which is enshrined in biodiversity conservation policy in many areas.

Maintain viable population sizes and facilitate increases of small populations where feasible

This management principle is based on the ecological principle that a species becomes vulnerable when its population drops below a critical level. It underpins strategies such as:

- reserve design
- the creation of buffer zones around reserves
- connecting habitats to link remnant populations of a species
- the translocation and reintroduction of desired species into ecosystems.

The overarching goal of increasing the population size of vulnerable species is still important in a changing climate, but some of the strategies and tools used to facilitate increases in population size may need to change because the climate is changing.

Maintain well-functioning ecosystems

This is the basic principle that underpins the maintenance of biodiversity and the provision of all ecosystem services. In addition, well-functioning ecosystems are the basis of resilience – the ability of ecosystems to absorb shocks, and maintain their structure and functioning. At regional and continental scales, a diversity of well-functioning ecosystems provides the broad range of habitats needed to maintain Australia's unique biodiversity and delivers the ecosystem services that are essential for the well-being of our society. This management principle, as currently applied, implicitly assumes that the climate fluctuates only within the limits of some natural variability (e.g. seasonal). Hence its influence on the location, composition and functioning of ecosystems does not change in a time frame of several centuries or more. This assumption is no longer valid, particularly if climate change this century reaches the predicted magnitude of glacial–interglacial transitions that occurred over thousands of years. Ecosystems will change in both location and composition as species respond differentially to a rapidly changing climate, almost surely affecting their functioning in ways that are difficult to predict a priori. It remains an open question as to whether current management and institutions can effectively apply the principle of maintaining well-functioning ecosystems under rapid climate change.

Remove or minimise existing stressors

This principle has become more prominent in informing management strategies in recent decades. Examples include:

- the control of land clearing in some regions
- control of fox, cat, goat and pig populations
- eradication of invasive species from islands
- the eradication or control of weeds in reserves.

This principle is a key component of a resilience-based strategy to enhance the adaptive capacity of ecosystems to deal with additional climate-related stressors. Unlike the previous two principles, this principle is immediately and directly relevant to the climate change challenge. Indeed, it will become even more important in future as climate change accelerates and its impacts on biodiversity become more severe.

Manage appropriate connectivity for species, landscapes, seascapes and ecosystem processes (see Boxes 3.3 and 6.2)

This principle has traditionally been associated with the conservation of highly mobile or migratory species, and with the need to connect remnant populations of threatened species or, at times, to isolate them (e.g. to protect island survivors from introduced predators). Some bird and mammal populations are managed with connectivity in mind, wherever land tenure permits this approach. It is obviously relevant to the climate change challenge because many species, especially those with long generation times, will need room and will require appropriate habitat in order to move (section 5.2.2). Thus this approach will become increasingly crucial in guiding policy and management responses to climate change in the next decades.

Protect a representative array of ecological systems

This principle recognises the fact that different environments support different biodiversity, and has been most obviously considered in the attempt to build representativeness into the 'comprehensive, adequate and representative' reserve system. However, it is also applied more locally in some reserves where a diversity of fire regimes (e.g. patch burning at Uluru Kata-Tjuta National Park and others) or seral stages of vegetation after disturbance are deliberately maintained in individual management units. Even though the individual species that characterise a particular location may change under climate change, continuing to encompass as much diversity as possible – at all scales, from patches to regions in areas managed for conservation – will remain a vital principle under climate change.

6.3 RELEVANCE OF MANAGEMENT PRINCIPLES TO ECOLOGICAL PRINCIPLES

The various management principles described in the previous section are relevant to the basic ecological principles underlying responses to disturbance or

Box 6.2 Regional vegetation networks, connectivity and biodiversity conservation in a rapidly changing climate. (David Lindenmayer)

One of the many kinds of impacts of rapid climate change will be range shifts among many elements of the biota (Parmesan 2006). Managing range shifts will be a far from straightforward task – particularly in landscapes and regions that have already been subject to substantial previous human modification, and where the existing cover of remnant native vegetation is fragmented (Opdam and Wascher 2004). One approach to facilitate range shifts in response to climate change has been to create ecological networks to maintain habitat, landscape and ecological connectivity (sensu Lindenmayer and Fischer 2007). Ecological networks have been defined as:

> 'A set of ecosystems of one type, linked into a spatially coherent system through flows of organisms, and interacting with the landscape matrix in which it is embedded' (Opdam *et al.* 2006, p. 324).

The concept of ecological networks has considerable appeal, particularly under a rapidly changing climate. Indeed, Bennett (1998, 2004) identified over 150 landscape-scale and regional-scale ecological networks around the world. Three high-profile ones in Australia are: the Alps to Atherton project, which has been renamed the Great Eastern Ranges Initiative (http://www.environment.nsw.gov.au) (Department of the Environment and Climate Change 2007); the Kosciuszko to Coast (http://www.k2c.org.au) in eastern Australia; and Gondwana Link in Western Australia (http://www.gondwanalink.org).

The Alps to Atherton project is designed to connect habitats and environments along the ridge of the Great Dividing Range and, in part, help facilitate latitudinal movements of species in response to changing climatic conditions. Kosciuszko to Coast creates an ecological network that spans a steep elevation gradient. It is designed, in part, to aid elevational shifts in species' ranges that might occur in response to rapid changes in climate, particularly temperature regimes. Both kinds of ecological networks also will be important for creating and/or maintaining large contiguous areas of relatively intact environments, which should promote the maintenance of populations of many species in situ. The Gondwana Link is attempting to reconnect a range of types of natural ecosystems in south-western Western Australia. The area targeted for linking spans over 1000 km between the karri forests in the far south-western part of Western Australia, and the mallee and woodland environments adjacent to the Nullarbor Plain. The Gondwana Link project aims to achieve this ambitious objective in several ways including purchasing land, halting land clearing, targeting large-scale revegetation efforts and working with key groups such as Landcare – which is responsible for improved rural land management (http://www.gondwanalink.org).

These kinds of ecological networks are important initiatives for maintaining and/or promoting various kinds of connectivity in a rapidly changing climate. However, their effectiveness remains largely unknown, and some deep thinking and carefully targeted monitoring will be required to quantify how successfully they contribute to the conservation of biodiversity. Part of the uncertainty around the long-term effectiveness of ecological networks centres around the fact that the physical linkages of habitats and environments as perceived by humans may not correspond to habitat connectivity for particular elements of the biota (Lindenmayer and Fischer 2006, 2007). That is, some species may not use ecological networks to move from one area to another (e.g. Sekercioglu *et al.* 2007). Given this, it appears that a range of strategies will be required to promote the various key kinds of connectivity needed for successful biodiversity conservation in landscapes and regions subject to rapid climate change. For example, Manning *et al.* (2009) championed the concept of landscape fluidity and the contributions to connectivity provided by various kinds of landscape elements in areas subject to multiple land uses (e.g. commodity production). They have also highlighted the connectivity roles that can be played by scattered paddock trees in landscapes dominated by agricultural production – both in Australia and around the world (Manning *et al.* 2006).

In summary, it is clear that a range of kinds of strategies will be important for maintaining or enhancing the various kinds of connectivity needed to conserve biodiversity under climate change. Ecological networks are one of these kinds of strategies.

change (Chapter 3). Table 6.1 examines the relative degree to which each of the current management principles are related to one or more of the ecological principles.

Current management principles for biodiversity conservation have different degrees of relevance to the ecological principles underlying responses to disturbance or change (Table 6.1). Some patterns emerge:

- In general, the management principles that are focused on individual species conservation (Management Principles 1–3) are very relevant to the ecological principles associated with individual species' responses to their environments (Ecological Principles 1–3), but do not address the other seven ecological principles.
- Maintaining well-functioning ecosystems (Management Principle 4) is highly relevant to principles of ecosystem functioning, and species interactions and roles (Ecological Principles 4–9), but have little direct relevance to ecological principles at the individual species level.
- Removing or minimising stressors on biodiversity (Management Principle 5) has relevance to almost all of the ecological principles (Ecological Principles 1–8), although other than reducing the stressor of land clearing, does not particularly address heterogeneity or connectivity (which give flexibility for movement of species across the landscape).
- Enhancing connectivity and maintaining large-scale landscape diversity (Management Principles 6 and 7) are highly relevant to nine of the ecological principles.

Maintaining large-scale landscape diversity provides the best overall scenario for biodiversity conservation in the face of disturbance or environmental change. This management principle encompasses the ecological principles supported by the individual species approaches while simultaneously addressing higher levels of biodiversity conservation. Managing for connectivity is also very important under a changing environment, but some vigilance is needed in using this approach, as connectivity per se can also be a conduit for movements by weedy plants or feral animals. Removing existing stressors and maintaining well-functioning ecosystems are also vitally important, as they address many ecological principles.

6.4 CURRENT CONSERVATION STRATEGIES AND TOOLS

The set of existing management principles outlined in the previous section and summarised in Table 6.1 form the foundation for strategies and tools currently used for biodiversity conservation. Strategies range from a focus on threatened species to a broader emphasis on landscapes and seascapes that have value in their entirety. The most important strategies and tools are protected area networks, off-reserve conservation, managing threatening processes, recovery planning and restoring ecosystems. These are explored in more detail as follows.

6.4.1 Protected area networks

Protected area networks form the cornerstone of Australia's conservation strategy for both terrestrial and marine ecosystems (Fig. 6.1 and Fig. 6.2). They respond most directly to the principles of ensuring species continue to exist in their historical habitats and maintaining well-functioning ecosystems. The protected area network aims to meet the criteria of comprehensive, adequate and representative (CAR). The National Reserve System (NRS) currently comprises over 8000 protected areas, including national parks, nature reserves, private conservation reserves and Indigenous Protected Areas. These areas together cover 88 million ha (11.5% of the continent) and represent the major terrestrial biodiversity conservation investment in Australia (Dunlop and Brown 2008). The reserve system provides the formal protection of viable samples of ecosystems, and long-term security of tenure and management. In 2004, the National Representative System of Marine Protected Areas comprised 7% of Australia's marine jurisdiction (excluding the Australian Antarctic Territory) covering over 64 million hectares. However, a recent review of the NRS (Sattler and Taylor 2008) found that Australia does not rank highly among the 17 mega-diverse countries regarding targets agreed under the Convention on Biological Diversity. Australia trails behind nations such as Colombia, China, United States, Peru, Ecuador, Venezuela and Malaysia in total percentage of land area protected (although management of protected areas in some of these countries remains ineffective, especially regarding illegal logging and poaching). Information from the Collaborative Australian Protected Areas Database shows that many of Australia's

Table 6.1 The relationship between management principles for conservation and the ecological principles of responses to environmental change – as described in Chapter 3 (Table 3.1), in terms of defining management actions that are needed for conservation.

	Current management principles for conservation						
Ecological principles relevant to environmental change	**1. Species should continue to live in their current locations**	**2. Offsets can help species to survive somewhere else**	**3. Maintain viable population sizes and facilitate increases where feasible**	**4. Maintain well-functioning ecosystems**	**5. Remove or minimise existing stressors**	**6. Manage appropriate connectivity for species, landscapes, seascapes and ecosystem processes**	**7. Protect a representative array of ecological systems**
1. Species differences: Species distributions and abundances reflect individual responses to the environment.	Species linked to locations	Same species in similar environments	Species-specific actions needed		Target stressors of particular species	Nature of connectivity is species-specific	Diverse landscapes support diverse species
2. Scale (time and space): Species distributions and abundances are responses to drivers at different scales.	Species linked to state of drivers	Same species where similar drivers	Species-specific management of drivers		Target scale of management of stressors for some species	Connectivity operates at multiple scales	Diverse drivers and scales support diverse species
3. Life histories: Life history attributes determine the ability of species to respond to change.	Life history attributes locally co-evolved	Life history attributes may help species fill any empty niches	Species-specific levels of resilience		Some stressors affect particular life history attributes	Importance of connectivity dependent on life history attributes	Diverse conditions support diverse life history attributes
4. Species interactions: No species exists in isolation from other species.				Must also manage above species level	Many stressors caused by, or affect, species interactions	Connectivity may depend on other taxa	Diverse options for species interactions enhance diversity
5. Species' roles: Some species affect ecological structure and processes more than others within communities and ecosystems.			Focus on critical species and others will follow	Some components of system more critical for management than others	Target stressors of key species for efficiency and greatest effect on ecological system	Prioritise connectivity of key species for efficiency and greatest effect on ecological system	Diverse key species enhance diverse dependent species
6. Trophic structures: Species are structured by their means of obtaining food into a larger trophic structure, or food web, of an ecosystem. The interaction of biota and physico-chemical environment yields ecosystem-level processes.				Species interactions determine ecosystem function	Target stressors that disrupt food webs for efficiency	Connectivity changes across trophic levels (i.e. multiple needs for whole web)	

Ecological principles relevant to environmental change	Current management principles for conservation						
	1. Species should continue to live in their current locations	**2. Offsets can help species to survive somewhere else**	**3. Maintain viable population sizes and facilitate increases where feasible**	**4. Maintain well-functioning ecosystems**	**5. Remove or minimise existing stressors**	**6. Manage appropriate connectivity for species, landscapes, seascapes and ecosystem processes**	**7. Protect a representative array of ecological systems**
7. Multiple drivers of environmental change: Communities and ecosystems change in response to many drivers, and these drivers themselves may interact.				Balance management of multiple drivers	Stressors interact often synergistically	Multiple approaches to connectivity needed	Create opportunities for a diversity of driver interactions
8. Non-linearity: Changes can be non-linear.			Population responses often non-linear	Ecosystem-scale responses sometimes non-linear	Effects of stressors can be non-linear (ensure stressors are below thresholds)	Effects of connectivity often non-linear (ensure thresholds are exceeded)	Non-linear relations between diversity of landscapes and species
9. Heterogeneity: Variations in time and space (heterogeneity) enhance biotic diversity.				Spatial diversity usually important to ecosystem function		Connectivity interacts with spatial heterogeneity	Manage to (reasonably) maximise spatial and temporal heterogeneity
10. Connectivity: Connectivity among resources, habitat, etc. shapes patterns of distribution, abundance and diversity.			Connectivity vital for some species			Manage for appropriate levels of connectivity (not always more)	Connectivity essential to link (or isolate) similar patches in diverse landscape

Key		
	Very highly relevant	Some relevance
	Highly relevant	Not particularly relevant

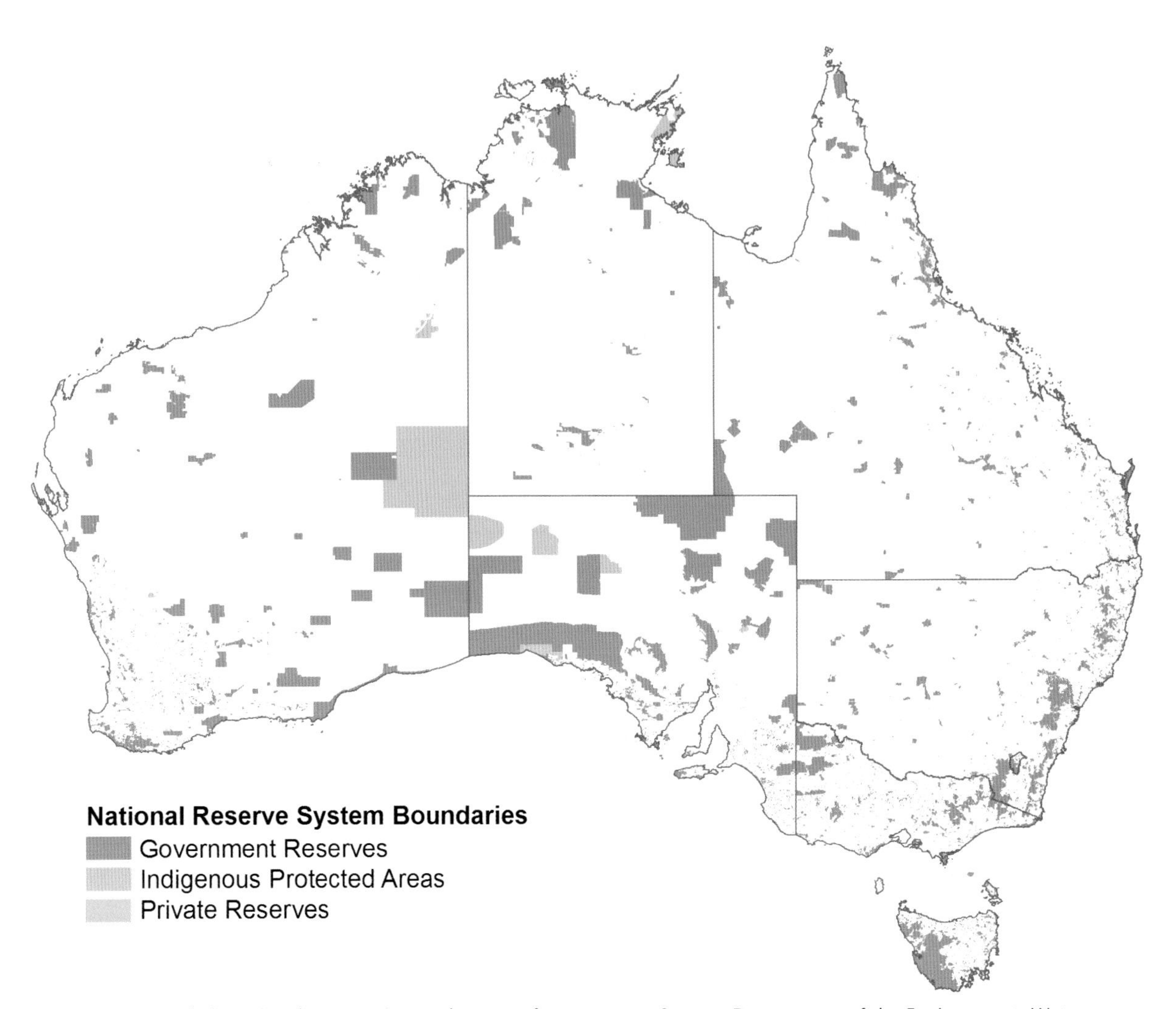

Figure 6.1 Australia's National Reserve System by type of governance. Source: Department of the Environment, Water, Heritage and the Arts.

bioregions still have less than 2% of their area included in conservation reserves, although the number of bioregions with less than 2% of their area reserved decreased from 26 to 16 between 1995 and 2005 (Gilligan 2006). Achieving a reserve system that is representative of the drivers of diversity will be as important under climate change as it is now, even though the precise forms of that diversity will be changing.

6.4.2 Off-reserve conservation

Off-reserve conservation evolved out of recognition that the protected area network, even if greatly enhanced, would not be adequate to meet many biodiversity conservation goals (Hale and Lamb 1997; Lindenmayer and Franklin 2002). Some species are found only in off-reserve areas (e.g. on private land), while others are migratory or nomadic and rely upon habitat found across entire regions. Some landholders have already been making important contributions to biodiversity conservation through revegetation and management of wildlife habitat. A range of local, state and national programs provides opportunities and support for off-reserve conservation of this kind. Expansion of these voluntary conservation efforts has become a major strategy in extending biodiversity conservation across private land. Off-reserve conservation supports the principles of facilitating movement of species across landscapes, facilitating increases in population size and

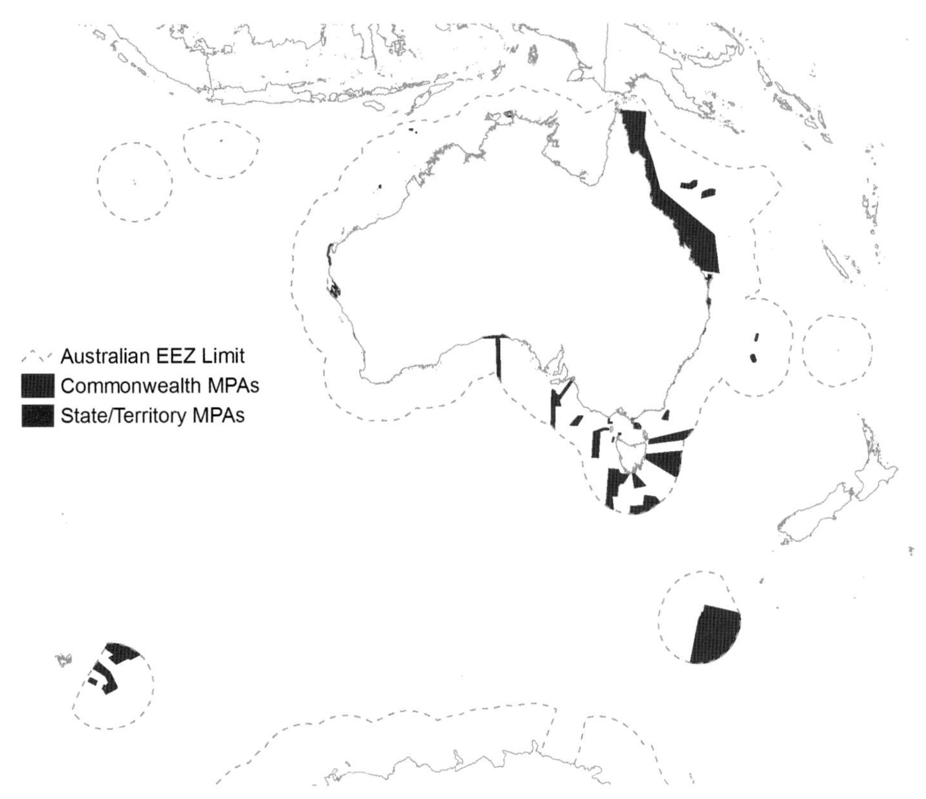

Figure 6.2 Australia's Marine Protected Areas. Source: Department of the Environment, Water, Heritage and the Arts.

maintaining well-functioning ecosystems. Taken together, protected areas and off-reserve conservation activities represent a progression towards landscape conservation, in which a range of land cover types, even if some are degraded or have less than optimal habitat, can deliver improved biodiversity outcomes and reduce extinction risk (Box 6.3). There are very few demonstrated examples of this type of landscape management anywhere in the world, and institutional and organisational arrangements to do this are yet to be identified and established (Lindenmayer and Hobbs 2007). However, this will become increasingly vital under climate change, due to the need for many species to move across the landscape matrix.

6.4.3 Managing threatening processes

Removal or minimisation of the very wide range of existing stressors on biodiversity is the key management principle addressed here. Threatening processes are recognised threats to biodiversity, due either directly or indirectly to humans. These include threats to the survival, abundance or evolutionary development of a species, population or ecological community. Reducing the impact of threatening processes can be an effective use of resources in that this addresses the cause of the problem, and can positively affect a range of species and communities, thus reducing the likelihood of future declines (Caughley and Gunn 1996). Vegetation clearance, modification and fragmentation comprise arguably the most serious threats to terrestrial biodiversity conservation in Australia, while fishing and pollution from land sources pose major threats to marine biodiversity. Invasive species – including exotic animals, weeds and diseases – significantly affect both production and environmental assets, and rank close to vegetation clearing as a driver of biodiversity loss. Hydrological disturbance has already had ecological consequences for Australian rivers, and is widely acknowledged to be a major cause of deteriorating conditions in many Australian river and floodplain ecosystems. The increase in areas affected by salinity due to rising water tables, caused by land clearing and irrigation, has been recognised for

Box 6.3 A hierarchical approach to protected area and off-reserve management in forests. (David Lindenmayer)

The conservation of forest biodiversity embodies a continuum of conservation approaches from the establishment of large ecological reserves through to an array of off-reserve conservation measures including the maintenance of individual forest structures at the smallest spatial scale.

Large ecological reserves are an essential part of all comprehensive biodiversity conservation plans and are critically important for many reasons (Lindenmayer and Franklin 2002). However, large ecological reserve systems are rarely comprehensive, adequate and representative for all elements of biodiversity. In other cases, past land management means there are few or no opportunities to set aside large ecological reserves. Hence, credible plans for forest biodiversity conservation must incorporate off-reserve approaches that complement reserve-based methods – that is, conservation strategies at the landscape and at the stand levels (Lindenmayer and Franklin 2002).

The five broad categories of approaches to landscape-level off-reserve forest management are:

- establishment of landscape-level goals for retention, maintenance or restoration of particular habitats or structures, as well as limits to specific problematic conditions (e.g. the amount of a forest landscape subject to prescribed burning)
- the design and subsequent management of transportation systems (generally a road network) to take account of impacts on species, critical habitats and ecological processes (Forman *et al.* 2002)
- the selection of the spatial and temporal pattern for harvest units or other management units (Franklin and Forman 1987)
- the application and/or management of appropriate disturbance regimes such as those involving fire or grazing
- the protection of aquatic ecosystems and networks (e.g. rivers, streams, lakes, ponds), specialised habitats (e.g. cliffs, caves), wildlife corridors, biological hotspots (e.g. spawning habitats, roosting areas for birds, camps for flying foxes, known threatened-species habitat), and remnants of late-successional or old-growth forest and disturbance refugia found within off-reserve forests.

It is important to distinguish between large ecological reserves and the protection of smaller areas within landscapes broadly designated for wood production.

The objective of off-reserve management at the stand level is to increase the contribution of harvest units to the conservation of biodiversity. The internal structure and composition of harvested units can have a significant influence on the degree to which a managed forest can sustain biodiversity and maintain ecosystem processes. Several broad types of strategies can contribute to the maintenance of structural complexity:

- structural retention at the time of regeneration harvest (e.g. large hollow trees and associated recruit trees, understorey thickets, large fallen logs)
- management of regenerated and existing stands to create specific structural conditions (e.g. through novel kinds of thinning activities). This may include the maintenance of open areas, as well as heath and grassland habitats within forests that can be critical for some key elements of biota
- long rotations or cutting cycles
- application of appropriate disturbance management regimes, such as prescribed burning to reduce fuel loads and reduce the risk of a high-intensity fire
- minimisation of linear disturbances, such as roads, which facilitate feral animal and weed movement.

Biodiversity is a multi-scaled concept. Therefore, attempts to conserve forest biodiversity must also be multi-scaled – with appropriate conservation strategies at the level of individual trees through to landscape and regional levels. Implementing an array of strategies at different scales is a risk-spreading approach. That is, if one strategy subsequently proves to be ineffective, others will be in place that might better conserve the entities targeted for management. Risk-spreading reduces the over-reliance on any one particular conservation strategy and attempts to deal with the considerable uncertainty regarding the effectiveness of current conservation management strategies. Risk-spreading is particularly appropriate for biodiversity conservation because it is often extremely difficult to accurately forecast the response of species to processes such as landscape modification, logging, prescribed fire and climate change (McCarty 2001).

many decades as a significant threat to biodiversity – particularly in the south-west of Western Australia and the Murray–Darling Basin. More generally, water resources in many catchments are already stressed due to intense competition for water supply among agriculture, urban areas, power generation, and environmental flows and other biodiversity requirements for water. Human modification of fire regimes affects the intensity, frequency and seasonality of fire, in turn affecting species populations, and the structure and composition of ecosystems in many regions (Bradstock *et al.* 2002).

Management of these threatening processes includes:

- controls on vegetation clearing
- improved allocation of water among competing users
- control of on-shore pollution sources
- restrictions on fishing
- modifications to various fire management techniques such as prescribed burning.

Current management approaches to deal with invasive species include biosecurity to prevent new species being introduced into Australia, and targeted action on established species including:

- eradication, biocontrol, chemical control and physical containment
- public and industry education
- changing management practices.

Management of threatening processes is becoming more difficult under a rapidly changing climate, as the direct impact of climate change on these threatening processes is undoubtedly important but very difficult to predict. Such management will be increasingly important under further climate change in order to maintain ecosystem function and resilience.

6.4.4 Recovery planning for threatened species and threatened ecological communities

Recovery planning for threatened species and communities has been effective in a few cases; for example, the noisy scrub-bird *Atrichornis clamosus* (Box 6.4; Danks *et al.* 1996) and the chuditch *Dasyurus geoffroii* (Orell and Morris 1994). Recovery plans seek to identify the threatening processes acting on a species or community, and then to set out the research and management actions necessary to stop the decline, and support the recovery, of the species or community. The aim of a recovery plan is to maximise the long-term survival in the wild of a threatened species or ecological community, but such plans now need to explicitly consider climate change. This approach most directly responds to the management principles of facilitating an increase in population size and/or addressing key threatening processes. However, while both Australian Government and state legislation promote the preparation and adoption of recovery plans, they do not require that funding to implement them be provided, hence recovery planning seldom leads to effective actions (Australian National Audit Office 2007). Planning and resourcing such long-term actions covering whole landscapes and seascapes will impose substantially increased demands on existing governance and management arrangements. This challenge is discussed further in Chapter 7.

6.4.5 Restoring ecosystems

Restoration ecology is a relatively new ecological science that aims to provide knowledge to restore or repair a considerably damaged, degraded or even destroyed ecosystem. The practice of ecological restoration is based on a wide range of activities including erosion control, control burning, removal of weeds, revegetation and reforestation, and the reintroduction of native species. Rehabilitation of mining sites is currently the most common type of ecosystem restoration in Australia and is technically advanced in many areas. Stream and riparian restoration has also been undertaken, with much research focused on measures to improve water quality, stream habitats and maintain geomorphic processes. Ecosystem restoration directly addresses the management principle of maintaining – or in this case, restoring – well-functioning ecosystems, but the implementation of this principle requires that climate change be taken into account when designing the nature of the ecosystem to be restored.

6.5 THE CURRENT POLICY AND INSTITUTIONAL LANDSCAPE

Biodiversity conservation over the past two decades has had a trend towards a redistribution of authority for managing natural resources – away from the traditional state and territory level, to both higher (national)

Box 6.4 Noisy scrub-bird *Atrichornis clamosus*. (Sarah Comer and Alan Danks)

The noisy scrub-bird was thought to be extinct until rediscovered at Two Peoples Bay, near Albany, Western Australia, in 1961. Management of the species is coordinated via the *South Coast Threatened Birds Recovery Plan*, which is implemented by the Western Australian Department of Environment and Conservation's South Coast Region with coordination by a Recovery Team comprising state government agency staff, scientists, the local natural resource management regional group and volunteers.

Two Peoples Bay Nature Reserve was created in 1967 to protect all noisy scrub-bird habitats known at that time. CSIRO and university researchers conducted research from 1966 onwards into the population genetics, diet and habitat (wet gullies) of the birds. Formal recovery planning started in 1980 and translocations began in 1983. From a single population of about 40 territorial males at the time of rediscovery, by 2001 the species had expanded to eight sub-populations with a total of 770 territorial males. A series of wildfires in the area between 2001 and 2005 dramatically reduced the population index, which stood at 368 in 2006. The number of wildfires from lightning strikes was unforeseen – there were more than in any other five-year period since the early 1960s.

Because fire can extirpate local populations and make habitat unsuitable for six to 15 years, the recovery plan has concentrated on establishing new populations elsewhere so that a single major fire, or even multiple fires over successive years, can never eliminate all birds. The strategy has been successful. Even though major wildfires significantly reduced total numbers, and the species is still listed as threatened, it occurs in much greater numbers and over a much greater area than was the case in 1961.

The noisy scrub-bird *Atrichornis clamosus* has recovered after successful implementation of a recovery plan.

Source: © Bert & Babs Wells/DEC.

As climate warming occurs in this area, predictions are for a general increase in the number of dry lightning storms and wildfires. Because the area of noisy scrub-bird habitat affected during the single five-year period 2001–2005 (more than 14 000 ha) was more than half the suitable noisy scrub-bird habitat in the area, a drying climate may place this species at severe risk, through both drying of its main habitat and increased fires.

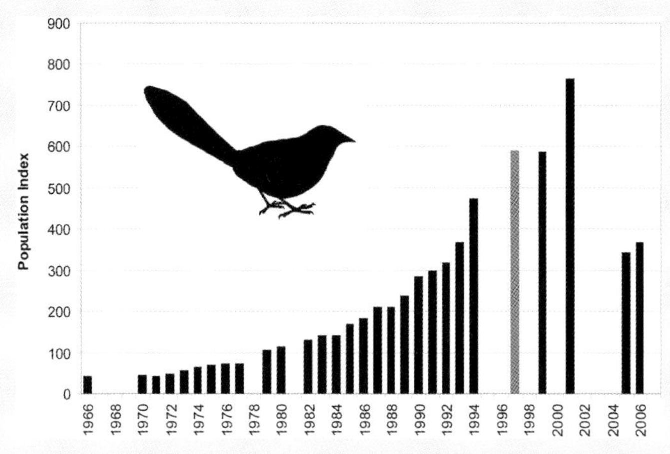

Population index (number of territorial singing males) for the noisy scrub-bird *Atrichornis clamosus*. Note: the 1997 index was based on an incomplete count and extrapolation.

Source: Sarah Comer.

Table 6.2 Examples of state and territory policies for biodiversity conservation.

Jurisdiction	Strategy	Date
Australian Capital Territory	*The A.C.T. Nature Conservation Strategy 1997*	1997
Victoria	*Victoria's Biodiversity – the Strategy* (replaces *Victorian Biodiversity Strategy 1997*)	2008
	Land and Biodiversity at a Time of Climate Change White Paper	2009
New South Wales	*NSW Biodiversity Strategy*	1999
	NSW Biodiversity and Climate Change Adaptation Framework	2007
	NSW Adaptation Strategy for Climate Change Impacts on Biodiversity	2007
	Adaptation Strategy for Climate Change Impacts on Biodiversity	2007
	A New Biodiversity Strategy for NSW: Discussion Paper	2008
Queensland	*Queensland Biodiversity Framework 2003*	2003
	Wet Tropics Conservation Strategy (2004)	2004
	Regional Nature Conservation Strategy for South East Queensland 2003–2008	2003
	ClimateSmart 2050 – Queensland Climate Change Strategy 2007: A Low-carbon Future	2007
	ClimateSmart Adaptation 2007–12	2007
	Issues Paper – Review of the Queensland Government Climate Change Strategy	2008
Tasmania	*Tasmania's Nature Conservation Strategy 2002–2006*	2003
	Draft Climate Change Strategy for Tasmania	2007
South Australia	*No Species Loss: A Nature Conservation Strategy for South Australia 2007–2017*	2007
Western Australia	*A 100-Year Biodiversity Conservation Strategy for Western Australia: Blueprint to the Bicentenary in 2029*	2008 Draft
Northern Territory	*Draft Northern Territory Parks and Conservation Masterplan*	2005 Draft

and lower (catchment or local government) levels. The Australian Government has used a number of powers to assume more influence over aspects of natural resource management, while at the same time there has been a strong move to devolve some policy and management responsibility to regional bodies and local governments. More recently, the rise of non-governmental organisations, including some with the financial resources to buy private land for conservation purposes, is adding a significant new element to the policy and institutional landscape. Society is thus already experimenting with novel modes of governance and management. While significant inertia remains in some quarters, the experiments now underway hold promise for building new institutions better adapted to dealing with the challenge of conserving biodiversity under a changing climate.

At the formation of the Australian Commonwealth in 1901, responsibility for and administration of natural resources was left virtually in its entirety with the states. All of the states and territories have biodiversity conservation legislation as well as strategies and policies developed over past decades (Table 6.2). However, the older policies and legislation predate the emergence of climate change as a significant issue for biodiversity conservation. Some of the more recent policies, including those under development or revision, give higher priority to climate change. These policies all have a number of common elements:

- consistency in regulation across all sectors that affect biodiversity
- recognition of the need for both on- and off-reserve actions
- partnerships with community and industry sectors
- long-term investment arrangements based on agreed outcomes
- flexible economic and other mechanisms – such as stewardship incentives, covenants, tenders and voluntary agreements
- underpinning programs.

Over time, the Australian Government has extended its influence in natural resource management through its taxation and funding powers, offshore marine fisheries responsibility, and use of external affairs powers – particularly as a result of entering into international treaties related to biodiversity conservation (Table 6.3). Australia has a long and relatively successful history of cooperation between state and Commonwealth governments. There has been increasing recognition that such 'cooperative federalism' has an increasingly important role to play in natural resource management through mechanisms such as the Council of Australian Governments (COAG) and various ministerial councils. The Natural Resource Management Ministerial Council in particular is the body responsible for taking forward high-level policy development for managing biodiversity under a changing climate.

Over the past decade or so, overarching biodiversity policy and strategies have been developed through joint Commonwealth–state processes and have been delivered through programs such as the Natural Heritage Trust, the National Action Plan for Salinity and Water Quality and, most recently, the Australian Government's Caring for our Country initiative. The first strategy that recognised climate change as a significant threat was the National Biodiversity and Climate Change Action Plan 2004–2007. This plan shifted the emphasis from the understanding and prevention of climate change impacts to understanding and adapting to impacts. The Plan also placed a greater emphasis on risk management as well as promoting adaptation and the enhancement of resilience.

Over the past two decades there has also been a trend to devolve much of the delivery of natural resource management (NRM) programs to the level of regional catchment management authorities and local Landcare groups. The benefits of this development are closer engagement with regional communities, 'landscape-wide' and whole-of-catchment approaches to management, and better targeting of Commonwealth and state government funding through integrated planning and investment strategies. There is significant variability across the range of regional bodies in terms of their organisational arrangements, capacity and performance, particularly in relation to delivery of biodiversity conservation outcomes. In addition, these bodies are limited in their ability to access and use new scientific results. This is becoming a serious problem under climate change, where information needs are evolving rapidly. However, it is likely that local or regional level of management will continue to be a key element in the institutional response to climate change (Keogh *et al.* 2006) (Chapter 7).

Local government undertakes a wide range of activities that affect NRM, such as land management, land use planning and fire control. These, however, have not yet been well integrated into biodiversity conservation and NRM activities at the state and national levels. In addition, local-level policies do not necessarily ascribe importance to the same species and ecological communities as occurs at the state and national levels. Local government has the potential to deliver on-ground actions, infrastructure support, regulation and incentive-based systems in support of biodiversity conservation. As yet, however, institutional arrangements for local government bodies to deliver outcomes consistent with state and national biodiversity objectives are not well developed. Furthermore, robust arrangements for effective collaboration between local government and other regional NRM bodies, such as catchment authorities, appear to be underdeveloped.

One of the most significant policy and institutional trends over the past two decades is the increasing importance of non-governmental organisations (NGOs) for biodiversity conservation. NGOs play three contrasting roles:

- NGOs such as the Wilderness Society, the Australian Conservation Foundation and the World Wide Fund for Nature (WWF) have traditionally focused their activities on political advocacy, community awareness and campaigns targeting specific sectors or high-profile issues. For example, WWF-Australia has produced reports about biodiversity and climate change, and also works with local communities to achieve conservation outcomes. Many smaller NGOs operate at national or local levels and influence particular aspects of conservation; for example, birds or local plant communities.
- Many NGOs actively promote research on biodiversity conservation through data collection and

Table 6.3 International conventions that directly influence Australian biodiversity conservation policy*

Convention	Brief description
Ramsar Convention on Wetlands (1971)	Provides for the listing of wetlands that meet specified significance criteria. There are currently 65 Australian wetlands listed under the Convention. The Convention promotes the concept of 'wise use', which was a precursor to the principle of sustainable development.
Convention on International Trade in Endangered Species (CITES) (1975)	Regulates the movement of endangered species and their products across international borders
International Whaling Convention (1946)	Regulates the management and conservation of the world's whale species, which were depleted from over-harvesting
Japan–Australia Migratory Bird Agreement (JAMBA) (1974) China–Australia Migratory Bird Agreement (CAMBA) (1986) Republic of Korea–Australia Migratory Bird Agreement (ROKAMBA) (2006)	Bilateral agreements that seek to conserve shorebirds (predominantly) that migrate annually between the northern and southern hemispheres. The agreements are supported by the protection and management of a network of important roosting and feeding sites, many of which are under threat from development.
World Heritage Convention (1972)	Provides for the identification, protection, conservation, presentation and transmission to future generations of cultural and natural heritage that is of outstanding value to the international community
Convention on Biological Diversity (1992)	Promotes the broader concepts of genetic, species and community biodiversity. The Convention introduced ecological sustainable development. It also included a program of action and promoted stronger community engagement through Agenda 21.
United Nations Framework Convention on Climate Change (UNFCCC) (1992)	Objective is to achieve stabilisation of greenhouse gas concentrations in the atmosphere at a level that would prevent dangerous anthropogenic interference with the climate system (Article 2)

* There are many more conventions, such as the World Trade Organization (1995), which may have major indirect effects on biodiversity.

monitoring. Birds Australia and Earthwatch are examples of such NGOs.

- More recently, organisations such as The Nature Conservancy and Australian Bush Heritage have supported the purchase of private land with high conservation value, in some cases returning it to the government estate while in others taking an active management role themselves.

The international aspects of biodiversity conservation are also increasingly influencing our national policy and institutional frameworks. Over the past 40 years the international community has worked to develop a number of global agreements in areas where coordinated efforts are required by nations to achieve biodiversity outcomes. Australia has been an active participant in the development of these agreements and has generally been a strong supporter of international cooperation. A number of agreements and conventions have been developed, signed and ratified to directly or indirectly address biodiversity conservation at global or regional scales. Table 6.3 describes the various international conventions that affect Australia's biodiversity conservation policy, and Box 6.5 highlights the role of national and international cooperative efforts to conserve migratory shorebirds and their habitat.

The trends described in this section have led to complicated institutional arrangements for managing conservation of Australian biodiversity, ranging from local to international scales. On the one hand, there has been greater apparent devolution of responsibility to regional and local levels. This has been driven in part by greater demand for public participation

Box 6.5 Trans-equatorial migrating shorebirds. (Andrew Burbidge)

Some 35 or so species of shorebirds, comprising about two million birds, are regular migrants to Australia. Most of the migrant species breed in northern China, Mongolia, Siberia and Alaska during June and July, and then migrate to Australia for the non-breeding season. The migration route used by these birds is known as the East Asian–Australasian Flyway.

Within Australia, 180 areas have been identified as being of international importance and a further 21 of national importance for these birds. In 1993, only 21% of these areas were in conservation reserves. The three most important are the south-eastern Gulf of Carpentaria in Queensland, and Roebuck Bay and Eighty Mile Beach in Western Australia (Watkins 1993).

The East Asian–Australasian Flyway crosses 20 countries. Shorebirds need to rest and 'refuel' during their migration between Australia and their breeding grounds. Along the way, they are subject to a number of threats – loss of feeding habitat and hunting are the major ones. Over the past three decades, Australia has entered into international agreements with Japan, China and the Republic of Korea, aimed at limiting hunting and protecting feeding habitat. Australia has also helped in developing the Asia-Pacific Migratory Waterbird Conservation Strategy. In addition, Australia is a signatory to the Ramsar Convention on Wetlands of International Importance, an intergovernmental treaty that provides the framework for national action and international cooperation for the conservation and wise use of wetlands and their resources. Currently, 65 Australian 'Wetlands of International Importance' are listed under the Ramsar Convention, covering approximately 7.5 million ha.

Migratory shorebirds continue to face a number of threats and there is increasing evidence that migratory shorebird populations are declining. Paramount among these threats is habitat loss through coastal development, especially in East Asia. Sea-level rise caused by climate change will further exacerbate habitat loss, since shorebirds feed in shallow water or on tidal flats exposed at low tide.

Mixed flock of trans-equatorial migratory shorebirds at Roebuck Bay, Western Australia.

Source: © Bert & Babs Wells/DEC.

because of reduced public trust in governments, science and experts since the 1970s (Ezrahi 1990), and also in part by the belief that local communities bring previously untapped knowledge and skills to the task. On the other hand, increased centralisation of responsibility has arisen in response to both the Commonwealth's increasing international responsibilities under relevant conventions (Table 6.3) – mostly recognised as Matters of National Environmental Significance under the *Environment Protection and Biodiversity Conservation Act 1999* – and the Commonwealth's growing fiscal role relative to states and territories, and a belief that efficient delivery of NRM programs is best achieved through purchaser–provider arrangements. In practice, this means funding is managed by agencies that are even further from on-ground actions than those of the states and territories, and therefore must impose onerous accountability requirements due to legal obligations under the *Financial Management and Accountability Act 1997* (Cwlth) and the *Commonwealth Authorities and Companies Act 1997* (Cwlth) regarding fiscal accountability. Many studies (e.g. Bäckstrand 2003; Fischer 2000; Norgaard 2004; Norgaard and Baer 2005; Plummer 2006) are now showing that centralised science and policy are poor at reflecting local differences, and that great benefits – at least through net reductions in transaction costs – can be obtained by linking local knowledge with science and policy in ways that capitalise on the best attributes of each. This creates challenges to develop the 'right' institutional design to foster the efficient establishment of accountability and trust.

The need to meet this challenge will become more acute as climate change causes regionally differentiated impacts that require local knowledge to manage effectively, but which require state and national support to resource this management and to coordinate between regions. There is thus a major challenge to incorporate community-based processes in our governance systems as efficiently as possible, by harnessing the potential of such processes to grow trust from the bottom up and the top down – thereby reducing the need for bureaucratic processes as a means of establishing this trust. There is also reason to think that meeting this challenge could result in more effective and efficient governance of natural resources. This point is discussed further in Chapter 7.

6.6 A PLATFORM FOR ADAPTATION TO CLIMATE CHANGE?

Which aspects of our past experience in biodiversity conservation can we carry over to the climate change challenge? As noted in sections 6.2 and 6.3, some of the existing principles that guide our efforts to conserve biodiversity are directly relevant to meeting the climate change challenge. Based on these principles, we already have some tools that can be used – often with reorientation and refinement – for the adaptation process. Five broad approaches aimed directly at the climate change challenge are evolving from the existing suite of tools and approaches (described in section 6.4 and considered in the following paragraphs). These five new approaches are:

- enhancing resilience in ecological systems at all levels (e.g. by representation, replication, restoration, refugia, relocation, and reducing other stressors)
- creating landscapes that maximise adaptation opportunities
- expanding and augmenting the reserve system
- undertaking specific in situ conservation actions based on the principles in sections 6.2 and 6.3
- undertaking ex situ conservation actions in some cases.

Enhance resilience

The concept of resilience traditionally meant the capacity for a system to recover to a previous state. In the context of biodiversity and climate change, resilience is considered as the extent to which species, ecosystems, landscapes and seascapes can undergo a shock without loss of functioning, structure, feedbacks and other valued characteristics (Walker and Salt 2006). For example, species might move rather than become globally extinct and ecosystems would continue to maintain functioning (e.g. nutrient cycling) as they change. To this extent, resilience may include the transformation of current ecological communities into new species mixes. It will be important, therefore,

Box 6.6 The use of surrogates in selecting areas for the National Reserve System. (David Lindenmayer)

Protected areas serve the needs of both species and ecosystem conservation, and they are a core component of conservation strategies worldwide. Three guiding principles – the CAR principles – underpin the design of a reserve system. This acronym stands for comprehensive, adequate and representative (Margules and Pressey 2000). *Comprehensiveness* refers to the need for a reserve system to capture the complete array of biodiversity ranging from species (and their associated genetic variation) to communities and ecosystems. *Adequacy* relates to the need for a reserve system to support populations of species that are viable in the long term. *Representativeness* means that a reserve system should sample the full range of species, forest types, communities and ecosystems from throughout their geographic ranges.

The design of a CAR reserve system will often be based on surrogates for overall levels of biodiversity. These surrogates are attributes thought to represent the distribution and abundance of species and assemblages. They can include forest types or other plant communities, particular climatic parameters, eco-regions or climate domains. Surrogates are essential in reserve design because it is impossible to document comprehensively all biodiversity (Lindenmayer and Burgman 2005). A region or landscape can be classified and mapped according to biodiversity surrogates, providing a spatial dataset against which reserve selection procedures can be applied (Margules and Pressey 2000).

to rethink the scale at which we assess resilience. At present it tends to be considered at the scale of individual parks and reserves. This will continue to be appropriate for ecosystem resilience, but the resilience of species and genetic diversity needs to be considered at a national or even international scale, as distributions shift and change.

Diversity of species is an important characteristic of resilient ecosystems; diverse ecosystems provide a range of functions and a range of alternative species to provide those functions when climate and disturbance regimes change. Actions such as appropriate fire management, control of invasive species and provision of environmental flows for aquatic ecosystems can build overall, general resilience of ecosystems against stressors and disturbances associated with climate change. Building resilience in ecosystems against specific threats arising from climate change (e.g. higher temperatures, drought or longer drying periods, more intense storms) could include enhancing connectivity along environmental gradients, and also providing better resourced and more effective management of invasive species, particularly those that may benefit from climate change.

The management and protection of refugia is an emerging strategy to enhance resilience. Low (2007) proposed the conservation of 'cool sites' – south-facing slopes, rock outcrops and deep gullies that offer relief from excessive temperatures – and aquatic refugia, particularly pools in streams that provide habitat for species during drought. Natural regeneration, particularly of eucalypts as well as other habitat-structuring species, is also likely to be an effective approach to recovering resilience. Natural regrowth selects for genotypes adapted to climate trends relative to parent stock, and may be better defined as 'reservoir' vegetation (a source for further adaptation) rather than 'remnant' vegetation (saved from environmental change) (Mansergh and Cheal 2007).

In practice, building resilience in complex systems like ecosystems is not a simple task, as the potential for increasing resilience in one dimension could lead to a reduction in resilience in another (Walker and Salt 2006). Furthermore, ecosystems need to be considered as complex social-ecological systems to truly build resilience that also incorporates the human aspect.

Create landscapes that maximise adaptation opportunities

Managing connectivity across landscapes is a major adaptive response to climate change (Boxes 3.3 and 6.2). Enhanced connectivity will enable mobile species in limited geographic ranges to:

- move as their habitat shifts (e.g. the Western Australian wheatbelt)
- provide more opportunities for breeding and survival

- link remnant populations
- increase the protection, and values, of refugia
- in theory promote genetic diversity and genetic adaptation.

In other cases, reducing connectivity can protect species that would be at risk. Achieving appropriate connectivity in terms of biodiversity and ecosystem functioning requires partnerships involving all types of tenures and including private landholders, all governments and regions, industry, and civil society. However, voluntary off-reserve activity by individuals or groups may not be sufficient to meet the objective of more appropriate connectivity. New market-based policy instruments, additional regulation, and more effective institutions and organisations, may be necessary. This issue is addressed in Chapter 7.

Expand and augment the reserve system

The expansion and maintenance of the network of protected areas – terrestrial, freshwater and marine – is a central element of a climate change adaptation strategy (Dunlop and Brown 2008; Box 6.6). However, as noted in section 6.2, the principles that underpin the design and maintenance of a reserve system will be different under climate change. Rather than providing habitat for species and ecosystems to be preserved where they currently are, reserves can provide the core elements of the more connected, dynamic landscapes and seascapes required to allow many species to move and reassemble in novel ways to form new ecosystems. This will need to build on the National Reserve System from the perspective of climate change, and focus on criteria such as key source populations and habitat, climate refugia, refugia from other stressors, migration corridors, stepping stones, and adaptation pathways. This approach also implies a reconsideration of the targets for protected areas taking more account of connectivity and likely changes in geographic ranges (Dunlop and Brown 2008). Continuing to augment the national reserve system with off-reserve conservation will be essential to manage the appropriate levels of connectivity. Even with an enhanced National Reserve System, off-reserve conservation is becoming more important as a strategy to cope with climate change (section 6.5).

Undertake in situ conservation actions based on basic principles

Specific in situ conservation actions based on ecological and management principles (sections 6.2 and 6.3) include:

- improving the robustness of ecosystems using natural processes such as recolonisation and natural selection, and taking into consideration a changing climate when selecting species for deliberate revegetation
- conserving specific environments already severely at risk as a high priority, such as grassy woodlands in south-eastern Australia
- removing or reducing existing threats such as invasive species (e.g. foxes, cats, rabbits, weeds), and identifying and addressing threats such as altered fire regimes (Williams *et al.* in prep.) that are compounded or enhanced by climate change (e.g. synergistic changes to fire regimes promoted by introduced gamba grass *Andropogon gayanus* in northern Australia)
- developing and implementing recovery plans for threatened species in the context of a changing climate
- investigating eco-engineering opportunities as appropriate, such as translocations of some species (Hunter 2007; Low 2007; Mansergh and Cheal 2007) and eco-engineering on a landscape basis (e.g. restoration). However, the large costs and low probabilities of success in some cases must also be considered
- preventing catastrophic collapse of systems in times of severe stress such as prolonged drought (e.g. ensure adequate water flow to aquatic ecosystems).

Undertake ex situ conservation actions

Ex situ conservation is becoming increasingly important as a means of conserving species that may not otherwise survive. The traditional collections of plants in botanic gardens and animals in zoos are now being augmented by germplasm banks (collections of genetic material – principally in the form of seeds – but also eggs and sperm, often stored cryogenically) and cooperative captive breeding programs based on stud books to minimise loss of genetic diversity. While traditionally viewed as a technique that should be used only as a last resort, ex situ conservation may become increasingly important with climate change as more species come under increasing threat of extinction in the wild. Even for species for which no suitable habitat will remain as climate changes, germplasm banks provide a hedge for the possibility that such habitat would re-emerge (Chapter 7).

6.7 SOME CHALLENGES TO INSTITUTIONAL CAPACITY

Tools and approaches for managing biodiversity under a changing climate are ultimately dependent on the organisations that develop and deploy them. As described earlier in this chapter, some organisations have built a solid base of experience in NRM – including biodiversity conservation – across a number of scales. Even without climate change, however, agencies and organisations are challenged to deliver effective, integrative management approaches and tools that meet several objectives simultaneously. The existing record of biodiversity loss and change, as described in Chapter 3, highlights the limited success of past and current approaches to conserve Australia's unique biodiversity. In most cases, the existing challenges to institutional capacity are being exacerbated by climate change. The most important of these challenges include:

- historical dominance of production industries over conservation
- resource limitations
- policy and jurisdictional differences
- key knowledge gaps.

6.7.1 Historical dominance of production industries over conservation

Agriculture, forestry, water provision, mining and other resource-oriented sectors have been important historically to the development of Australia and continue to play a large role in global markets. Because of their economic importance to Australia, these sectors have taken precedence over biodiversity conservation in terms of land use planning and resource allocation, yet all of them affect terrestrial and aquatic biodiversity. For example, much of the land now reserved in Australia's national system of protected areas is characterised by low levels of productivity (Lindenmayer and Burgman 2005; Pressey 1995), or is either too remote or too rugged to be of use to some resource-oriented sectors. Over the past few decades, this imbalance in emphasis has been redressed to some extent as environmental concerns grow in importance for the Australian public and the perceived economic value of protected areas increases. Nevertheless, there are legacy issues in legislation, land tenure and use, and in the institutional framework for NRM that reflect our historical emphasis on production and resource extraction. Furthermore, investment in biodiversity will be under greater, rather than less, pressure in the future due to global requirements for food and rising energy prices. The widespread changes now occurring across many Australian landscapes, however, provide opportunities for new approaches to combine production and conservation activities on the same landscape. The changes also provide opportunities for additional resources for biodiversity conservation through, for example, enhancement of commodity-based research and development corporations in biodiversity conservation, and direct investment by landholders in biodiversity conservation. These and other issues will be developed further in Chapter 7.

6.7.2 Resource limitations

Comparison of investment in biodiversity conservation with investment in physical capital such as roads and buildings – as well as in human and social capital through health, education and governance – would probably show that Australian investment in 'natural' capital is tiny. Investment in natural capital is literally at least two orders of magnitude smaller than that in physical capital – even when considering private investment by landowners in direct management and when NGOs involved in land acquisition and by industry research bodies is added to public sources of investment. There are several reasons for this, but reflect among other factors:

- the inability to establish markets for many ecosystem services (Chapter 2)
- as a consequence, the need for the public sector to ensure maintenance of ecosystem services
- the investment in the underlying natural resource base being driven by diverse priorities in the public sector.

For most societies, the historical flow of ecosystem services has been so abundant as to not cause concern. In many cases, they have been treated as free goods, which did not warrant priority treatment in state budgets. Consequently, much of the wealth generated by the consumption of ecosystem services has been invested in the provision of (non-ecosystem) services perceived to be scarce. As long as natural capital reserves were perceived to be sufficient to maintain the provisioning of ecosystem services that could not be substituted for by built, human or social capital, then investment in them would not have been a high public priority. Today, however, natural capital is struggling to continue to provide adequate delivery of these non-substitutable services (Chapter 3). In this context, a major reappraisal is required of the strategically appropriate level of public and private sector investment in natural capital.

6.7.3 Policy and jurisdictional differences

These include the following issues, which are developed further in Chapter 7.

Policies for resource-oriented and conservation-oriented sectors are usually developed in isolation from one another, leading to less than satisfactory outcomes for biodiversity. A good example is the need to develop rural industries and best practices that build sustainable livelihoods under climate change, while simultaneously developing and rewarding skills and capacities in managing land for biodiversity conservation.

Species do not respect jurisdictional boundaries, which creates management challenges even without climate change. As more species move at increasing rates in response to changing climate, cross-jurisdictional issues will become even more pervasive and complex. Furthermore, policy and management practices will need to adapt so that these species are not considered 'aliens' and potentially removed. In addition, the effects on resident species must be considered.

Policies in other areas of climate change can have impacts on biodiversity through so-called perverse incentives (Chapter 3) or unintended impacts. An example is the danger that land-based mitigation strategies (e.g. carbon sequestration, carbon offset schemes, biofuel plantings) can damage biodiversity through subsidies for, or promotion of, fast-growing monoculture-based production systems. On the other hand, well-conceived and implemented schemes can produce win-win situations in which both climate and biodiversity benefit.

One very specific area of interest is Indigenous land management, especially in relation to institutional arrangements, and to what extent the absence of private property rights will meet the challenges of rapid climate change.

Given the large uncertainties associated with the future trajectory of climate change and the equally large uncertainties associated with the ecological responses, adaptive management is the only way forward. Yet many management agencies are very risk averse, and are reluctant to use adaptive management for fear of it being perceived as a policy failure rather than as a learning-by-doing approach (Lindenmayer and Franklin 2002).

6.7.4 Key knowledge gaps

Climate change science, ecological theory, resource economics, management and institutions are all rapidly changing fields of scholarship. All of them are important for improved policy and management for biodiversity conservation under climate change, but they are often isolated from each other and from practitioners. Biodiversity management agencies need strong science groups within them to conduct high-priority research, taking care to increase research capacity internally while also taking advantage of research conducted by or in collaboration with others. They need also to drive active adaptive management programs to gauge the effectiveness of biodiversity management actions.

The relationships between biodiversity conservation and economic production are largely inseparable. What is required is new, management-relevant knowledge co-produced by policymakers, managers and researchers using participatory frameworks. New approaches to biodiversity conservation in the face of climate change are detailed in Chapter 7.

7 Securing Australia's biotic heritage

This chapter provides an overview and discussion of ways to secure Australia's biotic future. It focuses on the responses required to adapt natural ecosystems to enhance their resilience to climate change, discusses the implications of previous chapters and formulates recommendations. The chapter outlines the broad changes that are occurring in climate, as well as major socio-economic trends over the coming decades. It discusses management strategies and tools appropriate for biodiversity conservation under a changing climate, and proposes a substantially revamped institutional architecture based on the concepts of subsidiarity and polycentricity. As an innovative step towards implementation, the chapter proposes a systematic regional approach for biodiversity conservation – tailored to the characteristics and trends in particular regions – and builds towards an integrated response package encompassing management, education and governance. The chapter concludes by describing a set of challenges for resourcing the future.

7.1 BIODIVERSITY AT THE CROSSROADS

Australia's unique biotic heritage is at a crossroads. Over the past two centuries and particularly over the past 100 years, Australia has, for a variety of reasons, experienced high rates of biodiversity loss and ecosystem degradation (Chapter 3). The importation of European agricultural systems, forestry practices, pastoralism (open-range grazing), lifestyles and technologies has led to unintended and unanticipated yet profound changes to the biotic fabric of this ancient continent. Many of the stressors associated with this collision of culture and continent still operate on our ecosystems. More recently, a new global stressor – anthropogenic climate change – has arrived. We can already begin to see some of its impacts on species and ecosystems in Australia (Chapter 5). Although we are making some progress in understanding and managing our natural resources, including our biodiversity (Chapter 6), we continue to lose ground in the challenge to conserve our biodiversity in the longer term.

What is the future for Australia's biodiversity as climate change accelerates in the 21st century? This chapter examines the challenge that now confronts us in turning around the current rate of biodiversity loss, even with the increasing rate of biophysical and societal change. The following two sections describe the unprecedented climatic and socio-economic changes that are sweeping across the country. Sections 7.4 and 7.5 explore how innovative biodiversity management principles, strategies and tools can reverse the trends of species loss and ecosystem degradation. Section 7.6 examines transformations in institutions and governance systems that may be necessary to implement effective new approaches to biodiversity conservation in the 21st century. The chapter concludes by describing a systematic regional approach to biodiversity conservation, and identifying the resource requirements to meet the challenge. Roads of inaction, and even slow reaction, will lead to huge losses of biodiversity. A road of proactive planning and adaptation to alternative futures faces many challenges, but offers the hope of far better outcomes for Australia's biodiversity.

7.2 THE NATURE OF THE CLIMATE CHANGE THREAT IN THE 21ST CENTURY

Linking the magnitude and rate of climate change in the 21st century more directly to the implications for biodiversity is crucial for designing effective policy and management responses. As outlined in Chapter 4, warming of the climate system over the past century is unequivocal. It is very likely that anthropogenic greenhouse gas increases have caused most of the observed increase in globally averaged temperatures since the mid-20th century. Anthropogenic climate change is moving the Earth out of the envelope of natural variability that the world's ecosystems have experienced for at least the past two millennia, and probably much longer (Alverson *et al.* 2003; Steffen *et al.* 2004) (Fig. 4.6).

It is also very likely that the climate system will continue to warm through the rest of the 21st century and beyond, with changes to wind patterns, precipitation, sea level, extreme events, and many other aspects of weather and climate (IPCC 2007a). While there are uncertainties surrounding the magnitude and rate of further climate change (Chapter 4), all scenarios project significant change. In fact, current global emissions are tracking at or near the upper limit of the IPCC suite of projections (Raupach *et al.* 2007), implying a high likelihood of reaching a 2°C rise in global mean temperature this century and increasing the risk of much larger temperature rises by the end of the century (Anderson and Bows 2008). In terms of the vulnerability to climate change for key sectors in Australia, natural ecosystems were judged by the IPCC (2007b) as the most vulnerable sector, largely reflecting the very low adaptive capacity of natural ecosystems compared with other sectors (Fig. 7.1).

One of the greatest sources of uncertainty about the state of the world's climate by 2100 is the extent to which nations act to limit CO_2 emissions in the coming decades. The range of model projections of global mean temperature through the rest of the century, shown in Fig. 4.8, can therefore be schematised as three contrasting climate scenarios related to this degree of mitigation response. These scenarios (Fig. 7.2), which provide an essential basis for analysing responses, are:

- 'recovery', assuming that vigorous global mitigation efforts will succeed in limiting climate change to a temperature rise of 2°C or less, with a recovery towards the pre-industrial climate within 100–200 years
- 'stabilisation', assuming that the climate is eventually stabilised but at a significantly higher level than pre-industrial levels (about 3–4°C) and with a much longer time to return to pre-industrial values
- 'runaway', assuming unmitigated climate change, with accelerating change through this century and no stabilisation in sight. In this scenario the global mean temperature would be about 5–6°C higher than pre-industrial levels by 2100 and still rising.

The scenarios highlight that the Earth is committed to a further warming of at least 0.4°C regardless of human actions (Chapter 4), owing to the in-built momentum in the climate system. Thus, the trends for the three scenarios are identical up to 2030 or so. Beyond that, the rate and degree of global mitigation action will determine the alternative possible futures.

Two of the scenarios (stabilisation and runaway) would mean that over the next 100 years Australia could well experience changes in biome type and distribution at least as great as those associated with the 5000–10 000 year transition from the Last Glacial Maximum to the present. The current rate of climate change – a similar transition in temperature in only 100 years – is almost surely unprecedented for the past several million years (Steffen *et al.* 2004). This prompts assessments that the Earth will experience a massive wave of extinctions this century, with rates of species loss about 1000 times background levels (MA 2005). This is graphically illustrated in Fig. 7.3, which shows what the stabilisation and runaway scenarios might

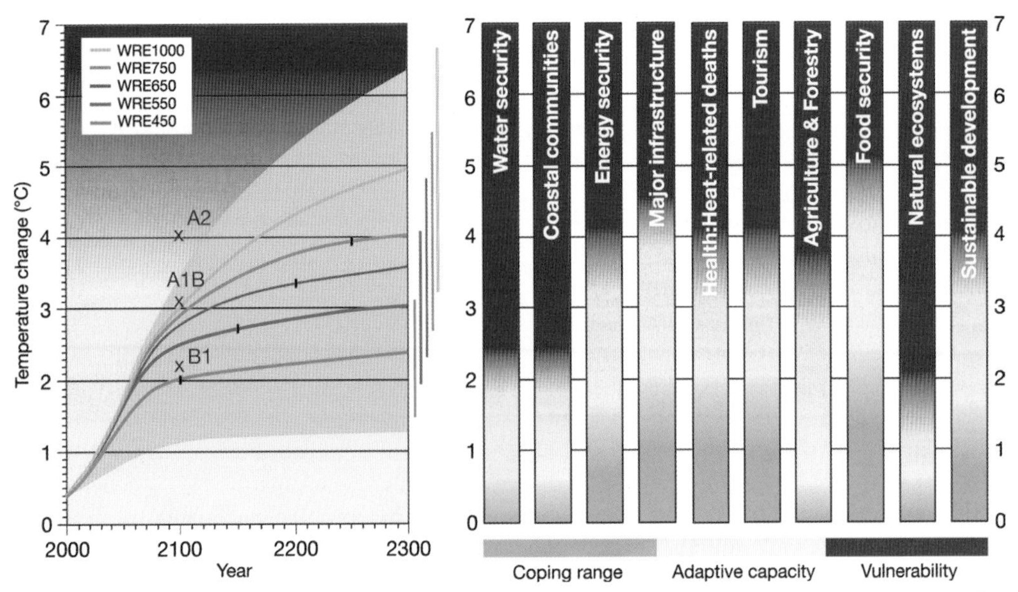

Figure 7.1 Vulnerability to climate change aggregated for key sectors in the Australia and New Zealand region, allowing for current coping range and adaptive capacity. The right-hand panel is a schematic diagram assessing relative coping range, adaptive capacity and vulnerability. The left-hand panel shows global temperature change taken from the IPCC Third Assessment Report: Synthesis Report. The coloured curves in the left panel represent temperature changes associated with stabilisation of CO_2 concentrations at 450 ppm, 550 ppm, 650 ppm, 750 ppm and 1000 ppm. Year of stabilisation is shown as black dots. Source: IPCC (2007b).

mean for Australia's biodiversity. It compares the equilibrium distribution of biomes at the Last Glacial Maximum to the present distribution, and then to the

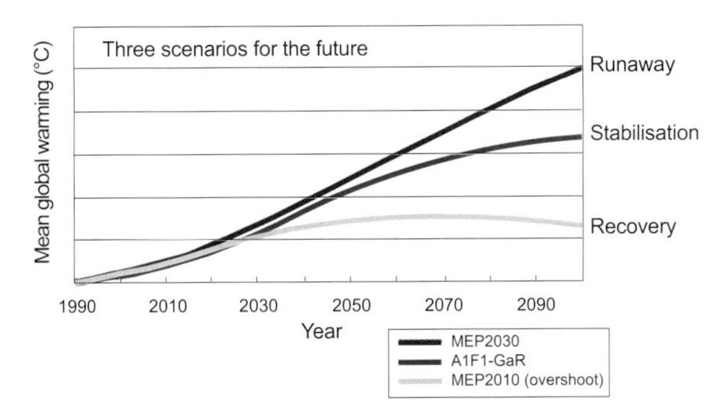

Figure 7.2 Three scenarios of future climate: recovery, stabilisation and runaway. Drawn using the IPCC suite of scenarios (IPCC 2007a) as driven by different assumptions about how humanity responds to the challenge of mitigating CO_2 emissions over the coming century (Box 4.1). The shorthand names range from the most optimistic 'recovery', through to the more realistic 'stabilisation' to the pessimistic 'runaway' (along which the world is currently tracking). These three scenarios inform risk management against an uncertain future when assessing management options for long-lived 'biological infrastructure' (Box 7.1). Source: Mark Stafford Smith.

simulated equilibrium biome distribution with a 2100 climate associated with a CO_2 concentration of 600 ppm. Among the many significant changes are an increase in the areal extent of desert and tropical xerophytic shrubland, and a reduction in temperate broadleaf forest (Fig. 7.3).

The preceding analysis provides the basis for an overall strategy for reducing the risk of climate change to Australia's biodiversity. The strategy has two major components:

- If Australian society wishes to minimise the risk of an unprecedented wave of extinctions over the next 100–200 years, mitigation of climate change – at the global scale – must be undertaken vigorously and rapidly to ensure that the rate and magnitude of change stay within the bounds of the recovery scenario. Mean temperature change above the 2°C level, with the associated changes in other aspects of climate, will lead to escalating loss of biodiversity with little or no chance of avoiding or reducing losses via adaptation (Hansen *et al.* 2008; IPCC 2007b; van Vliet and Leemans 2006).

Box 7.1 Climate scenarios, risk spreading and conservation strategies. (Mark Stafford Smith and David Lindenmayer)

In many ecosystems such as forests and woodlands, long-lived tree species have critical roles as key ecological structures (Manning *et al.* 2006). Therefore, the investments made today in the conservation management of these structures (and hence the large quantities of biodiversity associated with them) can have implications that span centuries (Lindenmayer and Franklin 2002), just as our investments in power stations and dams today create infrastructure legacies that last many decades. Yet, we cannot know what climatic conditions will be in 100–200 years time, because they depend on the mitigation that humans implement in response to the challenge of climate change. Therefore, we need to consider management strategies that hedge our bets against these different futures. We can do this by identifying the responses we should take to a variety of future climate scenarios; if all scenarios require the same responses, then this is a 'no regrets' option and we should get on with it. But if each scenario requires different actions, then we need to consider implementing multiple actions in the same region.

A hypothetical example comes from the ash-type forests of the Central Highlands of Victoria, south-eastern Australia. Large species of trees in these forests like mountain ash *Eucalyptus regnans* and shining gum *E. nitens* do not develop tree hollows suitable for occupancy for more than 40 species of vertebrates for at least 120–150 years (Lindenmayer *et al.* 1993). Exemplar species are the yellow-bellied glider *Petaurus australis* and yellow-tailed cockatoo *Calyptorhynchus funereus funereus*. A direct effect of a temperature rise of 2–3°C would not prevent the survival of such vertebrate species, as both have extensive distributions covering large parts of eastern Australia. However, we know they will decline rapidly if they do not have access to tree hollows for nesting and sheltering, as would happen if the mountain ash fails to recruit in the face of climate change. But our management needs to be different depending on how we think the future might play out.

Under the 'recovery' scenario (Fig. 7.2), we need to keep existing populations of tree species alive in situ. This is because by 2100 (well within the lifespan of some of the trees germinating today) the climate will begin to return towards conditions that the mountain ash can tolerate. For this scenario, we could identify some key refugia and work hard to keep them free from fire and other disturbances. This is preserving the *resilience* of the existing vegetation.

Under the 'stabilisation' scenario in Fig. 7.2 we need to establish tree species more typical of lower and warmer elevations, such as messmate *Eucalyptus obliqua* in areas where mountain ash lives today. When this species matures in 1–2 centuries, it will provide suitable tree hollows in those areas where mountain ash can no longer establish. Under this scenario, we might actively establish messmate (e.g. through deliberate planting or on-site releases of seed) in areas burned naturally over the coming decades. This approach is encouraging *directed transformation.*

Under the 'runaway' scenario in Fig. 7.2, the dominant tree species will continue changing over coming centuries, first to messmate then to yet other species characteristic of lower and warmer environments (e.g. red stringybark *E. macrorhyncha* and narrow-leaved peppermint *E. nicholii*). For this scenario, we might need to encourage the natural movement of species across the landscape to ensure a perpetual supply of large, long-lived trees species that can form hollows. This is promoting *self-organising transformation* of the vegetation community.

The key point is that we do not know which of these scenarios is actually going to happen, yet we need to begin this type of planning now. We can plan for such uncertainty by applying all three strategies in different places in the Central Highlands of Victoria. For example, we might recommend that one park follows the recovery strategy – particularly in areas with many good refugia (Mackey *et al.* 2002) – another specifically encourages the recruitment of messmate and yet another allows unrestrained recruitment of any long-lived tree species. In an active adaptive management approach, we can review this decision in 30–50 years as it becomes clearer which scenario is really playing out, and gradually refine the strategies.

In contrast, there will be conservation issues where the same management strategy is needed, irrespective of which future scenario manifests. For example, this may be the case for short-lived vegetation types, such as coastal heathlands, where managing disturbance and other threats may be appropriate regardless of the future (Woinarski 1999). In this case, the management strategy is one of *no regrets*. That is, we should get on with doing the same things in all places with this vegetation.

This kind of risk management using scenario-based thinking needs to become a key part of conservation planning, particularly as we will not know what the future holds until we are living it, yet we cannot wait until then to make decisions about some of our 'biodiversity infrastructure'.

- Under even the most modest climate change scenario (e.g. the recovery scenario), the potential impacts on biodiversity will increase through most of this century. Formation of novel ecosystems, abrupt changes in ecosystem structure and functioning, and surprising counterintuitive outcomes will become more common. Coupled with the existing stressors on biodiversity, these climate-related complications challenge the policy, management and governance communities to develop and implement innovative, adaptive and resilient (or transformative) regimes for

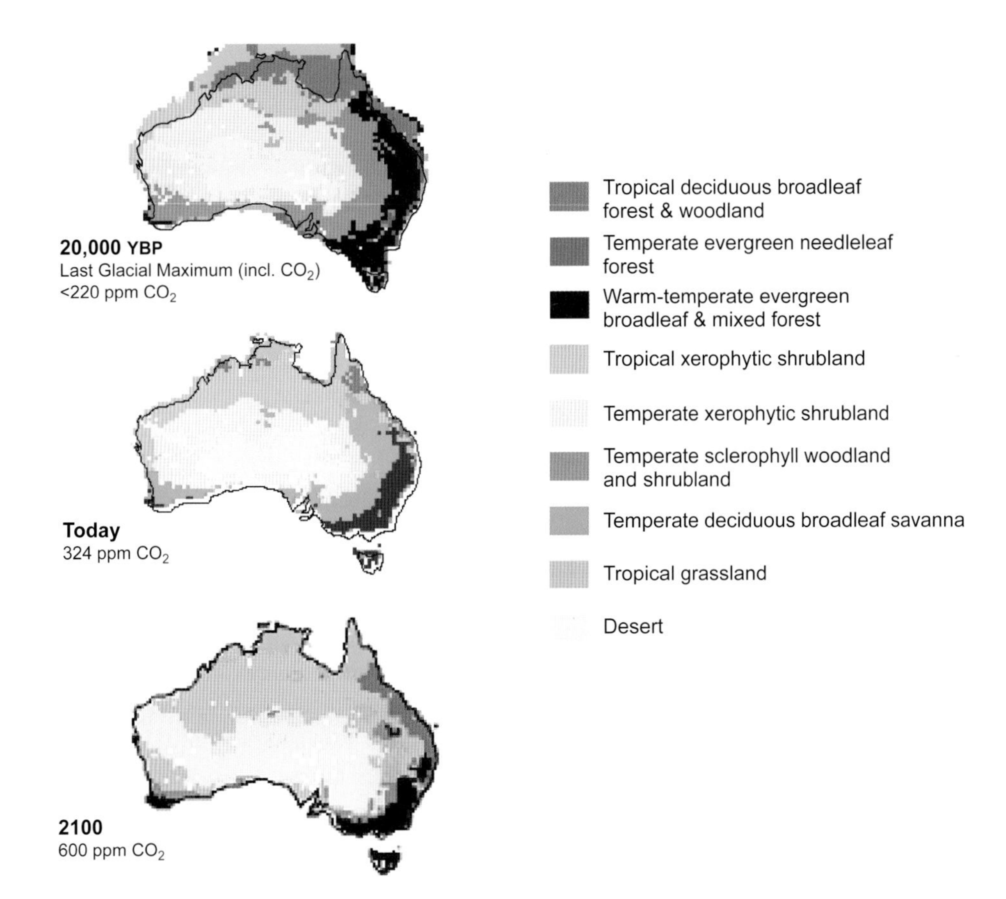

Figure 7.3 Historical, present and future Australian biome distributions. The maps show the distribution at the Last Glacial Maximum (approximately 200 ppm CO_2), the present distribution (ca. 325 ppm CO_2), and the future distribution at 2100 (ca. 600 ppm CO_2), all modelled using the dynamic global vegetation model BIOME 4.0 driven by the Hadley Centre climate model. The details of these simulated biome distributions are sensitive to the nature of the particular vegetation and climate models used; shown is a typical example of change. Source: Sandy Harrison, Bristol University, UK, personal communication.

the conservation of Australia's biodiversity under rapid change (sections 7.4 to 7.7). Thus, investing in adaptation is not an 'optional extra' but is central to *any* strategy to minimise the impacts of climate change on Australia's biodiversity.

In summary, the relationship between adaptation and mitigation is not an 'either-or' question but rather a 'both-and' imperative. That is, with agile, innovative and better-resourced conservation policy and management *now*, the current degradation of Australia's biodiversity and the impacts of modest climate change can be reduced. With rapid and effective global mitigation of climate change, the probability of more severe climate change and the associated risk of unavoidable, much higher rates of biodiversity loss in the coming decades and centuries can be significantly reduced.

7.3 SOCIO-ECONOMIC TRENDS THAT COULD AFFECT BIODIVERSITY

In addition to changes in climate, significant socio-economic changes are sweeping across Australia. Given the consistency in the underlying drivers of increasing population and individual affluence, most existing trends will continue or even intensify over the coming century – in some cases creating additional challenges but often offering new opportunities for dealing with climate change in the regions where they are occurring. Key socio-economic changes include:

- *decline of agriculture in marginal lands*. Marginal agricultural and rangeland regions are already experiencing declining human populations and land use diversification. Possible large-scale abandonment of agriculture in further large regions (e.g. northern part of the Western Australia wheatbelt, western New South Wales) would create opportunities for new management regimes that emphasise land stewardship and promote carbon storage, biodiversity and tourism markets in large areas of Australia
- *new landscapes*. Major new landscape uses, such as sequestration of carbon as offsets to fossil fuel emissions and the potential strong growth of a biofuels industry, are developing even in core agricultural regions. The associated financial markets and incentives currently take little heed of biodiversity but could readily redirect their priorities towards the joint supply of carbon and biodiversity services
- *high-density urban living*. There is considerable potential to manage inner-city, suburban and peri-urban areas differently for better biodiversity outcomes. As the form of Australian cities changes towards higher-density housing, the associated changes in land use offer opportunities for creative redevelopment that builds in more biologically diverse parklands, river corridors and green belts
- *'sea change' and 'tree change' movements*. Retirees (or escapees) from the cities often have a strong interest in nature and its preservation, and relatively high capacity for private investment in biodiversity outcomes. It is possible that the considerable investment in revegetation by these people (sometimes focused on mallee oil or blue gum plantations) could be better directed with respect to its potential value for biodiversity under climate change
- *expanding Indigenous estate*. With the advent of land rights in the 1970s and native title since the 1990s, an increasing proportion of land (particularly in more remote regions by area – now around 22% of the rangelands) has been returned to Aboriginal and Torres Strait Islander ownership (Bastin and ACRIS Management Committee 2008). The validity and value of Indigenous inputs to management have been recognised in other areas, particularly in joint management arrangements for parks, as well as weaker engagement around managing natural and cultural heritage on other tenures. In particular, the rapid expansion of Indigenous Protected Areas creates another new way of considering public/community partnerships for conservation
- *private sector conservation*. The continuing rise of private organisations like Bush Heritage Australia, The Nature Conservancy, Australian Wildlife Conservancy, Bush Tender and similar schemes for improving biodiversity outcomes on privately held lands offers additional opportunities for entraining more private investment in biodiversity conservation (Chapter 6).

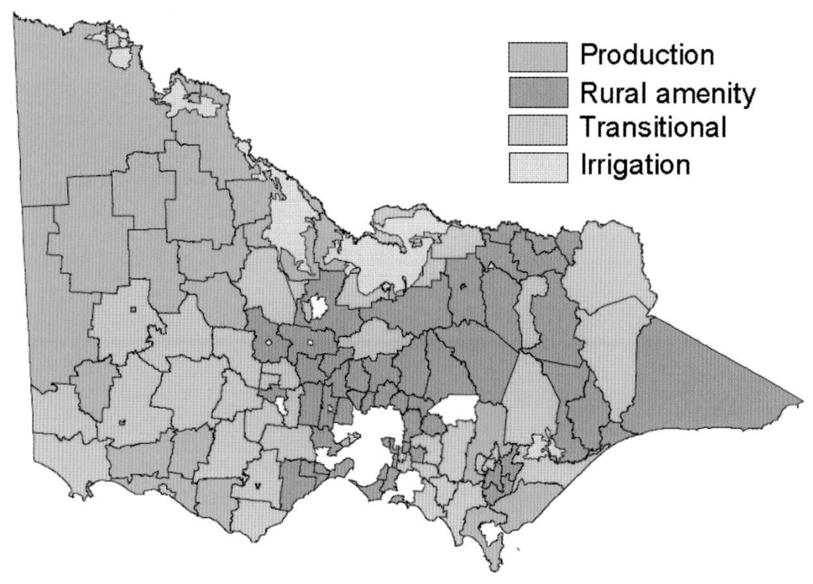

Figure 7.4 Stylised social landscapes of rural Victoria. Several types of social landscapes are shown based on production (including irrigation), conservation and amenity values. Source: Barr *et al.* (2005).

These trends have been characterised by Holmes (1997, 2006, 2009) as part of a worldwide movement among developed nations towards increasing non-agricultural production land uses. In these terms, Holmes sees regions being characterised in terms of changing balances of values – between production, conservation and 'amenity' values. In practice, many areas are moving towards complex mixes of all three. An example of the shifting socio-economic trends has been provided by the work of Barr *et al.* (2005), distinguishing agricultural production landscapes, rural amenity landscapes, rural transitional landscapes and irrigation landscapes – and their associated social, environmental and economic trends. Their work shows a complex set of changes within rural Victoria (Fig. 7.4).

These socio-economic trends are important inputs to the development of more innovative and integrated approaches to biodiversity conservation under a rapidly changing climate. In different ways they permit an increase in the investment in sustaining biodiversity that creates constructive partnerships between public and private investment, and which play out differently in different types of regions. The following three sections discuss the guiding principles and strategies/tools required to meet the climate change challenge, as well as possible implications for governance.

7.4 GUIDING PRINCIPLES AND APPROACHES FOR MANAGING BIODIVERSITY IN THE 21ST CENTURY

The legacy of past changes to biodiversity set the initial conditions for the future of Australia's biodiversity, while the nature of climatic and socio-economic changes over the next several decades will set the dynamic boundary conditions. The overarching goal remains to minimise the loss of biodiversity at a national level, though this no longer necessarily implies preserving all species in their current locations. This section explores some principles and approaches that can help guide the strategies and tools that need to be developed to meet the climate change challenge.

We present six principles and approaches for minimising loss of biodiversity. The first four are current management principles that are still very important under a changing climate, as described in Table 6.1, and the latter two are proactive approaches to biodiversity conservation under climate change.

- *Maintain well-functioning ecosystems.* With decades or centuries of projected climatic change that is significant in magnitude but uncertain in detail, the single most important principle guiding the management of biodiversity is the maintenance of well-functioning ecosystems. This is, in fact, a two-way principle. Maintenance of high levels of biodiversity is a good strategy to ensure the adequate functioning of an ecosystem. However, this is not a simple principle to implement under a changing climate. Maintaining or enhancing the resilience of ecosystems is crucial to ensure the continuation of adequate functioning; but when, under climate change, does maintenance of resilience of existing ecosystems become counterproductive and facilitation of transformation into new ecosystems become more appropriate? As transformation becomes more common later this century, monitoring the functioning of these new ecosystems and their ability to deliver the services on which society depends will be critical for implementing this principle.

- *Protect a representative array of ecological systems.* Not only is it important to have well functioning ecosystems, but also the full diversity of these systems needs to be included in areas managed for conservation. This basic principle of conservation needs renewed emphasis and reinterpretation under climate change. The principle of representativeness – representing all biodiversity in appropriately managed systems – remains essential under climate change. However, the purpose is now to represent as many different combinations of underlying environments and drivers, rather than specific arrays of current species. While the particular assemblages of species or genes in a single location may change, aiming to encompass diversity provides the best likelihood of having favourable conditions for all biodiversity somewhere. This applies at all scales from local through regional to national, in that a diversity of several stages of recovery since disturbance (e.g. time since fire) should be maintained locally. All environments should be represented in regional reserve systems, and a diversity of landscape architectures in terms of the arrangements of patches and connecting habitats should be represented in regional on- and off-reserve landscapes. This diversity should also be maintained at a national level.
- *Remove or minimise existing stressors.* The biggest threat to Australia's biodiversity continues to be a number of existing stressors, such as the direct human modification of ecosystems (e.g. land clearing) and the introduction of exotic species. These will continue to be important, but climate change presents a 'double whammy' – acting as an additional, direct stressor on species and ecosystems, as well as exacerbating the effects of many of the existing stressors. Thus, as a management principle, it will become even more important to minimise or remove existing stressors, with particular attention given to those stressors that might benefit from climate change (e.g. feral cats may benefit from climate change preferentially to other feral animals).
- *Manage appropriate connectivity of species, landscapes, seascapes and ecosystem processes.* With increasing pressure on species to migrate in response to a changing climate, and for ecosystems to disassemble and reassemble, there is a greater focus on achieving appropriate types of landscape and seascape connectivity to 'create space for nature to self-adapt' (Mansergh and Cheal 2007), while protecting some areas from disruption and invasion (Box 7.1). The concept of 'landscape fluidity', defined as the ebb and flow of organisms within a landscape (or seascape) through time (Manning *et al.* 2009), provides a more appropriately dynamic underpinning to biodiversity conservation in a rapidly changing world. Here, marine ecosystems may have some advantage, in that many (but not all) organisms may be able to change their geographical position in response to changes in the abiotic environment around them. Terrestrial ecosystems face more severe challenges, because most terrestrial organisms are less mobile, and are subject to more direct and pervasive human modification. Freshwater organisms may face the greatest barriers to dispersal, given that they may need to move between catchments.

Such an emphasis on the landscape as an integral part of biodiversity management indicates the need, in principle, to move from simplistic, polarised patterns of landscape structure and use – 'fortress agriculture' and 'fortress conservation' – to more fluid, multiple-use landscapes with space for self-adaptation of ecosystems in the landscape. Support for such adaptation must come from those who live in and on the landscape, and who must therefore be productively engaged in policy formulation and implementation. This principle implies the need to reverse the trend towards simplicity and efficiency (loss of diversity) in landscapes and in the coastal zone, and to build landscapes and ecosystems with more complexity, redundancy and resilience.

- *Eco-engineering may be needed to assist the transformation of some communities under climate change.* Driven somewhat by the growing interest and experience in restoration ecology, as well as the improved understanding of ecosystem structure and functioning, there will be cases where a passive 'let nature adapt' approach can and should be augmented by more proactive measures to conserve biodiversity (e.g. Hoegh-Guldberg *et al.* 2008). This approach invariably involves the direct and substantial modification of communities in directions consistent with the impacts of climate change. Eco-engineering has some major limitations that must be considered before it is applied; in particular, it is costly and not always successful (especially in a rapidly changing environment). Nevertheless, eco-engineering may constitute a necessary response principle in some cases. Eco-engineering should be focused on places where the best return on investment (both financially and in terms of biodiversity conserved) can be obtained. For example, Jones *et al.* (2008), using catchment modelling, found that the restoration of the riparian vegetation optimised carbon sequestration, water quality and biodiversity outcomes. Research is needed to identify similar critical intervention points in various landscapes in the context of climate change. Most likely, however, based on the ecological principles of how communities are structured (Chapter 3), the preservation or re-establishment of keystone or structuring species (or foundational species in aquatic systems) would give the best chance for an ecological system to self-organise in a way that would reduce total species loss and maintain ecosystem functioning.
- *Genetic preservation must be considered in some cases.* As a last resort approach, some species may need to be preserved outside of an ecosystem context, whether it is an existing or transformed natural ecosystem, or a human-engineered ecosystem. However, such last-resort ex situ methods should be seen in no way as substitutes for well-functioning ecosystems. Examples of ex situ conservation include:
  - the cryogenic seed bank for food species being established in Svalbard, Norway
  - refrigerated seed stores and cryogenic germplasm stores at Australian herbaria and botanic gardens
  - the potential role of zoos in conserving a very small number of charismatic and highly valued species
  - the breeding and maintenance of near-extinct species in isolated or quarantined areas, such as remote islands.

The above principles and approaches will almost surely achieve widespread, effective implementation only if they are supported by a set of more general principles that transform the way that societies think about and value the biotic world around them. The three most important of these general conceptual principles are:

- *Education and communication to bring the public along with change.* The public, as well as political and institutional leaders, must recognise that climate change is driving the natural world into uncharted territory in the 21st century. Furthermore, we are not starting with a 'clean slate'; the amount of biodiversity loss later in the century will depend to a large degree on what we do over the next decade or two to reduce the impact of current stressors. Innovative approaches to assisting nature to deal with the interacting set of new and existing stressors will be required. Social and political support is necessary for these new approaches to succeed. This may require re-examination of some strongly held views on biodiversity and its conservation (Hobbs *et al.* 2006). One such view is the need to preserve what we have now. Dunlop and Brown (2008) have already argued for a shift to minimising loss rather than 'preserving all'. Beyond even this, the public must learn to value new, unique, diverse ecosystems over individual species that may no longer inhabit them. In a rapidly changing abiotic environment, preservation strategies based on equilibrium dynamics will not work. Landscapes will change; some species will be lost and others will not persist in their current location. In general, the current emphasis on species will need to be balanced by an increasing focus on ecosystem services, processes and diversity. Managing for resilience of existing ecosystems (a preservation strategy) may work to a point, but we must also manage for transformation of ecosystems, landscapes, seascapes and perhaps even whole biomes. Such a wide-ranging remit for biodiversity management will, in turn, pose challenges to existing governance arrangements and administrative institutions. The increasing urbanisation of the Australian (and global) population also means that most of the public know less and less about the significance of biodiversity in providing services to their everyday lives. Engaging their interest in maintaining biodiversity is thus increasingly critical

- *Minimise threats and seize opportunities.* Biodiversity conservation must look towards new opportunities, and more creative strategies and tools. Many of the socio-economic trends described in section 7.3 offer opportunities for new conservation approaches and tools. Others, however, could easily turn into threats. For example, schemes to sequester carbon in landscapes or to produce biofuels to substitute for fossil fuels could easily lead to deleterious outcomes for biodiversity – especially if there is a trend towards highly simplified, industrial landscapes. However, with good research and astute policy development, these potentially perverse outcomes could also be turned into opportunities. Complex, biodiversity-rich ecosystems invariably store more carbon than simple monocultures (Mackey *et al.* 2008), and carefully designed biofuel systems can simultaneously lead to positive biodiversity outcomes (Foran and Poldy 2002). Creating such synergies among ecosystem services should become a central organising principle for all proposals to use landscapes for climate change mitigation
- *Apply a risk management approach to deal with irreducible uncertainties.* Significant uncertainties surround critical features of climate change science, such as how the hydrological cycle will change, and the consequences of these changes for water resources and water availability. Some of these issues, and the question of how much mitigation will be implemented over coming decades, represent irreducible uncertainties. Strategies and tools for biodiversity conservation under a changing climate must therefore embrace uncertainty as a central underpinning principle. A greater emphasis on risk management and adaptive management approaches is essential. The linear approach of research–policy–management–outcome needs to be replaced by an iterative, cyclical approach in which biodiversity outcomes are appraised, leading to new research, and adjusted policy and management (Box 7.2). Such an adaptive, cyclical approach needs high-quality information, based on monitoring and experimentation. Australian society needs to learn to accept some initial failures in policy and management approaches to deal with such a complex stressor as climate change. However, 'failures' are only true failures if management and policy fail to learn from them, adapt their approaches and do better the next time (Lindenmayer and Franklin 2002).

7.5 STRATEGIES AND TOOLS TO MEET THE CLIMATE CHANGE CHALLENGE

Climate change is a new threat to biodiversity in its own right, but it will also act through and exacerbate existing threats. Many existing management strategies and tools can be used to enhance biodiversity conservation, but some will need to be abandoned or revised and new ones will need to be developed. In this section, we assess current (stationary climate) and future (changing climate) management strategies at local, regional and continental scales. We begin in Table 7.1 by providing an overview of these strategies organised against ecological principles (Chapter 3).

The remainder of this section outlines some of the overarching strategies that will be needed to meet the

Box 7.2 Adaptive management, monitoring and climate change. (Roger Kitching and David Lindenmayer)

Adaptive management involves the integration of research, monitoring and management to improve the management of natural resources (Holling 1978). Adaptive management is often discussed by researchers and resource managers, but true adaptive management studies are extremely rare in practice (Bunnell *et al.* 2003; Haynes *et al.* 2006). Adaptive management is not management by trial and error. Rather, it is an iterative approach to management based on explicit, experimentally based tests of plausible management options. In true adaptive management, particular *management goals* are pursued in an iterative, cyclical fashion. Management procedures are put in place based on the best estimate of what will be appropriate to achieve the goals. The outcomes from these procedures are closely *monitored* using target measures that have been developed for the particular ecosystem. Management procedures are improved based on the closeness-of-fit between the state of the environment and the desired management goals, as indicated by the monitoring procedure. This *self-correcting procedure* is continued in a stepwise fashion until the management procedures in use are shown, through monitoring, to be producing the desired outcomes.

With rapid climate change, background ecosystem properties may change continually. An additional loop is required in the decision process for adaptive management, which has the effect of making the process open-ended because there is essentially no logical stopping point. This demands ongoing *adaptive monitoring* (Ringold *et al.* 1993); that is, the continuous re-appraisal of what are appropriate monitoring targets (Lindenmayer *et al.* 2003).

A consequence of the need for active adaptive management under a changing environment is that there will be an ongoing need for research to establish monitoring procedures and, of course, to conduct rigorous monitoring. However, on-ground adoption of adaptive management in a rapidly changing climate is a major challenge for resource management agencies. This is because examples of traditional adaptive management (in the absence of climate change) are remarkably uncommon (Haynes *et al.* 2006).

A rare exception comes from the wet ash eucalypt forests in the Central Highlands of Victoria, south-eastern Australia. These forests are important for the conservation of biodiversity, and also the production of timber and pulpwood. However, there are significant issues with the traditional forms of clearfell logging and their impacts on forest biodiversity in wood production ash forests. In 2003, a major adaptive management experiment was instigated that has been testing the effectiveness of altered logging practices for improved biodiversity conservation (Lindenmayer 2007b). The study has been established via a replicated and blocked experiment testing bird, mammal and vegetation responses to four experimental treatments. The experiment runs in parallel with a 25-year forest monitoring study (Lindenmayer *et al.* 2003), and together these projects explicitly link forest researchers, forest policy makers, forest managers and logging contractors as part of evolving best practice forest and biodiversity management in wet ash forests (Lindenmayer 2007b).

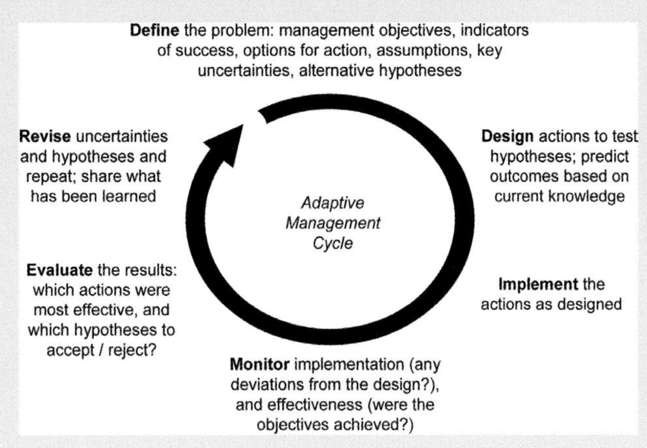

The adaptive management cycle.

Source: Murray and Marmorek (2004).

Table 7.1 Linking ecological principles (Table 3.1) to management strategies in response to a climate-changing world. Management strategies marked in italics will no longer be appropriate with climate change. In the second column, the term 'A stationary climate' refers to the climate before significant anthropogenic influence. The term 'A climate-changing world' refers to the most recent half-century and beyond, in which human influence on climate is significant. The abbreviation 'CC' refers to climate change.

		Management strategies		
Ecological principles	**Implications**	**Local**	**Regional**	**Continental**
1. Every species is different. Species distributions and abundances reflect individual responses to their environments.	**Stationary climate**: Species distributions relatively stable, distribution changes due to new disturbances, invasive species, or management regimes	• *Manage to maintain historical species and ecosystem composition.* • *Restore with local species.* • Control/eradicate exotic invaders.	• Manage hydrology on a catchment-wide scale. • Apply biosecurity to limit human-induced movement of organisms between major regions.	• Develop and manage a comprehensive, adequate and representative reserve system to capture maximum biodiversity. • Manage connectivity between regions. • Manage migratory species. • Apply border biosecurity.
	Climate-changing world: Species move individually in response to changing climate; some/many species will fail to adapt or move and will disappear.	• Reassess the purpose of management. • Allow native invaders that are moving in response to CC. • Choose plant species and genetic provenances for restoration with CC in mind. • Consider gene banking and ex situ conservation for species without solutions where later restoration may be possible.	• Integrate conservation reserves with off-reserve management. • Consider translocation of poorly dispersed and vulnerable species to suitable bioclimatic zones. • Develop multiple management strategies for any given environment to allow alternative species movements to occur.	• Integrate a comprehensive, adequate and representative reserve system with off-reserve conservation and management. • Integrate threatened species legislation, listings and management at the national scale. • Avoid 'one-size-fits-all' conservation management policies.
2. Species abundances and distributions are responses to drivers at different scales.	**Stationary climate**: Drivers are in balance except where human influences have changed them.	• Manage at multiple scales.	• Manage at multiple scales.	• Manage at multiple scales.
	Climate-changing world: Climate change is a new driver with potential novel feedback mechanisms.	• Monitor and manage distribution changes at local scales.	• Monitor and manage changes at larger scales.	• Monitor and manage changes at larger scales; expect, observe and adapt to new feedbacks between scales.

		Management strategies		
Ecological principles	**Implications**	**Local**	**Regional**	**Continental**
3. Life history attributes determine the ability of species to respond to change. Population genetic variability and breeding systems also determine ability to respond.	**Stationary climate**: Manage with understanding of species' life histories.	• Manage with understanding of species' life histories and reproductive strategies.	• Integrate local management strategies.	• Ensure management in different regions is appropriate to those environments.
	Climate-changing world: Some species will adapt better and/or be able to move better than others; successful combinations of life history attributes may change.	• Ensure appropriate response to changing species relative abundance. • Permit species and species combinations to 'experiment' (observe and adapt to results).	• Integrate local responses to changing species abundance • Recognise that 'picking winners' in ecosystem engineering is mostly impossible, so permit and learn from natural experiments.	• Coordinate management efforts and learning between regions and across multiple regions.
4. No species exists in isolation from other species.	**Stationary climate**: Species interactions are established over a long period of relative stability; an exception is the arrival of invasive species. Complex webs mean systems are normally resilient but pressures can cause very long-term changes in state.	• *Elimination of invasive taxa may return system to original state.*	• *Elimination of invasive species may return system to original state.*	• Implement continental-scale invasive species management strategies.
	Climate-changing world: Species interactions will be altered; arrival of 'new' species will affect existing species abundance and distribution, often in ways that cannot be predicted. Transformations of ecosystems will be common as resilience decreases.	• Plan for change; develop scenarios to assist managers make decisions about when to intervene and when not to intervene; allow monitored natural experimentation. • When losses are detected, expect consequences – some are only detectable in the long term.	• Plan for change; develop scenarios to assist managers make decisions about when to intervene and when not to intervene.	• Ensure policy facilitates flexible management responses in a changing world.

		Management strategies		
Ecological principles	**Implications**	**Local**	**Regional**	**Continental**
5. Some species affect ecological processes more than others within communities and ecosystems.	**Stationary climate**: Complex interactions and ecosystem stability are driven by a few species.	• Maintain keystone species, key structural species, foundational species, ecological engineers and top predators.	• Maintain keystone species, key structural species, foundational species, ecological engineers and top predators.	• Coordinate understanding of ecological processes and the response to change.
	Climate-changing world: Keystone and other important species may disappear from ecosystems with significant but hard-to-predict cascading effects.	• Identify species that will fill equivalent ecological roles. • Where not available, concentrate effort on existing important species, or breed new strains.	• Identify species that will fill equivalent ecological roles. • Consider/permit translocations beyond existing ranges (assisted colonisation).	• Implement national research into management of ecological processes in a changing world. • Review policy on translocations/assisted colonisation.
6. Species are structured by their means of obtaining food into a larger trophic structure, or food web, of an ecosystem. The interaction of biota and physico-chemical environment yields ecosystem-level processes.	**Stationary climate**: Changes at one level of a food web can affect other levels. Changes in physical environment, e.g. nutrient levels, can change whole ecosystems.	• *Manage to minimise change in species abundance.* • Prevent nutrient inflow and pollution.	• *Manage to minimise change in species abundance.* • Prevent nutrient inflow and pollution.	• *Management for maintenance of an existing mosaic of ecosystems and processes.*
	Climate-changing world: The loss of some species and arrival of others due to climate change may lead to novel ecosystems. CO_2 fertilisation will change ecosystem structure and composition.	• Expect novel ecosystems and respond accordingly.	• Expect novel ecosystems and respond accordingly.	• Review National Reserve System management policies. • Conduct research to better understand the effects of CO_2 fertilisation on Australian ecosystems, especially their interactions with a changing climate.
7. Communities and ecosystems change in response to many drivers, and these drivers themselves may interact.	**Stationary climate**: Humans drive changes in disturbance, mostly manageable at a local scale; most disturbances are local, some disturbances interact.	• Manage fire regimes. • Eradicate/control invasive species, preferably when they first appear. • Manage interactions between disturbances.	• *Limited regional fire management.* • Implement landscape scale management of invasives. • Manage interactions between disturbances.	• *Limited need for cross-jurisdictional fire management.* • Implement continental-scale management of widespread invasive species; biosecurity.

		Management strategies		
Ecological principles	**Implications**	**Local**	**Regional**	**Continental**
7. Communities and ecosystems change in response to many drivers, and these drivers themselves may interact.	**Climate-changing world**: Disturbance regimes will change at all scales, disturbances will interact in novel ways, invasive species will move into new areas, novel ecosystems will develop and ecological 'surprises' will occur.	• Plan for local changes in fire regimes. • Plan for altered disturbance regimes. • Identify priority areas in need of protection.	• Plan responses to more larger, hotter fires, and promote regional fire management. • Decide which areas have a high priority for protection. • Allow and learn from natural experiments.	• Undertake national integration of fire management, develop new fire control technologies and apply fire management across all biomes. • Coordinate management across institutions and jurisdictions. • Establish a national institute for invasive species control technology, including biocontrol.
8. Change can be non-linear.	**Static world**: Changes are relatively predictable, recovery from perturbation occurs in time (though some systems are in non-equilibrium states).	• Utilise traditional management approaches.	• Utilise traditional management approaches.	• Utilise traditional management approaches.
	Climate-changing world: Expect major surprises, and significant irreversible state changes; changes will be ongoing and permanent.	• Monitor local biodiversity. • Implement adaptive management at the local level.	• Monitor regional biodiversity. • Implement adaptive management at the regional level. • Prepare for transformation of some ecosystems and landscapes.	• Develop and implement national monitoring systems for biodiversity. • Research adaptive management techniques and protocols.
9. Variations in time and space (heterogeneity) enhance biotic diversity	**Stationary climate**: Local and regional diversity is substantially maintained by heterogeneity in environments and driving processes.	• Create reserves that encompass local heterogeneity. • Manage to maintain heterogeneity in processes, such as fire and soil moisture patchiness, throughout the landscape.	• Manage to maintain a diversity of landscapes at the regional scale.	• Develop a national comprehensive, adequate and representative reserve system.

		Management strategies		
Ecological principles	**Implications**	**Local**	**Regional**	**Continental**
9. Variations in time and space (heterogeneity) enhance biotic diversity	**Climate-changing world**: Combinations of heterogeneous environments and processes will change but heterogeneity itself will remain important.	• Reserves that encompass heterogeneity will be particularly important in permitting self-adaptation to occur. • Manage the entire mosaic and not only remnants of native vegetation cover to maintain heterogeneity in processes; monitor for/learn from species survival in unexpected combinations	• Consider the amount and configuration of habitat and other land cover types. • Coordinate management of reserve and off-reserve areas. • Do not apply the same management practices everywhere – a diversity of strategies reduces the chance of making the same mistake everywhere and allows natural experimentation.	• Develop state and national strategies to promote off-reserve conservation (e.g. stewardship schemes). • Build national commitment to biodiversity monitoring and adaptive management.
10. Connectivity among resources, habitat, etc. shapes patterns of distribution, abundance and diversity.	**Stationary climate**: Meta-population processes depend on maintaining connectivity for dispersal in most species; some species depend on isolation for survival; evolutionary processes require a dynamic mixture of isolation and connectivity.	• Apply principles of managing local or fragmented ecosystems.	• Develop corridors, 'stepping stones' and other strategies to promote connectivity at the regional scale.	• Develop linkages that promote connectivity between regions; limit risks of impaired continental-scale and intercontinental movements (e.g. migratory birds).
	Climate-changing world: Connectivity (or lack thereof) will lead to novel ecosystems: not all species can or will move at same rate; 'weedy' species are likely to disperse more successfully than: (i) long-lived keystone species and other long-lived species; and (ii) many species limited to specialist niches.	• Where appropriate, allow novel ecosystems to develop; translocate particular species where necessary. • Develop diversity in connectivity to permit alternative natural responses.	• Whole biomes may change – manage to maintain diversity; manage both species and ecosystems. • Consider cost/benefit of reducing connectivity around ecosystems that have no movement options.	• Develop policies that enable managers to respond to changing biodiversity patterns; educate the public about changing assemblages of species.

challenge of climate change. The description is not comprehensive; rather, it is intended to highlight some important existing strategies and some new approaches, and to inform further discussion on this issue. The aim is to address directly, and build on, several major management principles outlined in Table 6.1, and especially those:

- dealing with existing stressors and disturbances
- managing environments for appropriate connectivity via both the reserve system and off-reserve conservation
- involved in eco-engineering through landscape restoration
- capitalising on opportunities from climate mitigation strategies.

7.5.1 Dealing with existing stresses and disturbances under a changing climate

Important disturbances include land clearing and degradation; competing demands for freshwater in terms of quantity, quality and timing; invasive species; changing fire regimes; and changing climatic extremes (drought, floods and extreme temperatures). New or modified approaches to management are required to deal with each of these challenges. As noted earlier, active adaptive management, driven by collaboration between managers and researchers, must be the overriding management principle that underpins all of the issues and strategies described as follows:

- *Land clearing and degradation.* Land clearing has been the single greatest cause of biodiversity loss in Australia (Chapter 3) and even if no further clearing takes place, losses will continue because of habitat fragmentation and irrecoverably small and fragmented populations of some threatened species. The resulting loss of connectivity will exacerbate climate change effects on biodiversity, primarily as it prevents or limits species movement. Even where land is not cleared, its suitability for biodiversity conservation can be reduced in a number of ways. Apart from the negative impacts of invasive species and altered fire regimes (see further on), changes to biodiversity and to landscape function may be caused by altered hydrology, grazing, salinity and waterlogging, and also chemical application and pollution. With climate change, it is even more important that further land clearing not be permitted and that current legislation prohibiting clearing be enforced. Many current responses to degradation should be intensified, especially as they often benefit both production and conservation.
- *Invasive species.* Some invasive species will be 'winners' from climate change. For example, feral cats are known to have caused extinctions of indigenous mammals on arid islands – but not on high-rainfall ones – and hence changing rainfall patterns will most likely change the impact of cats on native mammals. Some 'sleeper' weeds will expand their range, and some new weeds may emerge. Some pathogens and their hosts will extend their range as temperatures increase. Increases in fuel loads due to introduced pasture species such as gamba grass *Andropogon gayanus* and buffel grass *Cenchrus ciliaris* alter fire regimes (Box 5.12). Key strategies for dealing with invasive species under a changing climate include improving quarantine and biosecurity, greatly increasing investment in invasive species control technologies (including biological control), improving bioclimatic modelling to predict the effects of climate change on distributions of invasive species, and managing connectivity carefully to minimise dispersal of invasive species into new areas.
- *Fire regimes.* Increases in temperature, longer and more intense dry periods, more extreme fire weather, more storms and lightning strikes, and introduced fire-cycle plants will affect fire regimes in all but the Top End – northern Australia (which is now at maximum fire frequency) (Williams *et al.* 2009). These changes will undoubtedly affect biodiversity. A recent synthesis of our current understanding of the consequences of climate change for fire regimes provides many important insights for biodiversity conservation strategies (Box 5.11). In terms of biodiversity, the most important features of managing fire regimes include adapting biodiversity management to account for changes in fire regimes, protection of refugia, and managing cumulative effects of fire and other stressors.
- *Water distribution.* One of the biggest existing stressors on native freshwater fish and aquatic

environments is the competing demand for water and the overallocation of water entitlements in some systems. Impacts stem from water quality issues to water flow and inundation patterns, both temporal and spatial. Removal of these types of stressors is complicated, often cross-jurisdictional, potentially expensive and at times highly political (e.g. Macquarie Marshes in New South Wales). However, dealing with these existing stressors is critical in strengthening the resilience and adaptation potential of native fish and many other aquatic species.

- *Climatic extremes.* Changes in climatic extremes will likely affect many species more than a change in mean temperature and rainfall (Fig. 5.3). Prolonged droughts, something seen recently in Australia, are likely to become more frequent and extreme. Higher temperature extremes are virtually certain, and other climatic extremes such as high rainfall events, floods, storms and high wind events are also more likely (IPCC 2007a). Arguably the most important of the climatic extremes for biodiversity is the likely change in water availability, implying a focus on areas where a combination of elevated temperatures and reduced rainfall will impose water stress on ecosystems and species. Appropriate responses for those species that are forced out of their tolerable environmental envelope by climatic extremes and are unable to migrate to suitable habitat could include translocation, captive breeding and germplasm storage. More generally, there is a need to plan and develop post-disturbance management actions, such as:
  - control of grazing pressure by domestic livestock after a drought
  - limitation of the extent and intensity of post-fire salvage logging (Lindenmayer *et al.* 2008b)
  - determination of which species are encouraged to re-establish after a disturbance.

7.5.2 Managing for appropriate connectivity via both the reserve system and off-reserve conservation

Given the very high uncertainties associated with many aspects of climate change that are important for biodiversity (especially at regional and local scales), managing to enhance the capability of ecosystems to self-adapt and to reorganise is a key underlying principle. Increasing connectivity is, in many cases, an appropriate response to achieve these goals. However, there is a need to define 'connectivity for what' and a need to determine why and where increased connectivity will be needed. In some cases increasing connectivity may be counterproductive (Chapter 3; Box 6.2). Appropriate connectivity is likely to be achieved by enhancing both the national reserve system and off-reserve conservation, and building on specific strategies to integrate them.

- *National Reserve System (NRS).* The NRS will remain a major strategy for conserving biodiversity under a changing climate (Dunlop and Brown 2008), but managing individual reserves as part of a reserve system will become even more important. Resources for managing protected areas are grossly inadequate even under current conditions, and will need significant increases to deal with the additional climate change stress. Additional strategies include:
  - expansion of the NRS towards meeting CAR objectives
  - integration of all types of protected areas into a single national system
  - promotion of larger reserves that provide larger buffers spatially and through larger populations
  - use of adaptive management within the NRS.
- *National Representative System of Marine Protected Areas (NRSMPA).* The marine equivalent of the NRS, the NRSMPA aims to set up a system of protected areas within the whole of Australia's marine jurisdiction. The NRSMPA is at an early stage of development and the opportunity exists for Australian governments to develop a world-class marine reserve system that is truly representative of our marine environments.
- *Off-reserve conservation.* While protected areas will remain a cornerstone of conservation policy and practice, they will, on their own, be inadequate to conserve biodiversity under climate change. Off-reserve conservation will assume growing importance as the 21st century progresses, and will

need to be enhanced by a range of strategies that include:
  - adequate funding or incentives for off-reserve conservation schemes including land and marine areas held under native title
  - establishment of extension-style education to promote voluntary self-management for biodiversity by landholders; for example, in the growing area of amenity landscapes (section 7.3)
  - use of regional-scale planning to integrate off-reserve conservation with the NRS and the NRSMPA.
- *Building connectivity.* Linking reserves with off-reserve conservation areas, and connecting landscapes and seascapes more generally, requires specific strategies to achieve the appropriate level of connectivity. These include the development of regional and continental-scale corridors and stepping stones, assisted migration and translocations and, where appropriate, the deliberate reduction of connectivity (e.g. in the alpine zone). Reducing connectivity may also be important to reduce the threat of aggressive invasive species overwhelming some ecosystems. In general, regions should have a diversity of levels of connectivity, provision for different rates of movement and provision for development of new combinations of species. This requires careful monitoring and adaptive learning to ensure that connectivity does not result in unforseen, perverse outcomes such as increased weediness and counterproductive species displacements.

7.5.3 Eco-engineering through ecosystem restoration

Planning for ecosystem restoration must now consider the effects of climate change, otherwise there may be failures in the medium to long term due to changing conditions. For example, the regeneration of forests after logging must use appropriate provenances and species for the anticipated climatic conditions. Within catchments, restoration of riparian zones provides particularly effective ecological returns (Jones *et al.* 2008). Translocations may become critical for island ecosystems, especially low-lying islands that are threatened by sea-level rise. Sea turtles, crocodiles and other reptiles will be particularly affected where the sex of hatchlings is determined by incubation temperature; translocations may be required to promote establishment of new sea turtle populations in cooler locations. For marine ecosystems, building structural features such as artificial reefs may facilitate dispersal as well as provide a key structural element around which new ecosystems may develop.

7.5.4 Capitalising on opportunities from climate mitigation strategies

Carbon trading and offset schemes, probably the most common climate mitigation approach in landscapes, offer an opportunity to promote sequestration in biomass while simultaneously benefiting biodiversity. On the other hand, there is the distinct possibility of perverse outcomes, depending on the design of the sequestration scheme. All biomass sequestration/carbon offset projects should be audited independently using credible full carbon accounting before they are approved, to ensure that such projects maximise biodiversity conservation and do not result in perverse outcomes. For example, the importance of old-growth forests for carbon storage needs to be recognised. Based on full carbon accounting, the conversion of old-growth forests to fast-growing plantations would result not only in significant losses to biodiversity, but also the acceleration of climate change through reduced carbon storage. In addition, long-lived tree species may be more resilient to climate change in the medium term than fast-growing exotic species. Beyond a thorough audit, a system of market-based instruments and other incentive approaches, such as biodiversity credits, should be established to avoid perverse outcomes and achieve positive outcomes. Such schemes could be linked directly to the carbon trading system so that incentives can be built in to deliver win-win outcomes for biodiversity conservation and climate mitigation.

7.6 BUILDING INNOVATIVE GOVERNANCE SYSTEMS

Many of the management initiatives discussed above will involve changes in the ways in which we do things – changes in rules, regulations and rights. These involve costs and benefits, and will create winners and losers. Significant social and economic issues will arise that will trigger political reactions from affected parties. Are our existing institutions robust enough to enable decisions to be made that are in the best current and future interests of society? How can we ensure efficient and equitable outcomes? Are more efficient governance institutions possible?

As emphasised in previous sections, biodiversity policy and management initiatives in response to climate change will involve the expanded use of some existing strategies and tools as well as new ones. This will further complicate an already complex policy scene. The increased demands on the institutions and organisations responsible for policy and management may lead to a lack of trust and confidence in these existing or future structures; this is already evident, for example, in the attitude of many farmers, who have been bruised in the past by the poor application of environmental policy (Productivity Commission Native Vegetation Inquiry 2004). The challenge of regaining and maintaining the public's trust will be made greater by the highly uncertain nature of climate change. The public will need to accept a changing policy and institutional landscape, just as they will need to accept novel, self-adapting communities and ecosystems rather than the preservation of existing communities and ecosystems in their present location (section 7.4).

The current policy and institutional landscape, which provides the base on which governance changes will need to build, is described in Chapter 6. The three most important features of that structure are:

- robust, cooperative arrangements between the Australian Government and the states, most recently implemented through COAG
- a more complex institutional architecture at the sub-state level, with local governments, regional NRM bodies and Landcare groups all playing particularly important roles in environmental management
- increasing fiscal and policy dominance of the Australian Government (Connell 2007), often expressed through purchaser–provider funding arrangements and the use of the private corporation model for public sector operation.

The need to integrate production, biodiversity, and cultural values and objectives in landscape management – within the context of accelerating climate change – presents unprecedented challenges to the institutional architecture developed since federation. Advocates of integrated catchment management have long argued for a more 'bottom-up' system (e.g. Burton 1985; Martin 1991) while, more recently, students of common pool resource management have been advocating alternatives to conventional 'top-down' governance (Marshall 2005). New approaches are possible and are required now.

The concepts of 'subsidiarity' and 'polycentricity' (Box 7.3) are central to building more effective and responsive environmental governance structures for the 21st century, particularly those focused on a regional scale. Embedding these concepts in a regionally oriented governance structure supports the retention of local ownership and cooperation (crucial for trust and legitimacy), while developing integrated solutions adapted to the particular knowledge, preferences, capacities, socio-economic trends and biophysical characteristics of a region (section 7.7). Such structures are likely to involve lower transaction costs and to be more effective than present and currently proposed arrangements (Connell 2007; Marshall 2008).

A regional governance system based on subsidiarity and polycentricity principles has the following benefits:

- capitalising on the local knowledge needed to identify community-owned environmental solutions
- responding in a more timely manner to local and regional issues
- gaining cooperation with governance decisions (or at least weakening opposition)
- allowing the reconfiguration of institutional arrangements and organisational structures to suit the different population densities in peri-urban through to remote regions, and as problems evolve
- motivating institutional learning and innovation (Marshall 2005, 2008).

The last point is crucially important in implementing an adaptive management approach, which is required to deal with the complexities and uncertainties of the climate change challenge.

These features will become increasingly important in delivering – with acceptable levels of transaction

Box 7.3 The concepts of 'subsidiarity' and 'polycentricity'. (Graham Marshall)

The concepts of subsidiarity and polycentric governance are increasingly appearing in discussions about environmental management in Australia (Marshall 2005, 2008). Instances of their use overseas are also reported in the literature (e.g. Acheson 1988; Ostrom 1990).

The principle of subsidiarity requires each governance function to be performed at the lowest level of governance with capacity to implement it effectively (Jordon 2000; Ribot 2004). This capacity involves both the representation of all parties with a substantive interest in the function of a body (McKean 2002), and the availability of sufficient physical, financial, human and social capital to perform that function (Marshall 2005, 2008).

A polycentric system of governance comprises multiple decision-making centres that retain considerable autonomy from one another. A monocentric governance system may comprise more than one organisation, but coordination occurs through a single integrated (often linear) command structure (Oakerson 1999; Ostrom *et al.* 1961, 1999). To meet both polycentric and subsidiarity design principles, then, a governance system needs not only to comprise multiple organisations and levels, but also to afford these multiple organisations real autonomy in how they perform the functions assigned to them (Andersson and Ostrom 2008; Berkes 2007; Ostrom 2005).

In practice, this would mean central government agencies permitting a much greater degree of autonomy in both the design and performance of local NRM bodies. In particular, information flow would be multidirectional rather than primarily top-down, and decision-making would be interactive and collaborative involving multiple centres in the system. Today's regional NRM system has taken a step in this direction, but remains largely monocentric with limited subsidiarity (Marshall 2008). Other NRM arrangements (e.g. many regional water planning processes; Smajgl 2009) similarly retain a strong element of central decision-making.

A future system might look more like a series of diverse, local, community-based NRM groups (perhaps growing from current NGOs) that send their chairpersons to a regional NRM body with greater autonomy than in most jurisdictions today. A hands-off central government role would focus on ensuring regional groups meet basic standards of local representativeness and accountability, facilitating learning between regions, and distributing central funding on the basis of conditions that constrain local autonomy as little as possible.

costs (section 6.5) – the new and modified strategies and tools required to support biodiversity conservation in a rapidly changing climate (section 7.5), where greater collaboration among policy, management and scientific communities is required. These have possible or perceived disadvantages, including:

- less 'tidy' governance arrangements than those of a monocentric 'one size fits all' design (Ostrom 2005)
- new kinds of transaction costs to which all players must become accustomed
- potential lower capacity to access scientific knowledge (Seymour *et al.* 2007).

Nevertheless, appropriate support from higher levels of government can mitigate the effects of all of these problems. Additionally, where an individual body within a polycentric governance system fails, this is likely to have only local effects rather than pervading the whole system.

In addition to a new type of institutional architecture, greater investment in biodiversity conservation (section 6.7) and much more effective collaboration among the policy, management and research communities are required. So too is the need to facilitate communication and collaboration between governments and NGOs involved in biodiversity conservation, so full advantage can be taken in policymaking and management of the flexibility, adaptiveness and innovation of NGOs.

One possible foundation on which a polycentric approach to purchaser–provider programs in NRM might be crafted builds on prior successful Australian exercises in cooperative federalism, and would involve partnership arrangements for sharing the costs of NRM investments among communities, governments, the private sector and other expected beneficiaries (Musgrave and Kingma 2003). This approach proposes a new national institution, which would review the status of the nation's natural resources and advise on progress in relation to targets. It should also advise the relevant bodies, from the Natural Resource Management Ministerial Council down, on the need to adjust targets and programs given growth in knowledge. The new institution would also audit the partnership arrangements between the Commonwealth, the states and community-based bodies to determine that they conform to the agreed partnership principles. Subsidiarity and polycentricity should be central to these principles. Lingering monocentric tendencies would be addressed by providing communities with an effective voice in how the commission is constituted and operates. Given the complexity of this undertaking – especially in building the ownership and trust of those whose on-ground cooperation is ultimately essential – the efficiency and effectiveness of this strategy would require it to be planned, assessed, reviewed and adapted in a decentralised and devolved way. The new institution would play a particularly crucial role in fostering polycentricity by ensuring that arrangements exist for accountability both upwards and downwards through the governance system. Regardless of the precise model, the roles outlined in this section need to be fulfilled in whatever system eventually emerges.

The new national institution, which could take the form of a national statutory biodiversity authority, is especially needed to lead and coordinate monitoring of the state of the nation's biodiversity (section 7.8). If we are to meet the challenge to our biodiversity caused by climate change, a necessary starting point is a comprehensive overview of its present status. Furthermore, that status must be monitored through time if we are to continue successfully to meet the challenge. The proposed body should lead preparation of the initial overview, and maintain oversight through an ongoing auditing and reporting process. Discharge of these functions will be essential to the efficiency and effectiveness of biodiversity conservation policy and management in the face of climate change.

Decentralised, polycentric governance systems offer many potential advantages over current governance arrangements in addressing the challenge of biodiversity conservation in a rapidly changing socio-economic and climatic environment. Two of the most important of these advantages – the ability to adapt the governance system to the particular characteristics of individual regions and the capacity to accommodate a wide range of funding sources – are the subjects of the next two sections.

7.7 TOWARDS A SYSTEMATIC REGIONAL APPROACH FOR BIODIVERSITY CONSERVATION IN THE 21ST CENTURY

On-the-ground application of general principles for biodiversity conservation under climate change requires consideration of the characteristics of particular areas or locales within their regional contexts – a 'place-based approach'. This could easily lead to a huge number of individual approaches, with a high risk of unnecessary duplication and loss of synergies, collaboration and learning. Such risks could be reduced by adopting a regional approach, with the regions defined by common socio-economic characteristics and trends, existing biomes and climate change regimes. Importantly, there are opportunities for synergies between existing socio-economic trends (section 7.3), which have their own drivers and momentum, and different strategies for biodiversity conservation. In addition, the new type of governance system described in the previous section is well suited to a regional approach that integrates biodiversity values, socio-economic trends and levels of government.

Table 7.2, Fig. 7.4 and Fig. 7.5 illustrate how one might think differently about regions that have varying current biodiversity value and that are on different socio-economic trajectories. Hobbs and McIntyre (2005) defined a series of agro-climatic zones that combine the underlying major biomes with broad-scale land use patterns (Fig. 7.5), and which Dunlop and Brown (2008) used to analyse how different impacts of climate change across the continent would have differentiated implications for the National Reserve System in each zone. Table 7.2 extends this analysis by disaggregating the zones in terms of Holmes' (2009) social trends towards production, amenity or protection-oriented landscape uses (see also Fig. 7.4 for Victorian example). The table shows how such a regional

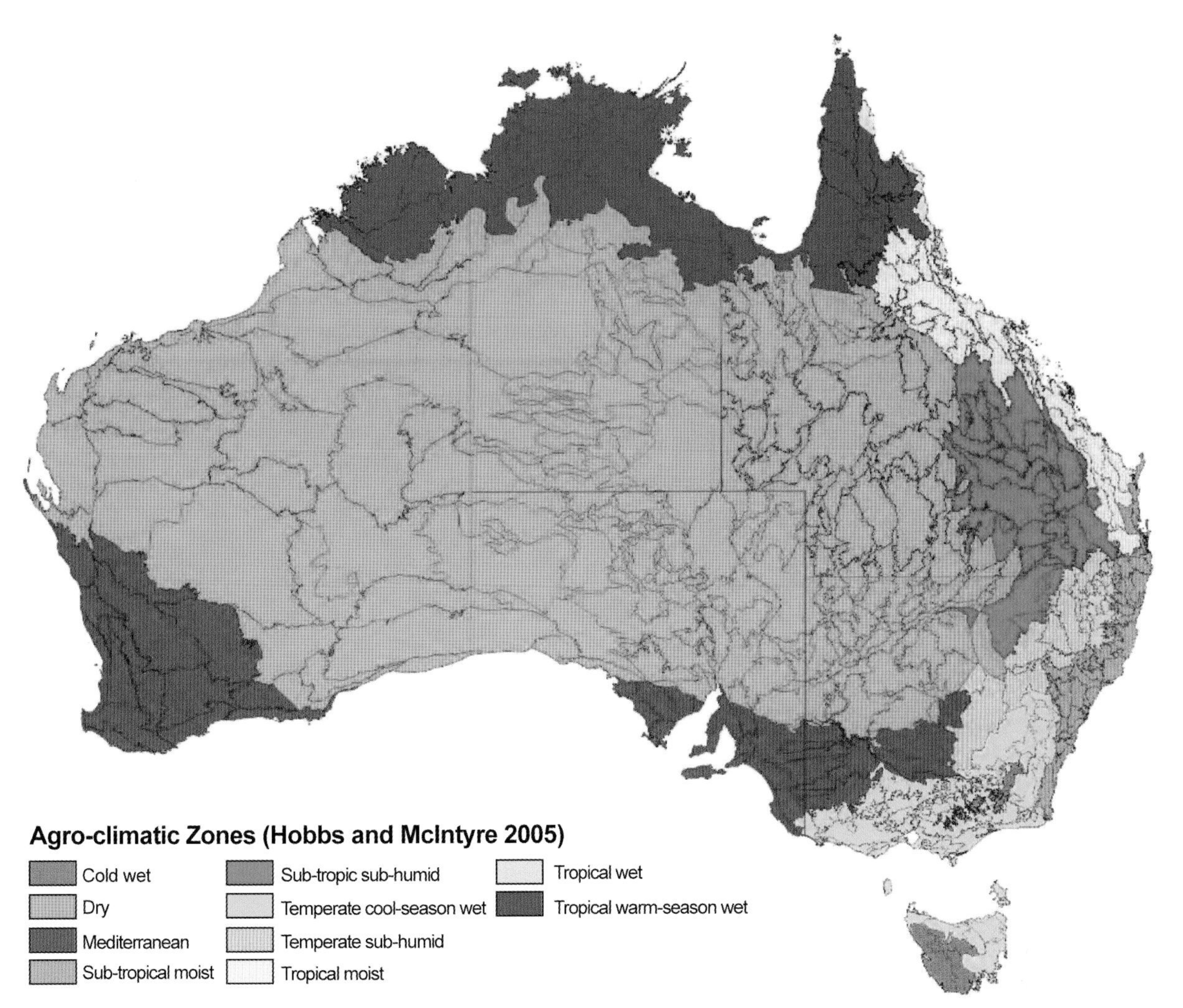

Figure 7.5 Map of agro-ecological zones in Australia overlaid on IBRA boundaries. Source: Hobbs and McIntyre (2005), as published in Dunlop and Brown (2008).

understanding might be applied to develop an integrated package of responses for each of these regional types to three climate change scenarios (Fig. 7.2), in terms of governance and investment sources, education, integrated response packages and 0–5 year action plans. *It must be emphasised that the table is meant only to demonstrate the approach, and is not meant to be complete, representative or thoroughly researched. It is a 'proof-of concept' exercise.* However, it is a critical illustration of the level of sophisticated analysis that is needed to understand the challenges created for biodiversity in each region from the combination of underlying environments, current and future socio-economic trends, and impacts of climate change. Furthermore, it identifies the opportunities that exist in recognising that different management responses are required in regions with different combinations of drivers. Much more work is required to implement the concept in practice, capturing local and regional knowledge in collaboration with policy and scientific inputs. In particular, more work is required to identify the potentially large array of organisations and institutions that could contribute to implementation of the concept. As the concept is developed further, the benefits of the approach can be articulated in more detail.

The analysis in Table 7.2 covers eight regions with contrasting socio-economic characteristics and trajectories. Each type of region has a specific area selected as an example, but other cases could have been given for most types. For each regional type, the analysis qualitatively indicates the state and trends over current and future decades in three important characteristics: population, agricultural and/or forestry production, and biodiversity. It also estimates the socio-economic trends that are at least partially responsible for changes in population, production and biodiversity. The analysis also lists the present biome/vegetation type and climatic regime. In this, it performs a very abbreviated version of the analysis carried out by Barr *et al.* (2005) for Victoria's rural regions (Fig. 7.6).

The subsequent columns work towards an integrated response package tailored to the characteristics and trends of the particular type of region. This is constructed from the perspective of enhancing biodiversity conservation under a changing climate in the context of regional socio-economic trends, while simultaneously hedging the response with respect to alternative potential future scenarios where these have different implications for actions (section 7.2). The climate change scenario column is particularly important – this briefly assesses whether different forms of action would be required in the different climate scenarios (recovery, stabilisation, runaway): if not, the actions are 'no regrets'; if so, then a mixed strategy may be needed to ensure preparation for all possible futures. This column is complemented with some comments on the education strategies and governance/institutional frameworks that will be required to support the integrated management package, which itself aims to build synergistically on the social trends of the region where possible. A significant strength of the approach is that the integrated response package (management, education, governance) is tailored to the characteristics of the particular region, where the region is considered as a linked social-ecological system. Table 7.2 also includes comments on a 0–5 year action plan for each regional type to suggest what can be done now to begin implementing the package.

A more complete systematic analysis aimed at such an approach would involve:

- collating the various regional socio-economic typologies already available for different regions
- running a qualitative expert knowledge-driven process for intersecting these with an appropriately resolved classification of environments (biome or biogeographic regions, or perhaps something like agro-ecological regions)
- identifying what climate change scenarios may be useful for planning in each type of region.

The results could then be taken into a series of regional workshops that deal with several geographically adjoining regions in practice. At a minimum level, the regions might be the agro-climatic zones of Hobbs and McIntyre (2005), intersected with trends towards maintaining core production (farming or forestry), taking land out of production in unsustainable marginal agricultural regions, or changing towards peri-urban intensification and amenity uses. It will also be useful to consider urban areas and marine regions separately. In the longer term, such regional differentiation should be taken up through regional NRM governance mechanisms (section 7.6) with ongoing local flexibility, adaptability and sensitivity to regional conditions.

7.8 RESOURCING THE FUTURE

The previous sections have outlined the strategies and tools, possible new institutional structures, and an indicative regional approach to support biodiversity conservation under a changing climate. These enhanced responses will place resource demands on society over and above those currently applied to biodiversity management. Of course, these will require direct financial resources, both public and private, but will also demand expertise and information that will not be instantly available. Short- and long-term strategies are needed to ensure these resources are available to meet needs as they become apparent.

7.8.1 Innovation, oversight and coordination

The most effective mechanisms by which independent expert advice has been provided to Australian governments in a disinterested and politically unbiased fashion have been through statutory bodies such as the Reserve Bank of Australia and the Australian Bureau of Statistics. In the broad environmental arena, the

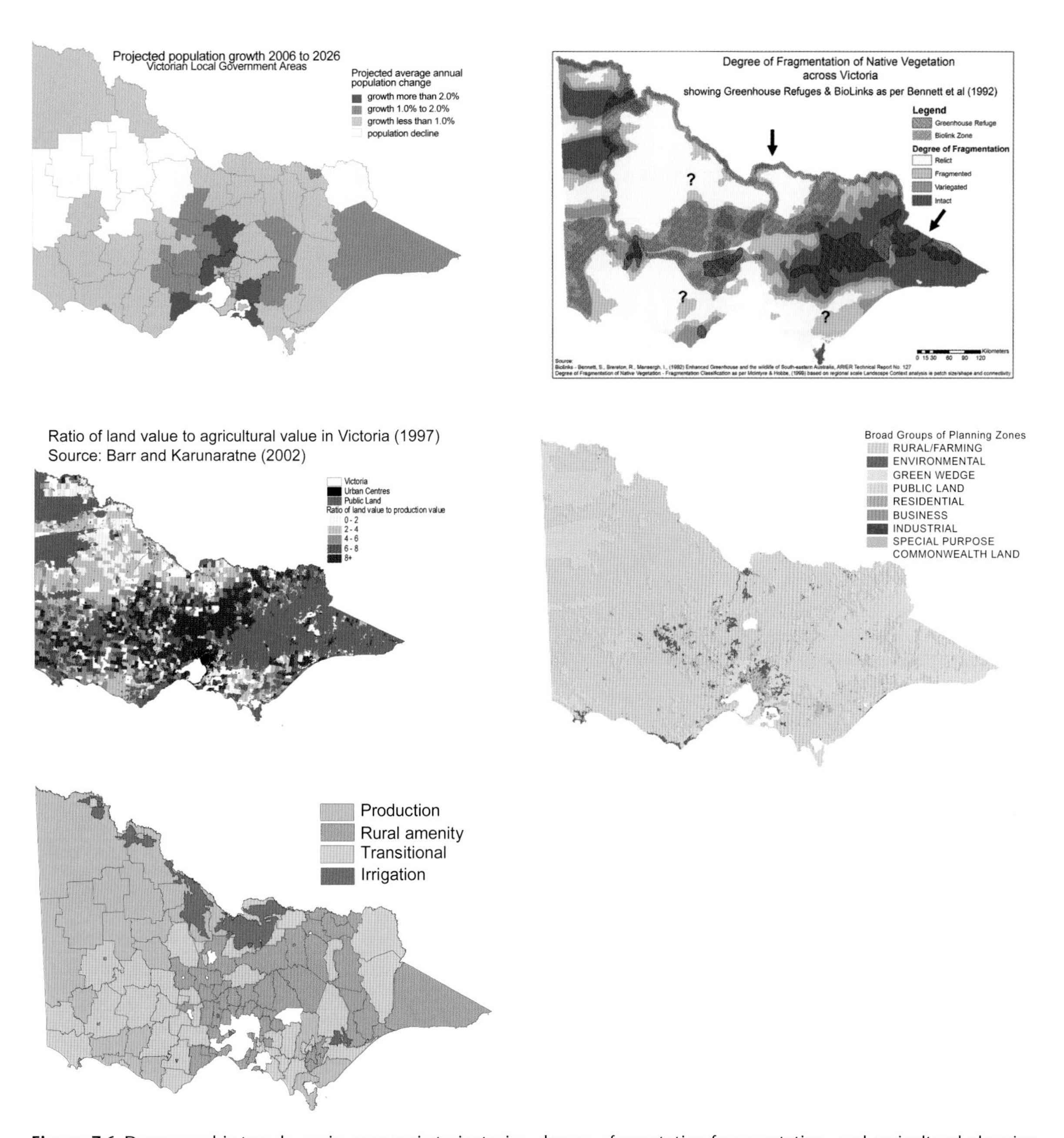

Figure 7.6 Demographic trends, socio-economic trajectories, degree of vegetation fragmentation, and agricultural planning and land values in Victoria. These are an example of the types of information needed to develop a systematic regional approach for biodiversity conservation. Source: Ian Mansergh; Neil Barr.

short-lived Resource Assessment Commission was a powerful and efficacious mechanism for providing advice on effective environmental decision-making.

The magnitude, seriousness and urgency of the likely impacts of climate change on biodiversity – acting directly and through the full range of other stressors – suggests the need for an ongoing independent body to oversee the monitoring of biodiversity-driven ecosystem services. The body should also advise – and if necessary, actively persuade – governments to tackle these problems effectively in an outcome-oriented and non-partisan manner. Whether this can be

Table 7.2 A systematic regional approach for biodiversity conservation in the 21st century. The table shows for various regional types: (i) the biomes and their climate trends; and (ii) the current state (low, medium, high) and trends of biodiversity (B), human population growth (H), economics ($), and agricultural production (P). The implications of the three climate scenarios (Fig. 7.2) are also shown (1 = 'Recovery' (recovers within 200 years), 2 = 'Stabilisation' (stabilises about 2100 but at ~3°C higher), 3 = 'Runaway' (keeps changing)), as well as approaches in governance and investment sources (off-reserve), education, integrated response packages, and a 0–5 year action plan.

Regional type (example)	Biome, climate, trend Agro-ecological zone	Trends in biodiversity (B), human population (H), economics ($), and agricultural production (P), and current state – low (L), medium (M) and high (H)				Future climate/ biodiversity (where known)	Implication of climate change scenarios*	Governance and investment sources (off-reserve)	Education	Integrated response package	0–5 year action plan (to begin now)
		B	H	$	P						
Amenity rural landscape (e.g. in eastern Victoria)	Temperate/ Mediterranean Drying with more extremes	L ↗	M ↗	M ↗	H ↘		1 ⇒ 3: Similar capacity to invest in assisting change in current vegetation, so some increase in severity/rates of change, but actions similar in all	Voluntary action based on extension and education Strong private investment from individual wealth with public guidance	Sophisticated; focused on solutions and actions; practical information on ecosystem management and appropriate species	Infusion of wealth Investment in environment Need information to avoid perverse incentives Possibility of big biodiversity gains	Vigorous education campaign Research on ecosystem replacement
Remote rural depopulating (e.g. Tanami Desert, central Northern Territory)	Desert savanna with warming trend Fire regime continues as for now	H ↘	L →	L ↘	L ↘		1 ⇒ 3: Increasing distance of movement by species across flat landscapes, so increasing need to facilitate. Refugia crucial for scenarios 1 and 2, but will be overwhelmed by 3	Stewardship contractual arrangements Mainly publicly funded conservation and Indigenous programs; carbon market may contribute	Encouraging the valuing of naturally functioning systems Practical information on how to conserve (training)	Modest stewardship payments and IPAs could give big biodiversity outcomes (and big ancillary benefits, e.g. cultural offsets issue)	Invest in remote area stewardship for multiple benefits
South-eastern temperate forest (e.g. central Victoria)	Temperate moist forest with a drying trend, which will include more fire and extreme events	H ↘	L →	M →	M →		1: Help key tree species persist 2: Replace key tree species with more climate tolerant equivalent 3: Facilitate ongoing shifts in key tree species composition	Government–industry agreements Opportunity to link climate change to production and biodiversity; carbon market is potentially a big player in public funding	Strong messages about managing fire, weeds and feral animals; encouragement to value new systems as communities change Targeted to forestry and conservation management	Fire management; carbon sequestration; water management; invest in new provenances Include biodiversity in management	Spread risk between 1 and 2 Replant some (but not all) areas with new species

Regional type (example)	Biome, climate, trend Agro-ecological zone	Trends in biodiversity (B), human population (H), economics ($), and agricultural production (P), and current state – low (L), medium (M) and high (H)				Future climate/ biodiversity (where known)	Implication of climate change scenarios*	Governance and investment sources (off-reserve)	Education	Integrated response package	0–5 year action plan (to begin now)
		B	H	$	P						
South-west Western Australia (world megadiversity hotspot)	Mediterranean: abrupt drying change in mid-1970s with a trend of increasing fire and drought	H ↓	M ↑	M ↑	M ↘	Drying trend, biodiversity loss	1: Invest in enhancing resilience for survival, focus on refugia 2: Invest in enhancing resilience, ex situ conservation, translocations and development of novel ecosystems 3: Will cause massive species loss; difficult choice between heavy 'engineering' investment or abandonment; main hope may be ex situ conservation	Public–private partnerships Potential for significant private mining and other production industry funding	Engage private industries/bodies about saving Gondwanan hotspot; information and action on minimising existing threats	Vegetation corridors and translocations Fire management Focus on refugia and south-west corner where rainfall should not decline as much; water demand and management	Urgent to engage mining leaders when mineral boom is strong Mitigate aggressively or 'Gondwana' is gone Manage to enhance resilience as much as possible
Core agriculture (e.g. northern Victoria, western slopes of New South Wales)	Modified temperate climate, drying with more extreme events Fire not important	L ↘	L →	M →	M ↗		1 ⇒ 3: Some increase in rate and severity; some adaptive action required at more severe scenarios	Strong contractual stewardship arrangements with Landcare-style support Driven by public spending to entrain behaviour of industry beyond its own immediate interests; carbon market should support this	Changing values from production only to multiple ecosystem services and income streams; extension of Landcare approach	Stewardship incentives; plantings to enhance connectivity and carbon uptake; limited potential for biodiversity outcomes	Trial and roll-out of stewardship payment system for carbon storage and biodiversity

Regional type (example)	Biome, climate, trend Agro-ecological zone	Trends in biodiversity (B), human population (H), economics ($), and agricultural production (P), and current state – low (L), medium (M) and high (H)				Future climate/ biodiversity (where known)	Implication of climate change scenarios*	Governance and investment sources (off-reserve)	Education	Integrated response package	0–5 year action plan (to begin now)
		B	H	$	P						
Developing tropical coastline (e.g. Great Barrier Reef and Wet Tropics of north Queensland)	Reef: sea-level rise, nutrient runoff, acidity, sea surface temperature, cyclones Tropical forest and woodland: more fires, harder to manage.	H ↘	M ↗	M ↗	M ↘		1: Invest heavily in enhancing resilience and protecting refugia Under 2 or 3, GBR is gone; Wet Tropics will retreat to small patches where engineering options may preserve species	GBRMPA, as an example agency, cooperating with local governments for marine areas; tourism industry–public sector alliances on land Private investment through tourist fees; public investment in stewardship	GBRMPA already running extensive education campaigns; need to convey sense of urgency and need to pull all relevant parties together	Increase reef resilience: particularly manage nutrient runoff, and tourism and fishing Rainforest: reduce existing stresses and manage fire on boundaries	Vigorous education in Australia and overseas – mitigate or lose the Reef Focus on reducing land use stresses
Heavily urbanising coastal zone (South-east Queensland)	Sub-tropical drying, sea-level rise, storms Fire not important	M ↘	M ↑	H ↗	M ↘		1 ⇒ 3: Some increase in severity/ rates but high investment possible. Low-lying areas (marshes, mangroves) vulnerable to sea-level rise at more severe scenarios	Regulation and buy-back on natural patches Public funds plus community private investment; little chance for significant industry investment	Persuade residents and developers to value nature and increase awareness of regulations. Engage volunteers.	Preserve habitat patches – wetlands, mangroves, etc. – but development pressure will limit possibilities; need to buy land for reserves; triage issues will arise	Vigorous education campaign Research on ecosystem replacement for patches of natural systems

Regional type (example)	Biome, climate, trend Agro-ecological zone	Trends in biodiversity (B), human population (H), economics ($), and agricultural production (P), and current state – low (L), medium (M) and high (H)				Future climate/ biodiversity (where known)	Implication of climate change scenarios*	Governance and investment sources (off-reserve)	Education	Integrated response package	0–5 year action plan (to begin now)
		B	H	$	P						
North Kimberley	Wet–dry tropical	H ↑	L ↗	L ↗	L ↗	Continuing high rainfall, possible slight increase.	For scenarios 1 or 2, preservation of region as wilderness, protection of refugia and good connectivity will minimise biodiversity loss. For scenario 3, some proactive measures may be necessary.	Oil and gas companies Mineral companies Ecotourism levy	Manage fire and feral animals; help Indigenous people remain in area and manage their land	Influx of tourists and dollars Oil, gas and mining companies investing in local area	Ensure recognition of one of the world's last large wilderness areas Better management of fire and feral animals Invest in Aboriginal management of land and sea for biodiversity

achieved by reviving the Resource Assessment Commission or establishing an entirely new statutory body, such as the commission proposed in section 7.6, is a matter for further debate. In any case, such a body should be established on a whole-of-government basis with high-level access as required.

An early task for such a body should be to run the first of a series of national forums, perhaps somewhat like the Australia 2020 Summit, on the current biodiversity crisis. We have identified earlier in this chapter that there must be an ongoing national discourse on the nature and purpose of biodiversity conservation in a rapidly changing world. Such issues failed to gain appropriate attention in the Australia 2020 Summit (19–20 April 2008; http://www.australia2020.gov.au). A further task that could well fall to this authority would be the independent audit of all proposals associated with emerging carbon and other environmental markets for potentially perverse outcomes. The body could also implement the first steps in iteratively improving the regional approach to biodiversity conservation outlined in the previous section.

7.8.2 Financial mechanisms

Current mechanisms for providing financial support for biodiversity conservation and management are inadequate. Direct funding from governments is generally offered with little, if any, long-term security. Projects often have no more than 12 months certainty and almost never more than three years. This short funding cycle may be appropriate for crisis management and for responding to short-term management demands. Climate change and biodiversity loss, however, are not short-term problems, and appropriate monitoring and management protocols need long-term financial commitments.

Providing estimates of the costs of alternative programs of action to pursue biodiversity outcomes is beyond the scope of this report. A reasonable observation, however, is that even modest aspirations will require budgets significantly greater than those of the present. If significant ecosystem and species losses are to be avoided, and desirable changes facilitated, the necessary budget increases will be substantial. Imaginative measures may be necessary if the biodiversity budget is to compete successfully with the alternative uses of available funds and be of a magnitude adequate for the task. Whatever measures are adopted, achieving substantial increases in financial resources requires broad support across Australian society, hence underscoring the importance of the national dialogue on the biodiversity crisis suggested previously.

Strategies are required to greatly boost public and private investment in environmental capital in general, and biodiversity conservation in particular. The existing suite of policy and program tools include the Caring for Our Country program, state and national investments in the National Reserve System and other conservation management, a variety of public–private partnership arrangements including Bush Tender, and support for non-government park purchases and tax arrangements. These all assist in meeting these needs, and must continue to be expanded imaginatively and with clear strategic direction. They nevertheless represent a piecemeal, small-scale approach to problem solving.

In the face of the additional stresses that are being caused by climate change, the nation cannot afford to continue to muddle along. The gap between investment in our common pool of environmental resources and that in our built, human and social capital is so large that serious national consideration must be given to a dramatically new approach to public and private resourcing of environmental management and biodiversity. One such approach could be an environmental levy, similar to the current Medicare Levy. A biodiversity credit scheme similar to carbon trading is in theory a possibility, but would be very difficult to design and achieve because there is no common currency for biodiversity, such as carbon dioxide for carbon emissions.

Another approach that merits consideration is the establishment of a trust fund, continuing the recent trend towards the establishment of trusts for major national priorities requiring significant resources. For example, the Building Australia Fund, Education Investment Fund, and the Health and Hospitals Fund (Budget Speech 2008–09). A national biodiversity trust could be established through collaborative state, territory and Commonwealth legislation, modelled on the following features of the Alaskan Permanent Fund (Barnes 2006). The trust would have the power to levy rents on all uses of biological resources in Australia and its waters, and the trustees would obey a charter to invest the financial capital from those levies to the best effect in sustaining the environmental capital of Australia. The trust beneficiaries would be defined as all Australians, who should be provided with incentives to recognise their benefits. The trust would be independent of government, except that trustees would be appointed by governments for terms that are long compared with the electoral cycle. The trustees would be empowered to set the resource rent rates. A variety of projects would be funded by the trust, aimed at reinvesting in natural capital, and related to NRM, and to the conservation of natural and cultural heritage. The trust could also invest its financial capital in companies that themselves act in favour of biodiversity. Note that the proposed trust would be vastly different from the short-term Natural Heritage Trust.

7.8.3 Expertise and people-power

Biodiversity responses to climate change will be complex and dynamic. Accordingly, there will be an ongoing demand for baseline information, monitoring and remedial action. A range of personnel will be required to fill these needs operating at a nested set of spatial scales. Some of these tasks will be in familiar areas with personnel such as restoration technicians, monitoring scientists, park rangers and research scientists. In other cases new roles will emerge, which are currently unfamiliar.

The expanded need for field environmental scientists, technicians and park rangers could probably be met with minimal retraining to add climate and biodiversity dimensions (as required) to current curricula. In other cases – such as the ongoing demand for taxonomic services for monitoring, restoration and fundamental understanding of ecosystems – there is a problem. There has been a well-documented long-term decline in the training and employment of taxonomists in Australian universities, museums, herbaria and other government institutions. In many instances, critical mass has already been lost and will only be regained slowly. Yet such scientists are crucial in all aspects of biodiversity science, not least in identifying and managing the impacts of climate change on biodiversity. Including taxonomic work among the priorities of funding bodies – such as the Australian Research Council, and the National Health and Medical Research Council – will help to reverse this downward trend, as will proper financial support for the Australian Biological Resources Study, and for scientific research within state museums, herbaria and other instrumentalities.

A new role will likely emerge as Australia moves to meet the challenge of managing the climate-driven changes in biodiversity. We have earlier identified the need for decentralised, regional bodies for on-the-ground management. In the dynamic situation that is arising as biodiversity changes in response to climate change, such regional bodies will only be effective if they are continually supplied with up-to-date scientific results. But scientific results alone will not be enough. The results themselves will need to be interpreted and incorporated into novel management actions under the guidance of a new kind of professional – biodiversity conservation facilitators. Such professionals would need to have scientific, pedagogic and people skills, a combination that seldom develops by accident. The training of such professionals needs to be developed against an agreed-upon curriculum that could be delivered as a coursework Masters program in any one of several educational modes (on campus, distance education, full-time, part-time, etc.). Such a scheme could be initiated by grants to an appropriate body (such as the National Climate Change Adaptation Research Facility) to train the first cohort of such professionals. The open educational marketplace may be expected to pick up on this training once it is proved viable.

7.8.4 Information demands

Coincident with the needs for maintaining capacity outlined above, there will be increased demands for education, problem-oriented research, technology transfer and capacity-building associated with these activities. Two aspects of education regarding biodiversity are in need of urgent action:

- In the popular terminology, 'biodiversity' is most often equated with species richness, and support for biodiversity conservation therefore still hinges on charismatic, usually vertebrate, species. This is despite the generally accepted and officially

Box 7.4 Broad research priorities for biodiversity. See also Chapter 5 for more detail of some of the biological and ecological research needs.

There are many elements of the national research effort that will generate results highly relevant to enhancing our abilities to cope with the impacts of climate change on biodiversity. Those endeavours, where either climate change is directly involved or where the urgent need to reverse current declines in effort requires changes in priorities, are summarised below.

1. While global climate modelling development (integrated atmosphere, ocean, cryosphere and land) proceeds and is relevant in Australia at the continental scale, there is urgent need for models and other tools that can quantify potential local impacts on particular locations, regions and ecosystems.
2. The role of monitoring is crucial in any attempt to measure and remedy impacts of climate on biodiversity: research on an ecosystem-to-ecosystem basis is needed to both identify key monitoring targets and design continental protocols for ongoing monitoring; however, this should not delay the early establishment of a national monitoring network.
3. Because many Australian plants and animals remain poorly known, monitoring and research into the impacts of climate change – or any other driver at the level of the ecological community – require vouchering and taxonomic research of those biodiversity elements being monitored or studied. The nation's biological collections in museums and herbaria provide the foundation for taxonomic research, and require better support and curation.
4. Research explicitly directed at the interaction between climate change and each of the other key drivers of ecosystem change – such as fire, salinity, disease, introduced predators, water extraction, grazing and clearing – should be enhanced. Modelling and data mining approaches are a likely first step in this direction.
5. There is a need for exemplary studies of how a range of key species (by function, taxon and habitat) will cope with projected change based on their life history characteristics.
6. Studies of community functioning, and in particular ecological cascades likely to emerge from loss of particular components of ecological communities, are needed for at least a subset of ecosystems.
7. Palaeogeographic and phylogeographic studies of past refugia under periods of climatic change are both highly informative and largely lacking in Australia; this needs to be rectified.
8. All else being equal, there is an imbalance in current knowledge in favour of terrestrial and/or marine systems. A priority area for future research should be freshwater ecological systems, which appear to have been neglected in studies of likely climate change impacts.
9. A program of research is required into social science issues, ranging through, for example, the political science of biodiversity policy, the socio-economics of the regional impacts of such policy, the institutions for biodiversity and natural resource governance, and market-based instruments as tools of biodiversity policy.

acknowledged definitions of biodiversity stretching from within-species genetic diversity to the ecosystem level (Chapter 2). As climate imposes change on the distributions and combinations of species across the landscape, so the focus of biodiversity conservation will shift to the services that ecosystems provide (Chapter 2) and the structural properties of such ecosystems rather than the identities of the particular species of which they are comprised. Both general and technical education needs to reflect this, and new curriculum materials and public education campaigns will be needed to make this shift. Of course, the role of individual charismatic species as conservation flagships will remain at least in the short to mid term.

- A high priority is overcoming the disconnection in knowledge of Australia's largely urban population with the biodiversity that is vital to our well-being. Public education aimed at urban populations on the importance of biodiversity to each and every citizen is vital. Appropriate public support for the extensive and essential – but inevitably expensive

– conservation measures must be ensured, again highlighting the urgent necessity for a national dialogue on the biodiversity crisis as well as a longer-term educational effort. Learning to appreciate the stresses upon ecosystem services and functioning will be a key principle. Maintenance of healthy water supplies, soils and local climate – together with pollination, tourism and amenity services – will form an appropriate foundation for this public education.

7.8.5 Research

There are many aspects of the climate–biodiversity interaction that are not yet well understood. Research to resolve issues, such as those set out in Box 7.4, must continue to be funded through the ARC and other granting schemes, as well as by state and territory conservation agencies. Biodiversity research currently suffers in public competition against other more fashionable areas of the life sciences, such as sub-cellular and molecular biology with all of their biomedical implications. Yet the understanding of biodiversity patterns and dynamics, as this publication has demonstrated, is at least as vital to the future well-being of Australians. Support for biodiversity must be given due priority and importance, with relevant funding earmarked for natural and social scientific research. Research to underpin active adaptive management of biodiversity in a changing climate is especially needed.

Of crucial importance is the development of appropriate, continent-wide, well-funded monitoring protocols. Much rhetoric has been expended on both the science and application of environmental monitoring. In general, both scientists and management agencies have avoided the imperative that effective monitoring must be based on continuing baseline surveys of biodiversity along major environmental gradients (e.g. degree of land use change, moisture availability). Affordable targets and the protocols required to monitor them effectively will differ from ecosystem to ecosystem and, under a changing climate, from time to time. Few Australian studies have confronted these issues head on and, at best, available studies only provide partial answers for a few ecosystem types and regions of the continent. There is an urgent and ongoing need for research on monitoring systems that will generate the continent-wide tools required to assess and modify management interventions undertaken within an adaptive framework.

7.8.6 Technology transfer and capacity-building

In a decentralised, regional approach to biodiversity management, such as that proposed in section 7.6, there will be an ongoing need for the results of research on issues such as adaptive management and monitoring to be transferred to the on-ground managers both promptly and effectively. This must be one of the key roles of biodiversity conservation facilitators.

8 Responding to the climate change challenge

Climate change will have an increasingly severe impact on Australia's biodiversity, the degree of impact depending on the rate and magnitude of temperature increases and other climatic changes. This chapter briefly summarises the key messages in this report and provides policy directions to assist the nation deal with the climate change challenge. The major outcomes of the assessment are presented within a framework of five key messages.

8.1 AUSTRALIA'S BIOTIC HERITAGE: VALUABLE AND WORTH CONSERVING

Australia has between 7% and 10% of all species on Earth. Its long isolation means that the majority of these species occur nowhere else. The structure and functioning of our terrestrial ecosystems differ markedly from analogous systems elsewhere. Unlike other continents, the lack of glaciation during the past few million years has left Australia with nutrient-poor soils, but also allowed the survival of many species-rich ecological communities. Aridity, high temperatures and sclerophyllous vegetation mean that fire plays an important role in shaping our landscapes. Freshwater is scarce, and many species have adaptations that reduce water use and loss. Aquatic species often have ways of surviving long droughts. A long coastline at the confluence of several of the world's great oceans supports a diverse array of marine species. The heritage values of biodiversity are important to Australians and to the rest of the world. They form part of our cultural identity and the direct contribution of indigenous biodiversity to the Australian economy is considerable. For example, tourists visiting Australia to see its species and landscapes contribute billions of dollars to the economy. Most importantly, Australia's natural biodiversity provides numerous 'ecosystem services' including food, fibre, fuel, genetic resources, fresh water, pollination, natural pest control and climate regulation. Living things are essential components of the indispensable life support system of our planet; if species loss and degradation of ecosystems continues, human society and the economy will be greatly affected. Estimates of the monetary value of biodiversity are very high. Indeed, biodiversity provides fundamental services that we would not be able to adequately replace through technology, no matter the cost, and which we require to survive. In short, these services are ultimately priceless.

Understanding why Australia is so species-rich, why our biodiversity is unique and why the conservation of our biodiversity is so important underpins our assessment. It is essential for understanding the profound implications of European settlement and the looming challenge of climate change, both of which drive change of such magnitude and rate that Australia's biosphere is undergoing rapid and continuing transformation.

8.2 AUSTRALIA'S BIODIVERSITY TODAY

The first humans – Indigenous Australians – arrived on the continent between 40 000 and 65 000 years ago. They came as hunters and gatherers, and had a relatively light ecological footprint, but they changed the landscape through hunting and by their use of fire.

European settlers arrived some 220 years ago and spread rapidly over the more productive parts of the country. They greatly altered many landscapes by clearing and farming, grazing rangelands, introducing exotic animals and plants, redirecting water resources, using fertiliser and other chemicals, changing fire regimes, urbanisation, mining, and harvesting resources such as fish and wood. Australia's biodiversity has experienced massive loss and changes over this period. Most drivers of these changes continue to operate in Australian ecosystems.

At the beginning of the 21st century, Australia's biodiversity is under considerable pressure. Our continent's rate of species extinctions is high in comparison with most other parts of the world, and many more species are on trajectories towards extinction. These changes in species diversity flow on to affect the structure and functioning of ecosystems in equally serious ways.

Although many are familiar with the historic record of change in Australia's biodiversity, the fact that these changes continue to unfold is fundamental to this assessment. The additional effects of climate change will exacerbate these current threats and cause unprecedented, additional stresses in their own right. Without an effective integration of knowledge about present stressors on biodiversity into climate change adaptation strategies and approaches, such efforts are virtually certain to fail.

8.3 THE CHALLENGE OF CLIMATE CHANGE

The Earth is warming rapidly – with human-driven increases in atmospheric greenhouse gas concentrations the primary cause – and climate change is likely to accelerate in the future. Australia is currently experiencing climate change consistent with the global pattern: higher temperatures, altered patterns of precipitation, sea-level rise, and changes in the magnitude and frequency of extreme events. Biodiversity is much less able to adapt to these and other climate changes than human systems.

Australia's plants, animals and micro-organisms are already responding to climate change and will continue to do so. Rising temperatures and greater extremes of temperature will significantly affect many species. Some are likely to become extinct. Reduced rainfall in parts of the country is affecting many species through reduced surface fresh water availability and lowered groundwater tables. Extensive, high-intensity fires are becoming more common. Sea temperatures and ocean acidity will continue to increase – leading to increased coral bleaching, coral reef loss and other changes – while sea-level rise and increased storm surges will affect coastal zones and low-lying islands.

The projected rate of temperature increase far exceeds that experienced during the past several million years. The transition from the Last Glacial Maximum (around 20 000 years b.p.) to the present (Holocene) climate took over 5000 years, compared with a potential recent change of similar magnitude in only 100 years. The degree to which Australia's biodiversity will be affected by climate change will depend on the magnitude and rate of climate change that is realised this century and beyond, which in turn depends on whether the world's nations can collaborate effectively to rapidly reduce greenhouse gas emissions.

Three future climate scenarios are considered in the assessment: a runaway scenario, a stabilisation scenario and a recovery scenario. Each of these scenarios has distinct and different implications for biodiversity conservation. Constructive and cost-effective adaptation approaches for vulnerable natural systems under a runaway climate scenario are not likely, and many extinctions and massive ecosystem change will result. A stabilisation scenario provides a focus for management actions to assist natural ecosystems adapt to some future, altered climate; however, the higher the level at which atmospheric greenhouse gas concentrations stabilise, the greater the change will be. Finally, a recovery scenario will require adaptation approaches that might be able to 'nurse' vulnerable systems through a period of increasing climate change, anticipating the prospect of better conditions at some future date, albeit centuries away. Planning for decisions with long-term implications needs to accommodate these possible futures until there is increasing certainty about future emission levels and about the degree of climate change that will result. However, for decisions with implications to 2050 only, the runaway scenario is the most appropriate to use for planning purposes, given current observations and the in-built momentum in the climate system.

Regardless of the level of global commitments to mitigation and regardless of whether stabilisation or recovery becomes the global goal, Australian managers and policymakers must prepare for a response time of centuries if substantial fractions of our biodiversity are to be conserved. In addition, our knowledge base is highly focused on the species level and on climate change as an isolated stressor, as this reflects the current state of the science. The most important impacts of climate change on biodiversity, however, will undoubtedly be the indirect ones at the community and ecosystem levels, together with the interactive effects with existing stressors. For example, warmer temperatures and declining snow cover in the Australian Alps are allowing greater numbers of introduced predators year-round at high elevations. This has many ecosystem-level effects and also threatens particular species such as the mountain pygmy-possum *Burramys parvus*, which is restricted to only 6 km^2 of the Alps, compounding other impacts of changing climate on the insulating snow cover and on post-hibernation food supplies. For the wetlands of Kakadu National Park, the major threats of climate change are not only the direct impacts on vulnerable species, but are also due to an intersection of effects; in this case, changing fire regimes, rising sea level and the resulting saltwater intrusion into freshwater wetlands, and the consequences of climate change for a suite of invasive weed and feral animal species. Such indirect effects highlight the many difficult but important issues that managers and policymakers face in dealing with the complex responses of our ecosystems to climate change – severe uncertainties, non-linearities, time lags, thresholds, feedbacks, rapid transformations, synergistic interactions and surprises. The current state of knowledge is not adequate for anything but the most general guidance on adaptation approaches.

8.4 BIODIVERSITY MANAGEMENT IN A CHANGING CLIMATE

Despite the lack of knowledge about system-level responses and the large uncertainties associated with the projections of future climate change, the situation is not hopeless. Concepts such as resilience and transformation provide positive, proactive avenues for reducing the vulnerability of biodiversity to climate change. The emphasis is on making space and opportunities for ecosystems to self-adapt and reorganise, and on the maintenance of fundamental ecosystem processes that underpin vital ecosystem services.

However, present institutional structures and levels of investment are already struggling to deal with current biodiversity change. In addition, a static view on climate is pervasively implicit in national, state and local government conservation strategies and many other policies, despite the best efforts of individual practitioners. For example, biodiversity conservation goals tend to be expressed in terms of maintaining species in their present locations. Current regulatory legislation subtly reinforces this approach. Consequently, the institutional architecture requires a transformation to implement revised policy and management tools with substantially increased resourcing.

Climate change should catalyse the transformation that is required to achieve a turnaround in the ongoing decline of Australia's biodiversity. Conservation of biodiversity is increasingly becoming a mainstream activity of governments, businesses, landowners, Indigenous Australians and community groups, leading to some notable conservation successes over the past couple of decades. Some of the ecological principles and strategies developed to conserve biodiversity in this period will be applicable in a future, climate-changing world. There is a relevant base of skill and experience on which to build, but it is significantly under-resourced.

The current policy and institutional landscape is changing rapidly, with simultaneous trends towards both centralisation and decentralisation. This state of flux provides opportunities to explore alternative institutional architectures and modes of policy delivery that can provide the flexibility needed to deal with a changing climate. Recognition in the community of the threat of climate change to biodiversity is growing rapidly, providing an opportunity for Australian society to re-examine its level of commitment to, and resourcing of, the conservation of the continent's unique biotic heritage in a rapidly changing world.

Innovative regional approaches to increasing adaptive capacity in biodiversity conservation under a changing climate can build on the major socio-economic trends sweeping across Australia. In the south-east, for example, abandonment of marginal agricultural areas, and an influx of retirees and escapees from urban areas, provide opportunities for integrating biodiversity values into these changing landscapes. More generally, integrated response packages in terms of governance, education, investment sources and action plans for biodiversity conservation can be tailored to the demographic and land use trajectories of specific regions around the country.

8.5 KEY MESSAGES AND POLICY DIRECTIONS

The impacts of climate change on Australia's biodiversity are now discernible at the genetic, species, community and ecosystem levels across the continent and in our coastal seas. The threat to our biodiversity is increasing sharply through the 21st century and beyond due to growing impacts of climate change, the range of existing stressors on our biodiversity and the complex interactions between them.

A business-as-usual approach to biodiversity conservation under a changing climate will fall short of meeting the challenge. A transformation is required in the way Australians think about biodiversity, its importance in the contemporary world, the threat presented by climate change, the strategies and tools needed to implement biodiversity conservation, the institutional arrangements that support these efforts, and the level of investment required to secure the biotic heritage of the continent.

The key following messages coming out of the assessment comprise an integrated set of actions. The order is arbitrary; they are highly interdependent and of similar priority. Taken together, they define a powerful way forward towards effective policy and management responses to the threat to biodiversity from climate change. The task is urgent. All key messages should be well towards full implementation within two years. Most need to be ongoing.

Reform our management of biodiversity

We need to adapt the way we manage biodiversity to meet existing and new threats – some existing policy and management tools remain effective, others need a major rethink, and new approaches need to be developed in order to enhance the resilience of our ecosystems.

As we are moving rapidly into a no-analogue state for our biodiversity and ecosystems, there is a need to transform our policy and management approaches to deal with this enormous challenge. Climate change presents a 'double whammy' – affecting species, ecosystems and ecosystem processes directly, as well as exacerbating the impacts of other stressors. Many effective management approaches already exist; the challenge is to accelerate, reorient and refine them to deal with climate change as a new and interacting complex stressor. The National Reserve System, the pillar of current biodiversity conservation efforts, needs to be enhanced substantially and integrated with more effective off-reserve conservation. Acceleration of actions to control and reduce existing stressors on Australian ecosystems and species is essential to increase resilience. However, there is a limit to how far enhancing resilience will be effective. Novel ecosystems will emerge and a wide range of unforeseen and surprising phenomena and interactions will appear. A more robust, long-term approach is to facilitate the self-adaptation of ecosystems across multiple pathways of adaptation that spread risk across alternative possible climatic and socio-economic futures. Active adaptive management – backed by research, monitoring and evaluation – can be an effective tool to support self-adaptation of ecosystems. An especially promising approach is to develop integrated regional biodiversity response strategies, tailored for regional differences in environments, climate change impacts and socio-economic trends.

Strengthen the national commitment to conserve Australia's biodiversity

Climate change has radical implications for how we think about conservation. We need wide public discussion to agree on a new national vision for Australia's biodiversity, and on the resources and institutions needed to implement it.

If the high rate of species loss and ecosystem degradation in Australia is to be slowed and eventually reversed, a more innovative and significantly strengthened approach to biodiversity conservation is needed. To meet this challenge – particularly under a rapidly changing climate – perceptions of the importance of biodiversity conservation and its implementation, in both the public and private sectors, must change fundamentally. A national discourse is therefore required on the nature, goals and importance of biodiversity conservation, leading to a major rethink of conservation policy, governance frameworks, resources for conservation activities and implementation strategies. The discourse should build a much broader and deeper base of support across Australian society for biodiversity conservation, and for goals that are appropriate in a changing climate. In particular, biodiversity education, policy and management should be reoriented from maintaining historical species distributions and abundances towards: (i) maintaining well-functioning ecosystems of sometimes novel composition that continue to deliver ecosystem services; and (ii) maximising native species and ecosystem diversity.

Invest in our life support system

We are pushing the limits of our natural life support system. Our environment has suffered low levels of capital reinvestment for decades. We must renew public and private investment in this capital.

There is as yet no widely accepted method – be it changes in natural capital, adjusted net savings or other indicators – to account for the impact of changes in Australia's biotic heritage due to human use. However, by any measure, Australia's natural capital has suffered from depletion and under-investment over the past two centuries. Climate change intensifies the need for an urgent and sustained increase in investment in the environment – in effect, in our own life support system. The challenge is to establish an enhanced, sustained and long-term resource base – from both public and private investment – for biodiversity conservation. In particular, significant new funding strongly focused towards on-ground biodiversity conservation work – carried out within an active adaptive management framework – is essential to enhance our adaptive capacity during a time of climate change. Monitoring the status of biodiversity is especially important, as without reliable, timely and rigorous information on changes in species and ecosystems, it is not possible to respond effectively to growing threats. An effective monitoring network would be best achieved via a national collaborative program with a commitment to ongoing, adequate resourcing.

Build innovative and flexible governance systems

Our current governance arrangements for conserving biodiversity are not designed to deal with the challenges of climate change. We need to build agile and innovative structures and approaches.

While primary responsibility for biodiversity conservation resides with each state and territory, over the past two decades many biodiversity conservation policies and approaches have been developed nationally through Commonwealth–state processes. There has also been a recent trend towards devolution of the delivery of NRM programs to the level of regional catchment management authorities and local Landcare groups. Dealing with the climate change threat will place further demands on our governance system, with a need to move towards strengthening and reforming governance at the regional level, and towards more flexibility and coherence nationally. Building on the strengths of current arrangements, a next step is to explore the potential for innovation based on the principles of: (i) strengthening national leadership to underpin the reform agenda required; (ii) devolving responsibilities and resources to the most local, competent level, and building capacity at that level; (iii) facilitating a mix of interacting regional governance arrangements sensitive to local conditions; and (iv) facilitating new partnerships with other groups and organisations, for example, with Indigenous and business entities. In addition, improved policy integration across climate change, environment protection and commercial natural resource use is required nationally, including across jurisdictional boundaries.

Meet the mitigation challenge

Australia's biodiversity has only so much capacity to adapt to climate change, and we are approaching that limit. Therefore, strong emissions mitigation action globally and in Australia is vital – and this must be carried out in ways that deliver both adaptation and mitigation benefits.

There is a limit above which biodiversity will become increasingly vulnerable to climate change even with the most effective adaptation measures possible. Global average temperature increases of 1.5 or 2.0°C above pre-industrial levels – and we are already committed to an increase of around 1.2 or 1.3°C – will likely lead to a massive loss of biodiversity worldwide. Thus, the mitigation issue is central to biodiversity conservation under climate change. To avoid an inevitable wave of extinctions in the second half of the century, deep cuts in global greenhouse gas emissions are required by 2020 at the latest. The more effectively the rate of climate change can be slowed and the sooner climate can be stabilised, the better are the prospects that biodiversity loss will be lessened. Societal responses to the mitigation challenge, however, could have significant negative consequences for biodiversity, over and above the effects of climate change itself. Examples include planting monocultures of fast-growing trees rather than establishing more complex ecosystems that both support more biodiversity and store more carbon, and inappropriate development of Australia's north in response to deteriorating climatic conditions in the south. However, with flexible, integrated approaches to mitigation and adaptation, many opportunities will arise for solutions that both deliver positive mitigation/adaptation outcomes and enhance biodiversity values.

Glossary

Italicised terms are defined elsewhere in the Glossary.

A

Acclimatisation/Acclimation: The changes in tolerance to temperature or another environmental factor by an *organism*. Acclimatisation usually occurs in a single organism over a short period of time.

Acidification: See *ocean acidification*.

Adaptation (evolutionary): Shift in the *gene pool* of a *population* as a result of *selection* for *traits* that give an advantage relative to an environmental factor.

Adaptation (human): Change in behaviour, *institutions*, etc. that confers advantage relative to an external factor such as the environment or a threat.

Adaptive management: Management practices that accommodate and respond to uncertain future events. A structured, iterative process (repetition of a process) of optimal decision-making in the face of uncertainty, with an aim of reducing uncertainty over time via monitoring. In this way, decision-making simultaneously maximises one or more resource management objectives and, either passively or actively, accrues information needed to improve future management. It is often characterised as 'learning by doing'.

Aestivation: A state of dormancy somewhat similar to hibernation. It takes place during times of heat and dryness, which is usually the summer months.

Alcohol dehydrogenase: An enzyme discovered in the mid-1960s in the fruit fly *Drosophila melanogaster*. It is responsible for catalysing alcohols into aldehydes and ketones, and vice versa, and is particularly sensitive to the environmental temperature in which the *organism* lives.

Alien (species): Not native (also referred to as *exotic*).

Allele: Diploid *organisms* (e.g. humans, most vertebrates) have paired homologous chromosomes in their somatic (non-reproductive) cells, and these contain two copies of each *gene*. An organism in which the two copies of the gene are identical – that is, have the same allele – is called homozygous for that gene. An organism in which the two copies of the gene are not identical is called heterozygous.

Alpha diversity: The biodiversity within a particular *community* or *ecosystem*. It is measured by counting the number of *taxa* within the ecosystem.

Angiosperm: Flowering plant with enclosed seeds. The flowering plants are the most widespread group of land plants. Angiosperms and *gymnosperms* comprise the two groups of seed plants.

Anthropocene: The current geological epoch in which humans and societies have become a geophysical force. It began around 1800 with the onset of industrialisation and an enormous expansion in the use of fossil fuels.

Anthropogenic: Resulting from or produced by human beings.

Apomixis (also called apogamy; adjective: apomyctic): Asexual reproduction, i.e. without fertilisation.

Archaea: A group of single-celled *micro-organisms* without a cell nucleus. They are similar to bacteria, but these two groups evolved differently and are classified as different domains in the three-domain system. Originally these organisms were named archaebacteria.

B

Beta diversity: The *species* diversity between ecosystems. This involves comparing the number of *taxa* that are unique to each of the ecosystems.

Biobank: A biobank is a repository for genetic material, with associated information describing or characterising the *organisms* to which the genetic material (usually DNA) belongs (see *germplasm*, *ex situ*).

Biodiversity: A word derived from **biological diversity**. The variety of all life forms: the different plants, animals and *micro-organisms*, their *genes*, and the *communities* and *ecosystems* of which they are part. Biodiversity is usually recognised at three levels: *genetic diversity*, *species* diversity and *ecosystem* diversity.

Biome: A climatically determined type of ecological community, often defined by the structure of the dominant life form and the geographic area (e.g. tropical rainforest, coral reef; scrub desert).

Bioregion (also called an ecoregion): An ecologically and geographically defined area.

Biosphere: The part of the Earth system comprising all *ecosystems* and living *organisms* in the atmosphere, on land (terrestrial biosphere) and in the oceans (marine biosphere). It includes derived dead organic matter, such as litter, soil organic matter and oceanic detritus.

Biota: All the plants, animals and *micro-organisms* of a particular region.

Bottleneck (genetic): An evolutionary event in which a significant percentage of a *population* or *species* is killed or otherwise prevented from reproducing, and the population is greatly reduced. Population bottlenecks increase *genetic drift*.

C

C3 grasses and **C4 grasses**: C3 grasses and C4 grasses employ different photosynthetic systems, with C3 being more common in the temperate zones, aquatic systems and woody plants, and C4 being more common in the tropics. Rising *carbon dioxide* (CO_2) levels are likely to alter the balance between C3 grasses and C4 grasses, with abundance of C4 *species* increasing in temperate areas – to the detriment of biodiversity and agricultural production.

C4 grasses: See *C3 grasses*.

CAR reserve system: CAR is an acronym for 'comprehensive, adequate and representative' and is defined as encompassing:

comprehensiveness – inclusion of the full range of *ecosystems* recognised at an appropriate scale within and across each *bioregion*

adequacy – the maintenance of the ecological viability and integrity of *populations*, *species* and *communities*

representativeness – the principle that those areas that are selected for inclusion in reserves reasonably reflect the biotic diversity of the *ecosystems* from which they are derived.

Carbon dioxide (CO_2): A chemical compound composed of two oxygen atoms bonded to a single carbon atom. It is a gas at standard temperature and pressure, and exists in Earth's atmosphere in this state. It is currently at a globally averaged concentration of approximately 383 parts per million (ppm) by volume in the Earth's atmosphere, although this varies by both location and time. Carbon dioxide is an important *greenhouse gas* because it transmits visible light but absorbs strongly in the infrared. Carbon dioxide is produced by all animals, plants, fungi and *micro-organisms* during *respiration*, and is used by plants during *photosynthesis*. It is, therefore, a major component of the carbon cycle. Carbon dioxide is generated as a by-product of the combustion of fossil fuels or vegetable matter, among other chemical processes. Carbon dioxide is also produced by volcanoes and other geothermal processes such as hot springs.

Carbon sequestration: The process of storing *carbon dioxide* in a solid material (*carbon sinks*) through biological or physical processes. It includes revegetation and pumping carbon dioxide into underground rock formations.

Carbon sink: A *carbon dioxide* reservoir that is increasing in size; the opposite of a carbon dioxide 'source'. The main natural sinks are oceans, and plants and other organisms that use *photosynthesis* to remove carbon from the atmosphere by incorporating it into biomass and releasing oxygen into the atmosphere. The concept of carbon dioxide sinks has become more widely known because the *Kyoto Protocol* allows the use of carbon dioxide sinks as a form of carbon offset.

Climate: The average weather in a region over a long period of time. Average weather may include average temperature, precipitation, wind patterns, ultraviolet (UV) levels and other physical measurements.

Climate change: Any long-term significant change in the 'average weather' that a given region experiences. In recent usage, the term 'climate change' often refers to changes in modern climate due to *global warming*.

CO_2: See *Carbon dioxide*.

Community or **Ecological community**: A naturally co-occurring biological assemblage of *species* that occurs in a particular type of habitat.

Connectivity: The extent to which particular *ecosystems* are joined with others of similar kind; the ease with which organisms can move across the landscape. Also applies to the extent to which populations of a species are able to interact with each other through *gene flow* (interbreeding.)

Coral: The term 'coral' has several meanings, but is usually the common name for the marine organisms from the class Anthozoa that exist as small sea anemone-like polyps, typically in colonies of many identical individuals. The group includes the important reef builders that are found in tropical oceans, which secrete calcium carbonate to form a hard skeleton.

Coral bleaching: The paling in colour that results if a *coral* loses its *symbiotic*, energy-providing *organisms* called *Zooxanthellae*. Coral bleaching is usually caused by high seawater temperatures, but can also be caused by pathogens, changes in water chemistry or increased sedimentation.

Coral reefs: Rock-like limestone (calcium carbonate) structures built by *coral* along ocean coasts (fringing reefs) or on top of shallow, submerged banks or shelves (barrier reefs, atolls); they are most conspicuous in tropical and sub-tropical oceans.

Corridor (for wildlife): A strip of habitat of varying width that facilitates animal movement between otherwise isolated patches of habitat.

Cryopreservation: A process where cells or whole tissues are preserved by cooling to sub-zero temperatures, such as –196°C (77 K), the boiling point of liquid nitrogen.

Cryosphere: The part of the world that is frozen. It includes the polar caps, the world's snowfields and glaciers. Although Himalayan glaciers have shrunk much less than those in Europe and North America, where 40% of the ice cover has disappeared in the past century, the *Intergovernmental Panel on Climate Change (IPCC)* is concerned that the Himalayan ice will disappear by 2035. More than 1.3 billion people depend directly on Himalayan ice for drinking water.

D

Demersal: The demersal zone is the part of the sea or ocean (or deep lake) comprising the *water column* that is near to (and is significantly affected by) the seabed and the *organisms* that live on, in or near the seabed.

Dispersal: The movement of organisms from one place to another. This differs from migration, which is a cyclical event due to seasonal changes in resources.

Disturbance (ecological): A temporary change in average environmental conditions that causes a pronounced change in an ecosystem. Outside disturbance forces often act quickly and with great effect, sometimes resulting in the removal of large amounts of biomass. Ecological disturbances include fires, flooding, windstorm, insect outbreaks, as well as anthropogenic disturbances such as forest clearing and the introduction of exotic species. Disturbances can have profound immediate effects on ecosystems and can, accordingly, greatly alter the natural community. Because of these and the impacts on populations, these effects can continue for an extended period of time.

Driver: A process that changes the trajectory of a *species* or *ecosystem*. Most ultimate drivers of biodiversity loss in Australia are human activities associated with consumption or development.

Dryas, younger (period): Also referred to as the 'big freeze', the Loch Loman Stadial or Greenland Stadial. A very cold period in the northern hemisphere from 12 700 to 11 500 years before present when average temperatures dropped within a decade by more than 5°C, remained low for 1300 years, and then warmed rapidly over a few decades to previous levels. The main cause of the 'big freeze' in Europe was the abrupt shutdown of the North Atlantic warm ocean current (the Gulf Stream, a *thermohaline current*) with the abrupt influx of fresh water from what is now the St. Lawrence Basin in Canada as the glacial ice front melted northward; rapid warming in Europe occurred with the subsequent restoration of the Gulf Stream. No climate change of this size, extent, or rapidity has occurred since. It is named after the tundra plant *Dryas octapetala*, which dispersed as far south as northern Spain during this period.

E

Ecological cascade: A chain, or cascade, of effects in an *ecological community* initiated by the removal of a *species* or addition of a new species, e.g. a series of secondary extinctions that is triggered by the primary extinction of a *keystone species* in an *ecosystem*.

Ecological community: see *Community*.

Ecological engineer: A *species* that has a profound effect on *ecosystem* dynamics by disturbing the physical environment, e.g. a burrowing/digging mammal such as the bilby, a potoroo or a bettong that affects soil water penetration, seed germination, fungal spore dispersal, etc., or a burrowing seabird that turns over soil and imports nutrients. The loss of such species from ecosystems can have profound long-term effects on species composition.

Ecological engineering (**Eco-engineering**): The design, construction and management of *ecosystems* that have value to both humans and the environment. This engineering discipline combines basic and applied science from engineering, ecology, economics and natural sciences for the restoration and construction of aquatic and terrestrial ecosystems. The field of ecological engineering is increasing in breadth and depth, as more opportunities to design and use ecosystems as interfaces between technology and environment are explored.

Ecological processes: Actions or events that shape *ecosystems*. Understanding ecological processes – whether they are natural disturbances like fire, or ongoing processes like nutrient cycling or *carbon sequestration* – is the key to the development and implementation of sustainable ecological management.

Ecological release: The process by which a *species* will expand its habitat and resource utilisation when a previous restriction on habitat and/or resource use is removed.

Ecological synergies: Synergy means working together and refers to the phenomenon in which two or more discrete influences or agents acting together create an effect greater than that predicted by knowing only the separate effects of the individual agents. Ecological synergy describes positive *symbiosis*.

Ecology: The scientific study of the distribution and abundance of life on Earth, and the interactions between *organisms* and their environment.

Ecosystem: An ecosystem is a dynamic complex of plant, animal and *micro-organism communities*, and the non-living (abiotic) environment (water, soil, climate, etc.), interacting as a functional unit. Humans can be an integral part of ecosystems.

Ecosystem engineering: Human intervention to promote certain outcomes in environmental management (see *Ecological engineering*).

Ecosystem processes: The physical, chemical and biological actions or events that link organisms and their environment. They include decomposition, production (of plant matter), nutrient cycling, and fluxes of nutrients and energy.

Ecosystem services: The benefits people obtain from *ecosystems* (e.g. food, renewable resources, water supply, recreational opportunities, oxygen, *carbon sequestration*, erosion control).

Ectomychorrhizal: Mycorrhizal describes a *symbiotic* (occasionally weakly pathogenic) association between a fungus and the roots of a plant, which helps the plant obtain nutrients. Ectomychorrizal fungi live on the outside of the root.

Ecotone: A transition area between two adjacent *ecological communities* or *ecosystems*. It may appear on the ground as a gradual blending of the two communities across a broad area, or it may manifest itself as a sharp boundary line.

El Niño–Southern Oscillation (ENSO): A global coupled ocean–atmosphere phenomenon. Although originally named for a local warming of the ocean near the coast of Peru in South America, 'El Niño' now refers to a sustained warming over a large part of the central and eastern tropical Pacific Ocean. Combined with this warming are changes in the atmosphere that affect weather patterns across much of the Pacific Basin, including Australia. These altered weather patterns often help promote further warming of the ocean because of the changes they cause in ocean currents.

Endemic (noun: endemism): Occurring only in the stated area.

Environmental envelope: The set of environments within which a species can persist, i.e. where its environmental requirements can be satisfied. The term is sometimes used synonymously with 'climate envelope' – the set of climates within which a species can survive.

Evapotranspiration: The sum of evaporation and plant *transpiration* from the Earth's land surface to the atmosphere. Evaporation accounts for the movement of water to the air from sources such as the soil, canopy interception and water bodies.

Evergreenness: Evergreen plants have leaves all year round. This contrasts with deciduous plants, which completely lose their foliage for part of the year.

Evolution: In biology, evolution is the process of change in the type and/or proportion of inherited *traits* of a population of organisms from one generation to the next. Though changes between generations are relatively minor, differences accumulate with each subsequent generation and can, over time, result in substantially different organisms in subsequent generations. Inherited traits are controlled by the *genes* that are passed onto offspring during reproduction.

Evolutionary radiation: A term used by biologists and palaeontologists to describe a dramatic and rapid (in geological terms) increase in the taxonomic diversity of a particular group of organisms. Typically, this diversification reveals a trend from a small number of similar ecological niches to a wide range of dissimilar ones. In other words, where a taxonomic group initially occupied one particular way of life, after a few million years it occupies many different ways of life (see *stem species*).

Exotic (species): Introduced (see *alien*).

Ex situ (conservation): Off-site conservation. This may involve translocation (moving animals or plants to a new site where threats are less), captive breeding or propagation. Endangered plants may also be preserved in part through a *seed bank*. Plants and animals can be preserved in *germplasm* banks (see *in situ*, *biobank*).

Extinction: The global disappearance of an entire *species* (as distinguished from *extirpation*).

Extinction debt: Describes the condition where a *threatening process*, e.g. *fragmentation* or *climate change*, leads to environmental conditions in which certain *species* will inevitably become extinct.

Extirpation: Local *extinction*.

F

Fire regime: The combination of fire frequency, intensity, interval and season. Different fire regimes can have different effects on *ecosystems*, e.g. frequent, low-intensity, cool-season fires can result in different combinations and abundances of plants and animals compared with infrequent, high-intensity, summer fires.

Fitness (genetic): Fitness is a central concept in evolutionary theory, and describes the capability of an individual of a certain *genotype* to produce viable offspring. If differences in individual genotypes affect fitness, then the frequencies of the genotypes will change over generations; the genotypes with higher fitness become more common. This process is called *natural selection*.

Fluidity (of an *ecosystem*): Being able to move *spatially* in response to environmental change.

Folivore: An herbivorous animal that specialises in eating leaves.

Food chain: The feeding relationships between species within an *ecosystem*. *Organisms* in a food chain are grouped into *trophic levels* based on how many links they are removed from the primary producers (plants). The pathways of the food and/or energy within the whole system are called a food chain or a *food web*.

Food web: See *Food chain*.

Founder effects: The loss of *genetic diversity* when a new colony is established by a very small number of individuals from a larger population.

Fragmentation: Removal (usually by land clearing) of large parts of a natural area, resulting in the retention of only small fragments (or remnants).

G

Gene: The unit of inheritance. The correct scientific term for this is *allele*. In cells, a gene consists of a long strand of deoxyribonucleic acid (DNA). Colloquially, the term gene is often used to refer to an inheritable *trait* that is usually accompanied by a *phenotype* (the expression of the gene in the organism's morphology or behaviour, as in 'brown eyes').

Gene flow: The transfer of *alleles* of *genes* from one population of a *species* to another.

Gene pool: The complete set of unique *alleles* in a *species* or *population*. A large gene pool indicates extensive *genetic diversity*, which is associated with robust populations that can survive bouts of intense *selection*. Low genetic diversity (see *inbreeding* and population *bottlenecks*) can cause reduced biological *fitness* and an increased chance of extinction.

Genera: See *genus*.

Genetic diversity: A level of *biodiversity* that refers to the total number of genetic characteristics in the genetic make-up of a *species*.

Genetic drift: The evolutionary process of change in the *gene* frequencies of a *population* from one generation to the next, due to the phenomenon of probability in which purely chance (*stochastic*) events determine which *alleles* (variants of a *gene*) within a reproductive *population* will be carried forward while others disappear.

Genotype: The entire genetic composition of an *organism*.

Genus (plural genera): A taxonomic category ranking below a family and above a *species*, and generally consisting of a group of species exhibiting similar characteristics. In taxonomic nomenclature the genus name is used – either alone or followed by a Latin adjective or epithet – to form the name of a species. The scientific name of a species is usually written 'genus name' then 'species name', e.g. the scientific name for humans is *Homo sapiens*; *Homo* is the name of the genus and *sapiens* is the specific epithet, i.e. it describes the species within that genus. There is only one living species of *Homo* (although some scientists have suggested that chimpanzees and bonobo should be included in *Homo*), while *Eucalyptus* includes several hundred species.

Germplasm: A term used to describe a collection of genetic resources of an *organism*. For plants, the germplasm may be stored as a seed collection, sometimes stored cryogenically (*cryopreservation*), or in a nursery or botanic garden. For animals, germplasm may be cryogenically stored as semen, eggs or fertilised early-stage embryos (see *biobank*).

Glacial-interglacial transitions: During the 2.5 million year span of the *Pleistocene* epoch, numerous glacials – or significant advances of continental ice sheets in North America and Europe – have occurred at intervals of approximately 40 000 to 100 000 years. These long glacial periods were separated by more temperate and shorter interglacials. During the interglacials, one of which we are experiencing now, the climate warmed to more or less present day temperatures and the tundra receded polewards following the ice sheets.

Global warming: The increase in the average temperature of the Earth's near-surface air and oceans since the mid-20th century, and its projected continuation. The *Fourth Assessment Report of the Intergovernmental Panel on Climate Change* concluded:
'Warming of the climate system is unequivocal.' 'Most of the observed increase in globally averaged temperatures since the mid-20th century is very likely due to the observed increase in anthropogenic greenhouse gas concentrations' ('very likely' in IPCC terminology means 'the assessed likelihood, using expert judgment, is over 90%').

Gondwana: The southern supercontinent that included most of the landmasses in today's southern hemisphere, including Antarctica, South America, Africa, Madagascar, Australia-New Guinea and New Zealand – as well as Arabia and the Indian subcontinent, which are in the northern hemisphere. Gondwana began to break up in the mid-Jurassic period (about 167 million years b.p.) with the break-up being complete about 33 million years b.p.

Gondwanan: Refers to patterns of evolutionary history and distribution of living *organisms*, typically when the organisms or close relatives of the organisms are restricted to two or more of the now discontinuous regions that were once part of Gondwana; e.g. the Proteaceae – a family of plants that is known only from Chile, South Africa and Australia – are considered to have a 'Gondwanan distribution'. Similar distribution patterns are seen in many plant and animal groups.

Greenhouse effect: The process in which the emission of infrared radiation by the atmosphere warms the Earth's surface. The name comes from an incorrect analogy with the warming of air inside a greenhouse compared with the air outside the greenhouse. The greenhouse effect was discovered by Joseph Fourier in 1824 and first investigated quantitatively by Svante Arrhenius in 1896. The Earth's average surface temperature of 14°C would, in the absence of the greenhouse effect, otherwise be about –19°C. *Global warming*, a recent warming of the Earth's lower atmosphere, is believed to be the result of an enhanced greenhouse effect due to increased concentrations of *greenhouse gases* in the atmosphere.

Greenhouse gas: Greenhouse gases are those gaseous constituents of the atmosphere, both natural and *anthropogenic*, that absorb and emit radiation at specific wavelengths within the spectrum of infrared radiation emitted by the Earth's surface, the atmosphere and clouds. This property causes the *greenhouse effect*. Water vapour (H_2O), *carbon dioxide* (CO_2), methane (CH_4), nitrous oxide (N_2O) and *ozone* (O_3) are the primary greenhouse gases in the Earth's atmosphere. As well as CO_2, N_2O, and CH_4, the Kyoto Protocol deals with the greenhouse gases sulphur hexafluoride (SF_6), hydrofluorocarbons (HFCs) and perfluorocarbons (PFCs).

Guild: A functional group of *species* that exploit the same class of environmental resources in a similar way and between which competition can be expected.

Gymnosperm: Seed-bearing plants with unenclosed seeds on the scales of a cone or similar structure. Gymnosperms include conifers (pines, firs, cypresses, etc.) and cycads (see *angiosperm*).

H

Habitat: The locality or natural home in which a particular plant, animal or group of closely associated organisms lives.

Holocene: The post-*Pleistocene* geological epoch, usually regarded as the past 10 000 years.

Homeostasis: The property of either an open system or a closed system, especially a living *organism*, that regulates its internal environment so as to maintain a stable, constant condition. Multiple dynamic equilibrium adjustments and regulation mechanisms make homeostasis possible.

Homeotherm: An animal that thermoregulates to maintain a stable internal body temperature regardless of external influence. Such animals (most mammals and birds) are often termed 'warm blooded', but this does provide an accurate description. Ectotherms ('cold blooded', which also is not a scientific or an accurate term) control body temperature through external means, such as the sun, or flowing air or water.

Hysteresis: The return to a previous state, once a *disturbance* or stress factor has been removed, that does not follow the same path of change that occurred when the original pressure was applied. In ecology, this can be due to time lags or differences in life history stages or life spans among interacting *organisms* and their environments.

I

Ice sheet: A mass of land ice that is sufficiently deep to cover most of the underlying bedrock topography. An ice sheet flows outwards from a high central plateau with a small average surface slope. The margins slope steeply, and the ice is discharged through fast-flowing ice streams or outlet glaciers, in some cases into the sea or into *ice sheets* floating in the sea. There are only two large ice sheets in the modern world – on Greenland and Antarctica. The Antarctic ice sheet is divided into east and west by the Transantarctic Mountains; during glacial periods there were others.

Inbreeding: Breeding between close relatives (or self-fertilisation within plants). If practised repeatedly, it usually leads to a reduction in *genetic diversity*.

Indigenous: Originating or occurring naturally in a particular locality; not introduced; native.

In situ (conservation): Conservation in the place where an entity occurs (see *ex situ*).

Institutions: Structures and mechanisms of social order and cooperation governing the behaviour of a set of individuals. Institutions are identified with a social purpose and permanence, transcending individual human lives and intentions, and with the making and enforcing of rules governing cooperative human behaviour.

Intergovernmental Panel on Climate Change (IPCC): The IPCC is a scientific intergovernmental body set up by the World Meteorological Organization (WMO) and the United Nations Environment Programme (UNEP). It constitutes: (i) the governments: the IPCC is open to all member countries of WMO and UNEP. Governments participate in plenary sessions of the IPCC where main decisions about the IPCC work program are taken and reports are accepted, adopted and approved. They also participate in the review of IPCC reports; (ii) the scientists: hundreds of scientists all over the world contribute to the work of the IPCC as authors, contributors and reviewers; and (iii) the people: as a United Nations body, the IPCC's work aims at the promotion of the United Nations human development goals.

The IPCC was established to provide decision-makers and others interested in climate change with an objective source of information about climate change. The IPCC does not conduct any research nor does it monitor climate-related data or parameters. Its role is to assess – on a comprehensive, objective, open and transparent basis – the latest scientific, technical and socio-economic literature produced worldwide that is relevant to the understanding of the risk of human-induced climate change, its observed and projected impacts, and options for adaptation and mitigation. IPCC reports should be neutral with respect to policy, although they need to deal objectively with policy-relevant scientific, technical and socio-economic factors. They should be of high scientific and technical standards, and aim to reflect a range of views, expertise and wide geographical coverage. In 2007, the IPCC released its Fourth Assessment Report (4AR). In December 2007, the IPCC, jointly with Al Gore Jr, was awarded the Nobel Peace Prize 'for their efforts to build up and disseminate greater knowledge about man-made climate change, and to lay the foundations for the measures that are needed to counteract such change.' (see http://www.ipcc.ch).

Introgression (genetic): *Gene flow* from one *species* into the *gene pool* of another by backcrossing an interspecific hybrid with one of its parents. Introgression is a long-term process; it may take many hybrid generations before the backcrossing occurs.

Invasive species: An *alien* species that adversely affects indigenous species or habitats. In Australia, animals (such as foxes, feral cats, rabbits and camels) and plants (such as bridal creeper, blackberry, buffel grass, gamba grass, lantana and bitou bush (boneseed)) are major invasive species.

Invertebrate: Any animal lacking a backbone.

IUCN: The International Union for the Conservation of Nature. The IUCN was founded in 1948 as the International Union for the Protection of Nature and in 1956 changed its name to the International Union for the Conservation of Nature and Natural Resources. In 1990 it became known informally as IUCN: the World Conservation Union, but since March 2008 that term is no longer used. The IUCN is the world's largest and oldest global environmental network. See http://www.iucn.org.

K

Keystone species: Species that have a disproportionately large influence on *community* and *ecosystems* in that their presence or absence changes the interactions among the species on which they depend.

Kyoto Protocol: A protocol to the international Framework Convention on Climate Change with the objective of reducing *greenhouse gases* that cause *climate change*. It was agreed to on 11 December 1997 at the 3rd Conference of the Parties to the Treaty when they met in Kyoto, and entered into force on 16 February 2005. Australia signed the Protocol in December 2007, and became a full member in 2008.

L

Last Glacial Maximum (LGM): The time of maximum extent of the *ice sheets* during the last *Pleistocene* glaciation (the Würm or Wisconsin glaciation), approximately 20 000 years b.p.

Life form: The characteristic morphology of a mature organism.

Life-history attribute: Any part of a *species*' life history (e.g. reproductive rate, longevity) that affects its ability to persist.

Limiting factor: A factor present in an environment that controls a process, particularly the growth, abundance or distribution of a *population* of *organisms* in an *ecosystem*. The availability of food, water, nutrients, shelter, and predation pressure are examples of factors limiting the growth of a population size.

M

Marsupials: One of three groups of living mammals, marsupials are separated from *monotremes* and *placental* (or eutherian) mammals by the female giving birth to live offspring and having a pouch (marsupium) in which the young is reared from a very early stage of development. There are also other morphological and physiological differences.

Mediterranean (climate): A *climate* that resembles that of the Mediterranean basin, having hot, dry summers and wet, cool winters. In Australia, the south-west of Western Australia and south-eastern South Australia have a Mediterranean climate.

Mesothermic: Climate found in the temperate zone where there are warm summers, and cool winters that are not cold enough to sustain snow cover.

Metapopulation: A group of spatially separated populations of the same species which interact at some level (see *Gene flow*). The term has been most broadly applied to species in naturally or artificially fragmented habitats (see *Fragmentation*).

Micro-organism: An *organism* that is microscopic (too small to be seen by the naked human eye). The study of micro-organisms is called microbiology. Micro-organisms are incredibly diverse and include bacteria, fungi, *archaea* and *protista* (two groups of single-celled *organisms*), as well as some microscopic plants and animals. They do not include viruses and prions, which are generally classified as non-living.

Millennium Ecosystem Assessment: See http://www.millenniumassessment.org. The Millennium Ecosystem Assessment (MA) was called for by the United Nations Secretary-General Kofi Annan in 2000. Initiated in 2001, the objective of the MA was to assess the consequences of *ecosystem* change for human well-being and the scientific basis for action needed to enhance the conservation and sustainable use of those systems and their contribution to human well-being. The MA has involved the work of more than 1360 experts worldwide. Their findings, contained in five technical volumes and six synthesis reports, provide a state-of-the-art scientific appraisal of the condition and trends in the world's ecosystems and the services (*ecosystem services*) they provide (e.g. clean water, food, forest products, flood control, natural resources), and the options to restore, conserve or enhance the sustainable use of ecosystems.

Mitigation: An *anthropogenic* intervention to reduce the anthropogenic forcing of the *climate* system; it includes strategies to reduce *greenhouse gas* sources and emissions, and to enhance greenhouse gas sinks.

Monitoring: Sampling and analysis designed to ascertain the extent of change from an expected or defined norm, or from past conditions.

Monotremes: One of three groups of living mammals, monotremes lay eggs rather than giving birth to live young. While fossil evidence suggests that the group was once more common and widespread, only five *species* exist today: two species in Australia (platypus and short-beaked echidna) and three species of long-beaked echidnas in New Guinea.

Mutualism: A biological interaction between individuals of two different *species*, where both individuals derive a *fitness* benefit; for example, increased survivorship.

Myrmecochorous: Seed *dispersal* by ants. About 1500 *species*, representing 87 *genera* and 24 families, of Australian plants are known to be myrmecochorous, i.e. regularly dispersed by ants because of ant-attracting structures (elaiosomes) on their seeds or fruits. Australian myrmecochorous plants are unexpectedly numerous and strikingly different from those known from the northern hemisphere. Most of them are *xerophytic* shrubs.

N

Natural selection: The process by which heritable *traits* that are favourable in a particular environment become more common in successive generations of a *population* of reproducing *organisms*, and unfavourable heritable *traits* become less common. Over time, this process can result in adaptations that specialise organisms for particular ecological *niches* and may eventually result in the emergence of new *species* (see *fitness*).

Naturalised: An *alien* (introduced) *species* that has become established in the wild.

Niche: The total range of conditions within which a *species* can survive, grow and produce viable offspring. The 'fundamental niche' defines the potential distribution of a species without any interactions with other organisms, while the 'realised niche' is the area actually occupied because of limitations due to other organisms such as competitors, predators, etc.

Nitrogen-fixing plant: Nitrogen fixation is the process by which nitrogen is taken from its natural, relatively inert molecular form (N_2) in the atmosphere and converted into nitrogen compounds (e.g. ammonia, nitrate, nitrogen dioxide). Some higher plants, such as legumes, contain *symbiotic* bacteria called rhizobia within nodules in their root systems, which produce nitrogen compounds that help the plant to grow and compete with other plants.

Novel ecosystem: New combinations of *species* and/or abiotic conditions. *Climate change* is expected to lead to development of numerous such *ecosystems*.

O

Ocean acidification: The name given to the ongoing decrease in the pH of the Earth's oceans, caused by their uptake of *anthropogenic carbon dioxide* from the atmosphere. Between 1751 and 1994, ocean surface pH is estimated to have decreased from approximately 8.179 to 8.104 (a change of –0.075). Although the natural absorption of CO_2 by the world's oceans helps *mitigate* the climatic effects of *anthropogenic* emissions of CO_2, it is likely that the resulting decrease in pH will have negative consequences, primarily for oceanic calcifying *organisms*. These use the calcite or aragonite polymorphs of calcium carbonate to construct cell coverings or skeletons. Calcifiers span the *food chain* and include organisms such as algae, *corals*, echinoderms, crustaceans and molluscs. While the full effects of ocean acidification are uncertain, it may have significant deleterious effects on many marine *species* and on coral reefs.

Off-reserve conservation: Biodiversity conservation conducted on land or water outside the formal conservation reserve system.

Organism: An individual form of life, such as a plant, animal, bacterium, *protist* or fungus.

Otoliths: Small, bony structures in the inner ear that stimulate hair cells when the head moves, providing information to the brain on orientation and acceleration. Fish otoliths accrete layers of calcium carbonate. The rate varies with the growth rate of the fish, resulting in the appearance of growth rings, which allows the determination of growth rate and age.

Ozone (O_3): A molecule consisting of three oxygen atoms. It is much less stable than O_2, the common form of oxygen in the atmosphere. Ground-level ozone is an air pollutant with harmful effects on the respiratory systems of animals. Ozone in the upper atmosphere filters potentially damaging ultraviolet light from reaching the Earth's surface. It is present in low concentrations throughout the Earth's atmosphere.

P

Palaeoecology: The use of data from fossils and *subfossils* to reconstruct (describe) the *ecosystems* of the past.

Permafrost: Soil permanently at or below 0°C. Most permafrost is located near the poles, but it may also occur at high elevations. The extent of permafrost can vary as the climate changes. Today, approximately 20% of the Earth's land mass is covered by permafrost. Measurement of the extent and depth of permafrost may be an indicator of climate change. Recently there has been significant thawing of permafrost in Alaska and Siberia. By the mid-21st century, the area of permafrost in the northern hemisphere is expected to decline by around 20% to 35%. The depth of thawing is likely to increase by 30% to a half of its current depth by 2080. Thawing of frozen peat leads to release of *greenhouse gases*, including methane, which exacerbates *global warming*.

Phenology: The study of the times of recurring natural phenomena. It has been concerned principally with the dates of first occurrence of natural events in their annual cycle. Examples include the date of emergence of leaves and flowers, the first flight of butterflies, the first appearance of migratory birds, and the dates of egg-laying of birds and amphibians. In the scientific literature on ecology, the term is used more generally to indicate the time frame for any seasonal phenomenon, including the dates of last appearance (e.g. the seasonal phenology of a *species* may be from April to September).

Phenotype (adjective phenotypic): Any observed quality of an *organism* – such as its morphology, development or behaviour – as opposed to its *genotype*, the inherited instructions it carries.

Phenotypic: See *phenotype*.

Photoperiod: The interval in a 24-hour period during which a plant or animal is exposed to light. Many flowering plants use a photoreceptor protein, such as phytochrome or cryptochrome, to sense seasonal changes in day length, which they take as signals to flower. Day length, and thus knowledge of the season of the year, is vital to many animals. A number of biological and behavioural changes are dependent on this knowledge. Together with temperature changes, photoperiod provokes changes in the colour of fur and feathers, migration, entry into hibernation, sexual behaviour and even the resizing of sexual organs.

Photosynthesis: The conversion of light energy into chemical energy by living *organisms*. Except for a few rare bacteria, the process requires chlorophyll, found only in algae and higher plants. The raw materials are *carbon dioxide* and water, the energy source is sunlight, and the end products are oxygen and energy-rich carbohydrates, e.g. fructose, glucose.

Phylogeography: The study of the historical processes that may be responsible for the contemporary geographic distributions of individuals. This is accomplished by considering the geographic distribution of individuals in light of the patterns associated with a *gene* genealogy.

Physiology: The study of the mechanical, physical and biochemical functions of living individual *organisms*.

Physiological: Characteristic of or consistent with the normal functioning of a living *organism*.

Phytophthora: A genus of water moulds, many *species* of which damage plants. *Phytophthora infestans* was the infective agent of the potato blight that caused the Great Irish Famine (1845–1849). Several species of *Phytophthora* have been introduced to Australia, the most damaging of which is *P. cinnamomi*, which causes root rot and which may cause the death of the plant due to water stress (the disease is sometimes referred to as dieback). It has caused, and is causing, massive plant deaths in the bush with concomitant loss of animal habitat – especially in south-western Australia but also in other southern states. *Phytophthora species* were formerly classified as fungi and are often still erroneously referred to as fungi.

Phytoplankton (see *Plankton*).

Placental (mammals): A derivation of 'placenta', an organ of the foetus of most mammals that attaches to the wall of the mother's uterus (womb), and provides for foetal nourishment and elimination of waste products. The other groups of mammals – *monotremes* (e.g. echidna, platypus) and *marsupials* – do not have a placenta.

Plankton: Any drifting *organism* that inhabits the pelagic zone (all but the bottom part of the *water column*) of oceans, seas or bodies of fresh water. It is a description of lifestyle rather than a genetic classification. They are widely considered to be some of the most important organisms on Earth, due to the food supply they provide to most aquatic life. Plankton are usually divided into 'phytoplankton' (primary producers with chlorophyll in their cells) and 'zooplankton' (planktonic *protista* and animals).

Pleistocene: A period of geological time beginning about 1.6 million years before present, immediately preceding the *Holocene* period.

Population (biological): The collection of individuals of a particular *species* in a stated area; they may or may not interact with other populations (see *gene flow*).

Protista: A group of micro-organisms that have cell nuclei but which are not fungi, animals or plants. Members of this group are closely related, but share common features such as being single-celled – or if they are multicellular, do not have specialised tissues.

R

Redundancy (ecological): Where one *species* appears to fulfil the same ecological role as another; the degree to which a given species can be substituted by another species to provide the same community functions or ecological services.

Refugium (plural: refugia): An area that has escaped or will escape changes occurring elsewhere and so provides a suitable *habitat* for *relict species*.

Relict species: An *organism* that at an earlier time was abundant in a large area but now occurs in only one or a few small areas. It can also refer to an ancient *species* that survives while related species become extinct.

Res nullius: A Latin term derived from Roman law whereby 'res' (objects in the legal sense – anything that can be owned, even slaves, but not subjects in law such as citizens) are not yet the object of rights of any specific subject. Such items are considered ownerless property and are usually free to be owned. Examples of 'res nullius' in the socio-economic sphere are wild animals or abandoned property. In English common law, for example, forest laws and game laws have specified which animals are 'res nullius' and when they become someone's property. Wild animals are regarded as 'res nullius', and as not being the subject of private property until reduced into possession by being killed or captured. A bird in the hand is owned; a bird in the bush is not. Bees do not become property until hived.

Resilience: The capacity of systems to absorb disturbance and reorganise while undergoing change so as to still retain function, structure, identity and feedbacks (Walker *et al.* 2004). In the context of climate change, resilience refers to the extent to which *ecosystems* can cope with a changing climate and continue to exist in their current state, in terms of composition, structure and functioning. The related term *transformation* refers to substantive changes in ecosystem composition, structure and functioning (i.e. a transition to a new state) in response to a changing climate. Application of these terms is scale dependent; i.e. transformation at one scale may be necessary to deliver resilience at higher scales.

Resistance: The degree to which a system does not respond to a shock (as opposed to *resilience*, which describes the extent to which it changes).

Respiration: The process whereby living *organisms* convert organic matter to *carbon dioxide*, and consume oxygen and release energy.

Restoration ecology: The study of renewing a degraded, damaged or destroyed *ecosystem* through active human intervention. Restoration ecology specifically refers to the area of scientific study that has evolved as recently as the 1980s.

Resprouters: Plants that survive fire and grow new shoots. Obligate seeders are killed by fire and can only regrow from seed.

S

Savanna: A tropical or subtropical woodland *biome*, the open canopy of which allows sufficient light to reach the ground to support an unbroken herbaceous layer consisting primarily of *C4 grasses*. The tree to grass ratio is very responsive to changes in precipitation, grazing, fire and land use management.

Sclerophylly (adjective: sclerophyllous): An *adaptation* in plants to arid conditions in which the leaves are rigid and have a thick, waxy cuticle. It is found in many plant groups, including some acacias, eucalypts and banksias.

Sea surface temperature (SST): The water temperature at the surface of the ocean. In practical terms, the exact meaning of 'surface' will vary according to the measurement method used. A satellite infrared radiometer indirectly measures the temperature of a very thin layer (about 10 micrometres thick) or skin of the ocean (leading to the phrase 'skin temperature') representing the top millimetre; a thermometer attached to a moored or drifting buoy in the ocean would measure the temperature at a specific depth. SST varies much more slowly than atmospheric temperatures due to the heat capacity of water. SST anomalies can be used as indicators of the phase of global climate fluctuations, such as the *El Niño–Southern Oscillation* (ENSO).

Sea-level rise: An increase in the mean level of the ocean. *Global warming* causes eustatic sea-level rise, due to an increase in the volume of the world's oceans as a result of thermal expansion and melting of ice. During the 20th century, the oceans have stored well over 80% of the heat that has warmed the Earth. The world average sea-level rise from 1950 to 2000 was about 1.8 mm/year. There are two long-term sea level records from Australia – at Fremantle (>90 years of data) and Fort Denison, Sydney (>80 years of data). Both show continuing rises in sea level during recent decades.

Seed bank: The collective name for seeds, often dormant, that are stored within the soil of many terrestrial ecosystems.

Sector (economics, human community): A part or section of a larger group.

Selection: see *Natural selection*.

Serotinous (noun: serotiny): Serotinous plants retain their seeds in a woody fruit or cone from which they are released after fire.

Sleeper weed: An alien plant established in the wild in low numbers that has the ability to increase its abundance and range dramatically under altered environmental conditions.

Spatial/Spatially (scale): Pertaining to area.

Species: A species is usually defined as a group of *organisms* capable of interbreeding and producing fertile offspring. While in many cases this definition is adequate, more precise or differing measures are often used, such as those based on similarity of morphology or DNA. Presence of locally adapted *traits* may further subdivide species into subspecies.

Stem species: An ancestral *species* from which related species diverged into different lineages. *Evolutionary radiation* often occurs from a survivor of mass extinctions or from the initial coloniser of a new area (such as Darwin's finches in the Galapagos Islands), or after *evolution* of a key innovation that allows expansion into a variety of *niches*.

Stochastic: Random; non-deterministic.

Stoma, stomate (plural: stomata): Small pores (openings) in leaves used for gas exchange. Two guard cells, either side of a stoma, regulate the size of the opening.

Stomata: See *stoma*.

Stomatal conductance: The control that leaf *stomata* exert on water vapour transpiration, carbon assimilation and *respiration* is expressed in terms of the stomatal conductance, which varies depending on several environmental factors, including CO_2 concentration in the atmosphere. Because *stomata* are a major conduit for the transfer of many gases between the terrestrial *biosphere* and the atmosphere, they play a major role in the biological control of the Earth's *climate* system and the chemistry of the atmosphere.

Stress (ecological): Factor(s) that reduce ability of an *organism* or *ecosystem* to thrive, e.g. drought, lack of nutrients, high temperature.

Stygofauna: Animals that live within groundwater systems, such as caves and aquifers. Usually they are small aquatic invertebrates, although stygofaunal vertebrates are known. Stygofauna can live within freshwater, brackish or saline aquifers; and within the pore spaces of limestone, calcrete or laterite. They are also found in marine caves and wells along coasts.

Subfossil: Remains whose fossilisation process is not complete, usually because of lack of time, or sometimes because the conditions in which they were buried were not optimal for complete fossilisation, i.e. mineralisation.

Symbiotic: A situation where two *organisms* (symbionts) live together in a close, mutually beneficial relationship.

T

Taxa: See *taxon*.

Taxon (plural: taxa): A taxonomic category or group, such as a phylum, order, family, *genus*, *species* or subspecies.

Temporal (scale): Pertaining to time.

Tetrapod: A vertebrate animal having four feet, legs or leg-like appendages. Amphibians, reptiles, birds and mammals are all tetrapods, and even snakes and legless lizards are tetrapods by descent.

Thermal stratification: Thermal layers in still water such as a lake. In seasonal environments, these have significant effects on *species* and *ecosystem* function. During the summer, when stratification occurs, there are usually three distinct levels based on temperature: the warm upper layer is mixed continually by wind and waves; the middle layer quickly cools with depth; and the cool bottom layer rarely exchanges water with the upper levels and is quite constant in temperature. As the body of water cools with colder weather, the upper level may eventually reach the same temperature as the lower levels, and the lake is said to 'turn over' and complete mixing occurs throughout the water column.

Thermohaline current: The thermohaline circulation (THC) is the large-scale ocean circulation pattern that transforms low-density upper ocean waters to high-density intermediate and deep waters, and returns those waters back to the upper ocean. Derivation is from 'thermo' for heat and 'haline' for salt, which together determine the density of sea water. Wind-driven surface currents in equatorial waters are driven polewards, only to cool, grow denser and sink at high latitudes. The dense water flows into the deep ocean basins where they are transmitted away from the poles, often to the basins of other large oceans. For example, the equatorial Atlantic produces the northbound Gulf Stream, one part of the THC, and is essentially responsible for the relatively warm northern European climate. The area also provides currents moving southward, upwelling eventually in the North Pacific with a transit time of around 1600 years. The water masses transport both energy (in the form of heat) and matter (solids, dissolved substances and gases) around the Earth and thus, the extensive mixing between the ocean basins makes the Earth's oceans a global system. The THC is sometimes called the ocean conveyor belt, the great ocean conveyer, the global conveyor belt or, most commonly, the meridional overturning circulation (often abbreviated as MOC).

Threatened (species): Likely to become extinct, threatened with extinction. A threatened *ecological community* is one that is likely to be destroyed. In the *IUCN* Red List of Threatened Species, threatened is a collective term including, from most to least threatened with extinction: critically endangered, endangered and vulnerable. This terminology is widely used in Australia.

Threatening process: Actions, either human or otherwise induced, that threaten the survival, abundance or evolutionary development of a *species*, *population* or *ecological community*, e.g. land clearing, introduced predators, weeds, pollution, fishing bycatch.

Tipping point (or critical threshold): A point where a small push away from one state has only small effects at first – but at some 'tipping point' the system can flip and go rapidly into another state, usually because of positive feedbacks. It has been applied in many fields from economics to human *ecology*. One example often considered in discussions of *climate change* is the possible abrupt reduction in the strength of the Gulf Stream of the North Atlantic due to changes in freshwater input near the Arctic, due in turn to an accelerating loss of floating sea ice near the North Pole, something that now may have passed a tipping point itself. A shutdown of the Gulf Stream would lead to decreased temperatures – even a mini-ice age, in northern Europe – and would affect all oceans (see *thermohaline current* and *Dryas, younger*).

Top predator (top-level predator, apex predator): A predator that, as an adult, is not normally preyed upon in the wild in significant parts of its range by creatures not of its own *species*. Being at the top of the *food chain*, top predators often shape the *ecological community* through their selective predation on a group of prey that also may be competitors with each other (see *keystone species*). Thus, they have a crucial role in maintaining the health of the community, i.e. they are *keystone species*. Loss of a top predator from a community can lead to significant changes in the abundance and variety of prey species. Top predators in Australia include dingoes, wedge-tailed eagles, feral cats, crocodiles, sharks and starfish.

Traits: Characteristics or properties of an entity. In biology it refers to a distinct *phenotypic* character of an *organism* that may be inherited, environmentally determined or somewhere in between.

Transformability: The capacity to create a fundamentally new system when the existing system is untenable.

Transformation: See *resilience*.

Transpiration: The evaporation of water from the surfaces of leaves through *stomata*. Transpiration enables flow of water, nutrients and minerals from the roots to the shoots and leaves.

Triage (*threatened species/ecological community* conservation): Allocation of priorities using a system similar to that used in prioritising medical emergencies for treatment. It infers not allocating resources in attempts to 'save' *species/communities* that have no or very little expectation of long-term survival (see *extinction debt*).

Trophic (level): A stage in a *food chain*, e.g. primary producer, herbivore, carnivore.

V

Vernalisation: The acquisition of the ability to flower in the spring by exposure to the prolonged cold of winter.

Vertebrate: An animal with a backbone (spinal column). A member of the subphylum Vertebrata of the phylum Chordata. Vertebrates comprise sharks and rays, bony fish, amphibians, reptiles, birds and mammals (including humans).

W

Water column: A conceptual column of water from the surface to bottom sediments. This concept is used chiefly for environmental studies that evaluate the *thermal stratification* or mixing (e.g. by wind-induced currents) of the thermal or chemically stratified layers in a lake, stream or ocean.

WWF: The World Wide Fund for Nature (formerly the World Wildlife Fund). WWF-Australia is the WWF national organisation in Australia.

X

Xerophile (adjective: xerophyllous): An *organism* living in or adapted to dry conditions.

Xerophyllic: Types of leaves adapted to dry conditions; often used to describe the plant having such leaves (see *xerophyte*).

Xerophyte (adjective: xerophytic): A plant that can grow and reproduce in conditions with a low availability of water.

Xerophytic: See *xerophyte*.

Z

Zooplankton: See *plankton*.

Zooxanthellae: The single-celled algae living in coral. Zooxanthellae can provide up to 90% of a coral's energy requirements via *photosynthesis*. In return, the coral provides the zooxanthellae with protection, shelter, nutrients (mostly waste material containing nitrogen and phosphorus) and a constant supply of the *carbon dioxide* required for photosynthesis (see *mutualism*). When corals are subjected to high environmental *stress*, e.g. high water temperatures, they can lose their zooxanthellae by either expulsion or digestion and die, a process known as *coral bleaching*.

Acronyms and abbreviations

ACRIS	Australian Collaborative Rangeland Information System
BIF	Banded ironstone formation
b.p.	Before present
CAMBA	China–Australia Migratory Bird Agreement
CAR	Comprehensive, adequate and representative
CITES	Convention on International Trade in Endangered Species
C–N	Carbon to nitrogen ratio
COAG	Council of Australian Governments
CRU	Climate Research Unit
CSIRO	Commonwealth Scientific and Industrial Research Organisation
CTBCC	Centre for Tropical Biodiversity and Climate Change
DCC	Department of Climate Change
DEST	Department of the Environment, Sport and Territories
EAC	East Australian Current
EAG	Expert Advisory Group
ENSO	El Niño–Southern Oscillation
EPA	Environmental Protection Authority
ESD	Environmental sex determination
ESRL	Earth System Research Laboratory
GBR	Great Barrier Reef
GBRMPA	Great Barrier Reef Marine Park Authority
GCM	General Circulation Model
IBRA	Interim Biogeographic Regionalisation for Australia
IPCC	Intergovernmental Panel on Climate Change
IUCN	International Union for the Conservation of Nature
JAMBA	Japan–Australia Migratory Bird Agreement
KNP	Kakadu National Park
NGO	Non-governmental organisation
NOAA	National Oceanic and Atmospheric Administration
NRM	Natural resource management
NRS	National Reserve System
NRSMPA	National Representative System of Marine Protected Areas
NSCABD	National Strategy for the Conservation of Australia's Biological Diversity
ppm	Parts per million
psu	Practical salinity unit
ROKAMBA	Republic of Korea–Australia Migratory Bird Agreement
SOI	Southern Oscillation Index
SRES	Special Report on Emissions Scenarios
UNFCCC	United Nations Framework Convention on Climate Change
WONS	Weeds of National Significance
WWF	World Wide Fund for Nature

References

ABRS (2008) *Australian Biological Resources Study*. Retrieved 17 September 2008, <http://www.environment.gov.au/biodiversity/abrs/>.

Access Economics (2007) *Measuring the economic and financial values of the Great Barrier Reef Marine Park 2005–06*. Research Publication 88. Great Barrier Reef Marine Park Authority, Townsville, Australia, <http://www.gbrmpa.gov.au/corp_site/info_services/publications/research_publications>.

Acheson JM (1988) *The lobster gangs of Maine*. University Press of New England, Hanover, NH.

Agriculture & Resource Management Council of Australia & New Zealand, Australian & New Zealand Environment & Conservation Council and Forestry Ministers (ARMCANZ) (2000) *Weeds of National Significance. Athel pine (Tamarix aphylla) Strategic Plan*. National Weeds Strategy Executive Committee, Launceston.

Alexander LV, Hope P, Collins D, Trewin B, Lynch A and Nicholls N (2007) Trends in Australia's climate means and extremes: a global context. *Australian Meterological Magazine* **56**: 1–18.

Alford RA and Richards SJ (1999) Global amphibian declines: a problem, in applied ecology. *Annual Review of Ecology and Systematics* **30**: 133–165.

Allen Consulting Group (2005) *Climate change risk and vulnerability: promoting an efficient adaptation response in Australia. Report to the Australian Greenhouse Office by the Allen Consulting Group*. Australian Greenhouse Office, Canberra.

Alley RB (2000) The Younger Dryas cold interval as viewed from central Greenland. *Quaternary Science Reviews* **19**: 213–226.

Alverson KD, Bradley RS and Pedersen TF (Eds) (2003) *Paleoclimate, global change, and the future*. IGBP Global Change Series. Springer-Verlag, Berlin.

Amos N, Kirkpatrick JB and Geise M (1993) *Conservation of biodiversity, ecological integrity and ecologically sustainable development*. Australian Conservation Foundation and World Wide Fund for Nature, Victoria.

Andersen AN, Cook GD, Corbett LK, Douglas MM, Eager RW, Russell-Smith J, Setterfield SA, Williams RJ and Woinarski JCZ (2005) Fire frequency and biodiversity conservation in Australian tropical savannas: implications from the Kapalga fire experiment. *Austral Ecology* **30**: 155–167.

Anderson AR, Hoffmann AA, McKechnie SW, Umina PA and Weeks AR (2005) The latitudinal cline in the *In(3R)Payne* inversion polymorphism has shifted in the last 20 years in Australian *Drosophila melanogaster* populations. *Molecular Ecology* **14**: 851–858.

Anderson K and Bows A (2008). Reframing the climate change challenge in light of post-2000 emission trends. *Philosophical Transactions. Series A, Mathematical, Physical, and Engineering Sciences* **366**: 3863–3882. doi:10.1098/rsta.2008.0138.

Andersson KP and Ostrom E (2008) Analyzing decentralized resource regimes from a polycentric perspective. *Policy Sciences* **41**: 71–93.

Andrewartha HG and Birch LC (1954). *The distribution and abundance of animals*. University of Chicago Press, Chicago.

Archer D (2006) *Global warming: understanding the forecast*. Blackwell Publishers, Oxford, UK.

Arnell NW (1999) Climate change and global water resources. *Global Environmental Change* **9**: S31–S49.

Aronson RB, Thatje S, Clarke A, Peck LS, Blake DB, Wilga CD and Seibel BA (2007) Climate change and invasibility of the antarctic benthos. *Annual Review of Ecology, Evolution, and Systematics* **38**: 129–154.

Atkins KJ (2006) *Declared rare and priority flora list for Western Australia*. Department of Environment and Conservation, Kensington, Perth.

Augee M and Fox M (2000) *Biology of Australia and New Zealand*. Pearson Education, Sydney.

Australian and New Zealand Environment and Conservation Council (ANZECC) (2001) *Review of the National Strategy for the Conservation of Australia's Biological Diversity*. ANZECC, Australia.

Australian Bureau of Statistics (2008) Retrieved 17 September 2008, <http://www.abs.gov.au/>.

Australian National Audit Office (2003) Performance audit: referrals, assessments and approvals under the *Environment Protection and Biodiversity Conservation Act* 1999. Audit Report No. 38 2002–03.

Australian State of the Environment Committee (2001) *Australia State of the Environment 2001.* CSIRO Publishing, Melbourne, on behalf of the Department of the Environment and Heritage.

Bäckstrand K (2003) Civic science for sustainability: Reframing the role of experts, policy-makers and citizens in environmental governance. *Global Environmental Politics* **3**: 24–41.

Banfai DS and Bowman DMJS (2005) Dynamics of a savanna-forest mosaic in the Australian monsoon tropics inferred from stand structures and historical aerial photography. *Australian Journal of Botany* **53**: 185–194.

Banfai DS, Brook BW and Bowman DMJS (2007) Multiscale modelling of the drivers of rainforest boundary dynamics in Kakadu National Park, northern Australia. *Diversity and Distributions* **13**: 680–691.

Barbier E and Cox M (2002) Economic and demographic factors affecting mangrove loss in the coastal provinces of Thailand, 1979–1996. Special issue of *Ambio* **31**: 351–357.

Barnes P (2006) *Capitalism 3.0: a guide to reclaiming the commons.* Berrett-Koehler Publishers, San Francisco, CA.

Barr NR (2005) *The changing social landscape of rural Victoria.* Department of Primary Industries, Melbourne.

Barr N and Karunaratne K (2002) *Victoria's small farms.* CLPR Research Report No. 10. Department of Natural Resources and Environment, Bendigo.

Barrett G, Silcocks A, Barry S, Cunningham R and Poulter R (2003) *New atlas of Australian birds.* Birds Australia, Melbourne.

Bastin G and ACRIS Management Committee (2008) *Rangelands 2008 – taking the pulse.* National Land and Water Resources Audit, Canberra.

Bates BC, Hope P, Ryan B, Smith I and Charles S (2008) Key findings from the Indian Ocean Climate Initiative and their impact on policy development in Australia. *Climatic Change* **89**: 339–354.

Bayliss B, Brennan K, Eliot I, Finlayson M, Hall R, House T, Pidgeon B, Walden D and Waterman P (1997) Vulnerability assessment of predicted climate change and sea level rise in the Alligator Rivers Region, Northern Territory Australia. *Supervising Scientist Report 123*, Supervising Scientist, Canberra.

Beattie AJ and Ehrlich PR (2001) *Wild solutions.* Melbourne University Press, Melbourne.

Beaumont LJ, McAllan IAW and Hughes L (2006) A matter of timing: changes in the first date of arrival and last date of departure of Australian migratory birds. *Global Change Biology* **12**: 1339–1354.

Beer T and Williams AAJ (1995) Estimating Australian forest fire danger under conditions of doubled carbon dioxide concentrations. *Climatic Change* **29**: 169–188.

Beeton RJS, Buckley KI, Jones GJ, Morgan D, Reichelt RE and Trewin D (2006) *Australia State of the Environment 2006*, Independent report to the Australian Government Minister for the Environment and Heritage, Department of the Environment and Heritage, Canberra. Retrieved 17 September 2008, <http://www.environment.gov.au/soe/2006/index.html>.

Bell PRF (1992). Eutrophication and coral reefs – some examples in the Great Barrier Reef lagoon. *Water Research* **26**: 553–568.

Bell PRF and Elmetri I (1995) Ecological indicators of large-scale eutrophication in the Great Barrier Reef lagoon. *Ambio* **24**: 208–215.

Bennett AF (1998) *Linkages in the landscape: the role of corridors and connectivity in wildlife conservation.* IUCN, Gland, Switzerland.

Bennett G (2004) *Integrating biodiversity conservation and sustainable use. Lessons learned from ecological networks.* IUCN Gland, Switzerland and Cambridge, UK.

Bennett S, Brereton R and Mansergh I (1992) *Enhanced greenhouse and the wildlife of south eastern Australia.* ARIER Technical Report No. 127. Arthur Rylah Institute for Environmental Research, Melbourne.

Bennett S, Kazemi S, Kelly S, Marsack P, Nelson N and Hosking J (2007) The possible effects of projected sea-level rise. Supplement to *Wingspan* **17**:17.

Berg RY (1975) Myrmecochorous plants in Australia and their dispersal by ants. *Australian Journal of Botany* **23**: 475–508.

Bergkamp G and Orland B (1999) *Wetlands and climate change: exploring collaboration between the Convention on Wetlands (Ramsar, Iran, 1971) and the UN Framework Convention on Climate Change*. International Union for the Conservation of Nature, <http://www.ramsar.org/key_unfccc_bkgd.htm>.

Bergstrom D (2003) Impact of climate change on terrestrial antarctic and subantarctic biodiversity. In *Climate change impacts on biodiversity in Australia* (Eds M Howden, L Hughes, M Dunlop, I Zethoven, D Hilbert and C Chilcott) pp. 55–57. Biological Diversity Advisory Council, CSIRO Sustainable Ecosystems, Canberra.

Berkelmans R and Oliver JK (1999) Large-scale bleaching of corals on the Great Barrier Reef. *Coral Reefs* **18**: 55–60.

Berkes F (2007) Community-based conservation in a globalized world. *Proceedings of the National Academy of Sciences of the United States of America* **104**: 15188–15193.

Berry SL and Roderick ML (2002) CO_2 and land-use effects on Australian vegetation over the last two centuries. *Australian Journal of Botany* **50**: 511–531.

Bishop MJ and Kelaher BP (2007) Impacts of detrital enrichment on estuarine assemblages: disentangling effects of frequency and intensity of disturbance. *Marine Ecology – Progress Series* **341**: 25–36.

Blackburn S (2005) Coccolithophorid species and morphotypes – indicators of climate change. In *Regional impacts of climate change and variability in south-east Australia*. Report of a joint review by CSIRO Marine Research and CSIRO Atmospheric Research. pp. 14–15. Hobart, Australia.

Blakers M, Davies SJJF and Reilly PN (1984) *The atlas of Australian birds*. Melbourne University Press, Melbourne.

Boland CRJ (2004) Introduced cane toads *Bufo marinus* are active nest predators and competitors of rainbow bee-eaters *Merops ornatus*: observational and experimental evidence. *Biological Conservation* **120**: 53–62.

Bond WJ and Keeley JE (2005) Fire as a global 'herbivore': the ecology and evolution of flammable ecosystems. *Trends in Ecology & Evolution* **20**: 387–394.

Bowman DMJS (2000) *Australian rainforests: islands of green in a land of fire*. Cambridge University Press, Cambridge.

Bowman DMJS (2003) Bushfires: a Darwinian perspective. In *Australia burning: fire ecology, policy and management issues* (Eds G Cary, DB Lindenmayer and S Dovers) pp. 3–14. CSIRO Publishing, Melbourne.

Bowman DMJS, Boggs GS and Prior LD (2008) Fire maintains an *Acacia aneura* shrubland–*Triodia* grassland mosaic in central Australia. *Journal of Arid Environments* **72**: 34–47.

Bowman DMJS and Dingle JK (2006) Late 20th century landscape-wide expansion of *Allosyncarpia ternata* (Myrtaceae) forests in Kakadu National Park, northern Australia. *Australian Journal of Botany* **54**: 707–715.

Bowman DMJS and Prior LD (2005) Turner Review No. 10 Why do evergreen trees dominate the Australian seasonal tropics? *Australian Journal of Botany* **53**: 379–399.

Bowman DMJS, Walsh A and Milne DJ (2001) Forest expansion and grassland contraction within a eucalytpus savanna matrix between 1941 and 1994 at Litchfield National Park in the Australian monsoon tropics. *Global Ecology and Biogeography* **10**: 535–548.

Braby MF (2000) *Butterflies of Australia: their identification, biology and distribution*. CSIRO Publishing, Melbourne.

Bradshaw CJA, Field IC, Bowman DJMS, Haynes C and Brook BW (2007) Current and future threats from non-indigenous animal species in northern Australia: a spotlight on World Heritage Area Kakadu National Park. *Wildlife Research* **34**: 419–436.

Bradshaw RHW and Holzapfel CM (2006) Evolutionary response to rapid climate change. *Science* **312**: 147–148.

Bradstock RA, Williams JE and Gill MA (Eds) (2002). *Flammable Australia: the fire regimes and biodiversity of a continent*. Cambridge University Press, Melbourne.

Braithwaite LW, Belbin L, Ive J and Austin M (1993) Land use allocation and biological conservation in the Batemans Bay forests of New South Wales. *Australian Forestry* **56**: 4–21.

Breed B and Ford F (2007) *Native mice and rats.* CSIRO, Melbourne.

Brodie J (1997) Nutrients in the Great Barrier Reef Region. In *Nutrients in marine and estuarine environments. Australia: State of the Environment Technical Paper Series (Estuaries and the sea).* (Ed. P Cosser) pp. 7–28. Department of the Environment and Heritage, Canberra.

Brook BW (2008) Synergies between climate change, extinctions and invasive vertebrates. *Wildlife Research* **35**: 249–252.

Brook BW, Sodhi NS and Bradshaw CJA (2008) Synergies among extinction drivers under global change. *Trends in Ecology & Evolution* **23**: 453–460.

Brook BW and Whitehead PJ (2006) The fragile millions of the tropical north. *Australasian Science* **27**: 29–30.

Brown C and Grace B (2006) Athel pine in the Northern Territory: a strategic approach to eradication. *15th Australian Weeds Conference Proceedings: managing weeds in a changing climate.* (Eds C Preston, JH Watts and ND Crossman). Weed Management Society of SA Inc, Adelaide.

Bruno JF, Selig ER, Casey KS, Page CA, Willis BL, Harvell CD, Sweatman H and Melendy AM (2007) Thermal stress and coral cover as drivers of coral disease outbreaks. *PLoS Biology* **5**(6): doi:10.1371/journal.pbio.0050124.

Buddemeier RW and Smith SV (1988) Coral-reef growth in an era of rapidly rising sea-level-predictions and suggestions for long-term research. *Coral Reefs* **7**: 51–56.

Bull CM and Burzacott D (2002) Changes in climate and in the timing of pairing of the Australian lizard, *Tiliqua rugosa*: a 15 year study. *Journal of Zoology* **256**: 383–387.

Bunn SE and Arthington AH (2002) Basic principles and ecological consequences of altered flow regimes for aquatic biodiversity. *Environmental Management* **30**: 492–507.

Bunnell FL (1998) Evading paralysis by complexity when establishing operational goals for biodiversity. *Journal of Sustainable Forestry* **7**: 145–164.

Bunnell F, Dunsworth G, Huggard D and Kremsater L (2003) *Learning to sustain biological diversity on Weyerhauser's coastal tenure.* Weyerhauser Company, Vancouver.

Burbidge AA (2004) *Threatened animals of Western Australia.* Department of Conservation and Land Management, Kensington, Western Australia.

Burbidge AA, Johnson KA, Fuller PJ and Southgate RI (1988) Aboriginal knowledge of the mammals of the central deserts of Australia. *Australian Wildlife Research* **15**: 9–39.

Burbidge AA and Morris KD (2002) Introduced mammal eradications for nature conservation on Western Australian islands: a review. In *Turning the tide: the eradication of invasive species, Auckland.* (Eds CR Veitch and MN Clout). IUCN SSC Invasive Species Specialist Group.

Burton JR (1985) *Development and implementation of Total Catchment Management policy in New South Wales. A discussion paper.* NSW Government, Sydney.

Byrne M (2007) Phylogeography provides an evolutionary context for the conservation of a diverse and ancient flora. *Australian Journal of Botany* **55**: 316–325.

Byrne M, Elliott C, Yates C and Coates D (2008) Extensive pollen dispersal in *Eucalyptus wandoo*, a dominant tree of the fragmented agricultural region in Western Australia. *Conservation Genetics* **9**: 97–105.

Byrne M, Morrice MG and Wolf B (1997) Introduction of the northern Pacific asteroid *Asterias amurensis* to Tasmania: Reproduction and current distribution. *Marine Biology* **127**: 673–685.

Byrne M, Yeates DK, Joseph L, Kearney M, Bowler J, Williams MA, Cooper S, Donnellan SC, Keogh JS, Leys R, Melville J, Murphy DJ, Porch N and Wyrwoll K-H (2008) Birth of a biome: insights into the assembly and maintenance of the Australian arid zone biota. *Molecular Ecology* **17**: 4398–4417.

Cai W (2006) Antarctic ozone depletion causes an intensification of the Southern Ocean super-gyre circulation. *Geophysical Research Letters* **33**: L03712.

Campbell BD, Stafford Smith DM and McKeon GM (1997) Elevated CO_2 and water supply interactions in grasslands: a pastures and rangelands management perspective. *Global Change Biology* **3**: 177–187.

Canadell JG, Le Quéré C, Raupach MR, Field CB, Buitenhuis ET, Ciais P, Conway TJ, Gillett NP, Houghton RA and Marland G (2007) Contributions to accelerating atmospheric CO_2 growth from economic activity, carbon intensity, and efficiency of natural sinks. *Proceedings of the National Academy of Sciences of the United States of America* **104**: 18866–18870.

Cantrell BK, Chadwick B and Cahill A (2002) *Fruit fly fighters: eradication of the papaya fruit fly.* CSIRO Publishing, Canberra.

Carey C and Alexander MA (2003) Climate change and amphibian declines: is there a link? *Diversity and Distributions* **9**: 111–121.

Cary GJ (2002) Importance of a changing climate for fire regimes in Australia. In *Flammable Australia: the fire regimes and biodiversity of a continent.* (Eds RA Bradstock, JE Williams and AM Gill) pp. 26–48. Cambridge University Press, Melbourne.

Cary G, Lindenmayer DB and Dovers S (Eds) (2003) *Australia burning: fire ecology, policy and management issues.* CSIRO Publishing, Melbourne.

Castellano MA and Bougher NL (1994) Consideration of the taxonomy and biodiversity of Australian ectomycorrhizal fungi. *Plant and Soil* **159**: 37–46.

Castles I (1992) *Australia's environment: issues and facts.* Australian Bureau of Statistics, Canberra.

Caughley G and Gunn A (1996) *Conservation biology in theory and practice.* Blackwell Science, London.

Caughley G, Grigg GC, Caughley J and Hill GJE (1980) Does dingo predation control the densities of kangaroos and emus? *Australian Wildlife Research* **7**: 1–12.

Cazenave A and Nerem RS (2004) Present-day sea level change: observations and causes. *Reviews of Geophysics* **42**: 139–150.

Chambers LE (2005) Migration dates at Eyre Bird Observatory: links with climate change? *Climate Research* **29**: 157–165.

Chambers LE (2007) Observed and projected climate change impacts. State of Australia's birds 2007: birds in a changing climate. Supplement to *Wingspan* **17**.

Chambers LE (2008) Trends in timing and migration of south-western Australian birds and their relationship to climate. *Emu* **108**: 1–14.

Chambers LE and Griffiths GM (2008) The changing nature of temperature extremes in Australia and New Zealand. *Australian Meteorological Magazine* **57**: 13–35.

Chambers LE, Hughes L and Weston MA (2005) Climate change and its impact on Australia's avifauna. *Emu* **105**: 1–20.

Chapman AD (2005) *Numbers of living species in Australia and the world.* Report for the Department of the Environment and Heritage, Canberra.

Chapman JW and Carlton JT (1994) Predicted discoveries of the introduced isopod *Synidotea laevidorsalis* (Miers, 1881). *Journal of Crustacean Biology* **14**: 700–714.

Chappell J, Chivas A, Wallensky E, Polach HA and Aharon P (1983) Holocene paleo-environmental changes, central to north Great Barrier Reef inner zone. *BMR Journal of Australian Geology and Geophysics* **8**: 223–235.

Chappell J and Woodroffe CD (Eds) (1985) Morphodynamics of Northern Territory tidal rivers and floodplains. In *Coasts and tidal wetlands of the Australian monsoon region*, Mangrove Monograph No. 1. Australian National University North Australia Research Unit, Darwin.

Chiew FHS and McMahon TA (2002) Modelling the impacts of climate change on Australian streamflow. *Hydrological Processes* **16**: 1235–1245.

Church JA, Hunter JR, McInnes KL and White NJ (2006) Sea-level rise around the Australian coastline and the changing frequency of extreme sea-level events. *Australian Meterological Magazine* **55**: 253–260.

Churchill S (1998) *Australian bats.* Reed New Holland, Sydney.

Climatic Research Unit (CRU) (2009) *1: Global temperature record.* University of East Anglia, Norwich, UK.

Coates DJ (2000) Defining conservation units in a rich and fragmented flora: implications for the management of genetic resources and evolutionary processes in south-west Australian plants. *Australian Journal of Botany* **48**: 329–339.

Cogger HG, Cameron EE, Sadlier RA and Eggler P (1993) *The Action Plan for Australian Reptiles.* Australian Nature Conservation Agency, Canberra.

Commonwealth Scientific and Industrial Research Organisation (CSIRO) (2001) *Climate change: projections for Australia.* CSIRO Climate Impacts Group, Melbourne.

Commonwealth Scientific and Industrial Research Organisation and the Australian Bureau of Meteorology (CSIRO and BOM) (2007) *Climate change in Australia: observed changes and projections*. Technical Report. CSIRO and Bureau of Meteorology, Melbourne.

Connell D (2007) *Water politics in the Murray-Darling Basin*. Federation Press, Annandale, Australia.

Corbett LK (2001) *The dingo in Australia and Asia*. JB Books, Adelaide.

Cork S, Stoneham G and Lowe K (2008) Ecosystem services and Australian natural resource management (NRM) futures. Paper to the Natural Resource Policies and Programs Committee (NRPPC) and the Natural Resource Management Standing Committee (NRMSC). Department of the Environment, Water, Heritage and the Arts, Canberra.

Corlett RT and Primack RB (2005) Tropical rainforests and the need for cross-continental comparisons. *Trends in Ecology & Evolution* **21**: 104–110.

Costanza R, d'Arge R, de Groot R, Farberk S, Grasso M, Hannon B, Limburg K, Naeem S, O'Neill RV, Paruelo J, Raskin RG, Sutton P and van den Belt M (1997) The values of the world's ecosystem services and natural capital. *Nature* **387**: 253–260.

Cowie ID and Werner PA (1993) Alien plant species invasive in Kakadu National Park, Tropical Northern Australia. *Biological Conservation* **63**: 127–135.

Cowling RM, Witkowski ETF, Milewski AV and Newbey KR (1994) Taxonomic, edaphic and biological aspects of narrow plant endemism on matched sites in Mediterranean South Africa and Australia. *Journal of Biogeography* **21**: 651–664.

Cox CB and Moore PD (1993) *Biogeography: an ecological and evolutionary approach*. Blackwell Science Ltd, Oxford, UK.

Crisp MD, West JG and Linder HP (1999) Biogeography of the terrestrial flora. In *Flora of Australia Volume 1. Introduction*, 2nd edn. (Eds AE Orchard and HS Thompson) pp. 321–367. CSIRO, Melbourne.

Crome FH (1994) Tropical rainforest fragmentation: some conceptual and methodological issues. In: *Conservation biology in Australia and Oceania*. (Eds C Moritz and J Kikkawa) pp. 61–76. Surrey Beatty, Sydney.

Crossland MR (2001) Ability of predatory native Australian fishes to learn to avoid toxic larvae of the introduced *Bufo marinus*. *Journal of Fish Biology* **59**: 319–329.

Crowley G, Garnett S and Shephard S (2009) Impact of storm-burning on *Melaleuca viridiflora* invasion of grasslands and grassy woodlands on Cape York Peninsula, Australia. *Austral Ecology* **34**: 196–209.

Daily GC (Ed.) (1997) *Nature's services: societal dependence on natural ecosystems*. Island Press, Washington, DC.

Daniels MJ and Corbet L (2003) Redefining introgressed protected mammals: when is a wildcat a wildcat and when is a dingo a wild dog. *Wildlife Research* **30**: 213–218.

Danks A, Burbidge AA, Burbidge AH and Smith GT (1996) *Noisy scrub-bird recovery plan. Wildlife Management Program No. 12*. Department of Conservation and Land Management, Como, Western Australia.

D'Antonio CM and Vitousek PM (1992) Biological invasions by exotic grasses, the grass/fire cycle, and global change. *Annual Review of Ecology and Systematics* **23**: 63–87.

Davis MB (1984) Climate instability, time lags, and community disequilibrium. In *Community ecology* (Eds J Diamond and TJ Case) pp. 269–284. Harper and Row, New York.

Delcourt HR and Delcourt PA (1991) *Quaternary ecology: a paleological perspective*. Chapman and Hall, London.

Delong SC (1996) Defining biodiversity. *Wildlife Society Bulletin* **24**: 738–749.

Department of Environment and Climate Change (2007) Alps to Atherton initiative: a continental-scale lifeline to engage people with nature. In *The Great Eastern Ranges Initiative*. (NSW Department of Environment and Climate Change. http://www.environment.nsw.gov.au/ger/index.htm) pp. 1–54. Department of Environment and Climate Change NSW, Sydney.

Department of Planning and Community Development (2008) *Victoria in Future 2008:Victorian State Government Population and Household Projections 2006–2036*. Victorian Government report.

Department of the Environment and Heritage (2005) *National Vegetation Information System (NVIS) Stage 1, Version 3.0 Major Vegetation Groups*, <http://www.deh.gov.au/erin/nvis/mvg/index.html>.

Department of the Environment and Heritage (2007) *No species loss: a nature conservation strategy for South Australia 2007–2017*, <http://www.environment.sa.gov.au/biodiversity/pdfs/nsl_strategy.pdf>.

Department of the Environment, Sport and Territories (DEST) (1996) *National Strategy for the Conservation of Australia's Biological Diversity 1996*. DEST, Canberra.

Deutsch CA, Tewksbury JJ, Huey RB, Sheldon KS, Ghalambour CK, Haak DC and Martin PR (2008) Impacts of climate warming on terrestrial ectotherms across latitude. *Proceedings of the National Academy of Sciences of the United States of America* **105**: 6668–6672.

Devney C and Congdon B (2007) Demographic and reproductive impacts on seabirds? Supplement to *Wingspan* **17**: 14–15.

Diaz RJ and Rosenberg R (2008) Spreading dead zones and consequences for marine ecosystems. *Science* **321**: 926–929.

Dickman CR and Woodford Ganf R (2007) *A fragile balance: the extraordinary story of Australian marsupials*. Thames & Hudson, Fishermens Bend, Victoria.

Dingle H, Rochester WA and Zalucki MP (2000) Relationships among climate, latitude and migration: Australian butterflies are not temperate-zone birds. *Oecologia* **124**: 196–207.

Dixon KW, Roche S and Pate, JS (1995) The promotive effect of smoke derived from burnt native vegetation on seed germination of Western Australian plants. *Oecologia* **101**: 185–192.

Done TJ (1999) Coral community adaptability to environmental change at the scales of regions, reefs and reef zones. *American Zoologist* **39**: 66–79.

Done TP, Whetton P, Jones R, Berkelmans RWC, Lough JM, Skirving WJ and Wooldridge SA (2003) *Global climate change and coral bleaching on the Great Barrier Reef*. Final report to the State of Queensland Greenhouse Taskforce through the Department of Natural Resources and Mines. Australian Institute of Marine Science, Townsville.

Dow DD (1977) Indiscriminate interspecific aggression leading to almost sole occupancy of space by a single species of bird. *Emu* **77**: 115–121.

Dow DD (1979) Agonistic and spacing behavior of the noisy miner *Manorina melanocephala*, a communally breeding honey-eater. *Ibis* **121**: 423–436.

Dunlop JN (2007) Climate change signals in the population dynamics and range expansion of seabirds breeding off south-western Australia. Supplement to *Wingspan* **17**: 16.

Dunlop N (2001) Sea-change and fisheries: a bird's-eye view. *Western Fisheries*: 11–14.

Dunlop M and Brown PR (2008) *Implications of climate change for Australia's National Reserve System: a preliminary assessment*. Report to the Department of Climate Change. Department of Climate Change, and the Department of the Environment, Water, Heritage and the Arts, Canberra

Easterling DR, Meehl GA, Parmesan C, Changnon SA, Karl TR and Mearns LO (2000) Climate extremes: observations, modeling, and impacts. *Science* **289**: 2068–2074.

Eby P (2000) The results of four synchronous assessments of relative distribution and abundance of grey-headed flying-fox *Pteropus poliocephalus*. In *Proceedings of a workshop to assess the status of the grey-headed flying-fox*. (Eds G Richards and L Hall) pp. 66–77. Australasian Bat Society, Canberra.

Edmonds T, Lunt I, Roshier DA and Louis J (2006) Annual variation in the distribution of summer snowdrifts in the Kosciuszko alpine area, Australia, and its effect on the composition and structure of alpine vegetation. *Austral Ecology* **31**: 837–848.

Eliot I, Finlayson CM and Waterman P (1999) Predicted climate change, sea level rise and wetland management in the Australian wet-dry tropics. *Wetlands Ecology and Management* **7**: 63–81.

Ellis EC and Ramankutty N (2008) Putting people in the map: anthropogenic biomes of the world. *Frontiers in Ecology and the Environment* **6**: 439–447.

Emison WB (1996) Use of supplementary nest hollows by an endangered subspecies of Red-tailed Black Cockatoo. *Victorian Naturalist* **113**: 262–263.

English V, Jasinska E and Blyth J (2003) *Aquatic root mat community of caves of the Swan Coastal Plain, and the Crystal Cave Crangonyctoid Interim Recovery Plan 2003–2008*. Department of Conservation and Land Management, Perth.

Environmental Protection Authority (EPA) (2007) *State of the environment report: Western Australia 2007*. Environmental Protection Authority, Perth.

Estades CF and Temple SA (1999) Deciduous-forest bird communities in a fragmented landscape dominated by exotic pine plantations. *Ecological Applications* **9**: 573–585.

Everist SL (1974) *Poisonous plants of Australia*. Angus & Robertson, Sydney.

Ezrahi Y (1990) *The descent of Icarus: science and the transformation of contemporary democracy*. Harvard University Press, Cambridge, MA.

Fenner F and Fantini B (1999) *Biological control of vertebrate pests: the history of myxomatosis in Australia – an experiment in evolution*. CABI Publishing, Wallingford.

Fischer F (2000) *Citizens, experts, and the environment: the politics of local knowledge*. Duke University Press, Durham, NC.

Fischer J, Lindenmayer DB, Blomberg SP, Montague-Drake R, Felton A and Stein JA (2007) Functional richness and relative resilience of bird communities in regions with different land use intensities. *Ecosystems* **10**: 964–974.

Fitzpatrick MC, Gove AD, Sanders NJ and Dunn RR (2008) Climate change, plant migration, and range collapse in a global biodiversity hotspot: the *Banksia* (Proteaceae) of Western Australia. *Global Change Biology* **14**: 1–16.

Flematti GR, Ghisalberti EL, Dixon KW and Trengove RD (2004) A compound from smoke that promotes seed germination. *Science* **305**: 977.

Folke C, Carpenter S, Walker B, Scheffer M, Elmqvist T, Gunderson L and Holling CS (2004) Regime shifts, resilience, and biodiversity in ecosystem management. *Annual Review of Ecology, Evolution, and Systematics* **35**: 557–581.

Foran B and Poldy F (2002) *Future dilemmas: options to 2050 for Australia's population, technology, resources and environment*. Working paper 02/01. CSIRO Sustainable Ecosystems, Canberra.

Ford HA (1979) Birds. In *Natural history of Kangaroo Island* (Eds MJ Tyler, CR Twidale and JK Ling) pp. 103–114. Royal Society of South Australia, Adelaide.

Forman RT, Sperling D, Bissonette JA, Clevenger AP, Cutshall CD, Dale VH, Fahrig L, France R, Goldman CR, Heanue K, Jones JA, Swanson FJ, Turrentine T and Winter TC (Eds) (2002) *Road ecology: science and solutions*. Island Press, Washington, DC.

Franklin JF and Forman RT (1987) Creating landscape patterns by forest cutting: ecological consequences and principles. *Landscape Ecology* **1**: 5–18.

Galloway RW (1988) The potential impact of climate changes on the Australian ski fields. In *Greenhouse, planning for climate change*. (Ed. GI Pearman) pp. 428–437. CSIRO Publishing, Melbourne.

Garnett ST and Brook BW (2007) Modelling to forestall extinction of Australian tropical birds. *Journal of Ornithology* **148**: 311–320.

Garnett ST and Crowley GM (2000) *The Action Plan for Australian Birds 2000*. Environment Australia, Canberra.

Gascon C, Lovejoy TE, Bierregaard RO Jr., Malcolm JR, Stouffer PC, Vasconcelos HL, Laurance WF, Zimmerman B, Tocher M and Borges SH (1999) Matrix habitat and species richness in tropical forest remnants. *Biological Conservation* **91**: 1–7.

Gedney N, Cox PM, Betts RA, Boucher O, Huntingford C and Stott PA (2006) Detection of a direct carbon dioxide effect in continental river runoff records. *Nature* **439**: 835–838.

German Advisory Council on Climate Change (2006) *The future oceans – warming up, rising high, turning sour*. WBGU, Berlin.

Ghannoum O, von Caemmerer S and Conroy JP (2001) Plant water use efficiency of 17 Australian NAD-ME and NADP-ME C_4 grasses at ambient and elevated CO_2 partial pressure. *Australian Journal of Plant Physiology* **28**: 1207–1217.

Gibbons P and Lindenmayer DB (2007) The use of offsets to regulate land clearing: no net loss or the tail wagging the dog? *Ecological Management and Restoration* **8**: 26–31.

Gienapp P, Teplitsky C, Alho JS, Mills JA and Merilä J (2008) Climate change and evolution: disentangling environmental and genetic responses. *Molecular Ecology* **17**: 167–178.

Gifford RM and Howden M (2001) Vegetation thickening in an ecological perspective: significance to national greenhouse gas inventories. *Environmental Science and Policy* **4**: 59–72.

Gill AM (1975) Fire and the Australian flora: a review. *Australian Forestry* **38**: 4–25.

Gilligan B (2006) *The National Reserve System Programme – 2006 evaluation*. Commonwealth of Australia, Canberra.

Gleason SM, Williams LJ, Read J, Metcalfe DJ and Baker PJ (2008) Cyclone effects on the structure and production of a tropical upland rainforest: implications for life-history tradeoffs. *Ecosystems* **11**: 1277–1290.

Glen AS and Dickman CR (2005) Complex interactions among mammalian carnivores in Australia, and their implications for wildlife management. *Biological Reviews* **80**: 387–401.

Global Amphibian Assessment (2007). Retrieved 3 August 2008, <http://www.globalamphibians.org/>.

Government of Western Australia (2002) *A Biodiversity Conservation Act for Western Australia. Consultation paper*, <http://www.dec.wa.gov.au/our-environment/biodiversity/proposed-biodiversity-conservation-act.html>.

Graham CH, Moritz C and Williams SE (2006) Habitat history improves prediction of biodiversity in rainforest fauna. *Proceedings of the National Academy of Sciences of the United States of America* **103**: 632–636.

Grantham BA, Eckert GL and Shanks AL (2003) Dispersal potential of marine invertebrates in diverse habitats. *Ecological Applications* **13**: 108–116.

Great Barrier Reef Marine Park Authority and Australian Greenhouse Office (2007a) *Climate change and the Great Barrier Reef: a vulnerability assessment*. (Eds JE Johnson and PA Marshall). Great Barrier Reef Marine Park Authority, Townsville.

Great Barrier Reef Marine Park Authority and Australian Greenhouse Office (2007b). *Great Barrier Reef climate change action plan 2007–2012*. Great Barrier Reef Marine Park Authority, Townsville.

Green K (1998) Introduction. In *Snow: a natural history; an uncertain future*. (Ed. K Green) pp. xiii-xix. Australian Alps Liaison Committee, Canberra.

Green K (2006) Effect of variation in snowpack on timing of bird migration in the Snowy Mountains of south-eastern Australia. *Emu* **106**: 187–192.

Green K and Osborne WS (1994) *Wildlife of the Australian snow-country: a comprehensive guide to alpine fauna*. Reed, Sydney.

Green K and Pickering CM (2002) A scenario for mammal and bird diversity in the Australian Snowy Mountains in relation to climate change. In *Mountain biodiversity: a global assessment*. (Eds C Koerner and EM Spehn) pp. 241–249. Parthenon Publishing, London.

Greenstein BJ and Pandolfi JM (2008) Escaping the heat: range shifts of reef coral taxa in coastal Western Australia. *Global Change Biology* **14**: 513–528.

Grey MJ, Clarke MF and Loyn RH (1998) Influence of the noisy miner *Manorina melanocephala* on avian diversity and abundance in remnant grey box woodland. *Pacific Conservation Biology* **4**: 55–69.

Griffin GF and Friedel MH (1985) Discontinuous change in central Australia: some implications of major ecological events for land management. *Journal of Arid Environments* **9**: 63–80.

Griffin GF, Morton SR, Stafford Smith DM, Allan GE, Masters KA and Preece N (1989) Status and implications of the invasion of tamarisk (*Tamarix aphylla*) on the Finke River, Northern Territory, Australia. *Journal of Environmental Management* **29**: 297–315.

Groffman PM, Baron JS, Blett T, Gold AJ, Goodman I, Gunderson LH, Levinson BM, Palmer MA, Paerl HW, Peterson GD, Leroy Poff N, Rejeski DW, Reynolds JF, Turner MG, Weathers KC and Wiens J (2006) Ecological thresholds: the key to successful environmental management or an important concept with no practical application? *Ecosystems* **9**: 1–13.

Groves RH, Hosking JR, Natianoff GN, Cooke DA, Cowie ID, Johnson RW, Keighery GJ, Lepschi BJ, Mitchell AA, Moerkerk M, Randall RP, Rozefelds AC, Walsh NG and Waterhouse BM (2003) *Weed categories for natural and agricultural ecosystem management*. Department of Agriculture, Fisheries and Forestry, Canberra.

Hale P and Lamb D (Eds) (1997) *Conservation outside nature reserves*. Centre for Conservation Biology, University of Queensland, Brisbane.

Halse SA, Burbidge AA, Lane JAK, Haberley B, Pearson GB and Clarke A (1995) Size of the Cape Barren Goose population in Western Australia. *Emu* **95**: 77–83.

Hannon SJ and Schmiegelow FKA (2002) Corridors may not improve the conservation value of small reserves for most boreal birds. *Ecological Applications* **12**: 1457–1468.

Hansen J, Sato M, Kharecha P, Beerling D, Berner R, Masson-Delmotte V, Pagani M, Raymo M, Royer DL and Zachos JC (2008) Target atmospheric CO_2: where should humanity aim? *Open Atmospheric Science Journal* **2**: 217–231. doi: 10.2174/1874282300802010217.

Harper MJ, McCarthy MA and van der Ree R (2005) The use of nest boxes in urban natural vegetation remnants by vertebrate fauna. *Wildlife Research* **32**: 509–516.

Harrington GN and Sanderson KD (1994) Recent contraction of wet sclerophyll forest in the wet tropics of Queensland due to invasion by rainforest. *Pacific Conservation Biology* **1**: 319–327.

Haynes RW, Bormann BT and Martin JR (Eds) (2006) *Northwest Forest Plan – The first 10 years (1993–2003): Synthesis of monitoring and research results*. General Technical Report PNW-GTR-651. USDA Forest Service, Pacific Northwest Research Station, Portland, Oregon.

Hays GC, Richardson AJ and Robinson C (2005) Climate change and marine plankton. *Trends in Ecology & Evolution* **20**: 337–344.

Heath L (2008) *The vulnerability of the world heritage values of Australia's world heritage properties to climate change impacts*. Report to Australian Greenhouse Office. The Australian National University, Canberra.

Heatwole H (1987) Major components and distribution of the terrestrial fauna. In *Fauna of Australia, Vol 1A: General Articles*. (Eds GR Dyne and DW Walton). AGPS, Canberra.

Hennessy K, Fawcett R, Kirono D, Mpelasoka F, Jones D, Bathols J, Whetton P, Stafford Smith M, Howden M, Mitchell C and Plummer N (2008) *An assessment of the impact of climate change on the nature and frequency of exceptional climatic events*. CSIRO and Australian Bureau of Meteorology.

Hennessy K, Fitzharris B, Bates BC, Harvey N, Howden SM, Hughes L, Salinger J and Warrick R (2007) Australia and New Zealand. In *Climate change 2007: impacts, adaptation and vulnerability. Contribution of Working Group II to the Fourth Assessment Report*. (Eds ML Parry, OF Canziani, JP Palutikof, PJH van der Linden and CE Hanson) pp 507–540. Cambridge University Press, Cambridge, UK.

Hennessy KJ, Lucas C, Nicholls N, Bathols JM, Suppiah R and Ricketts JH (2005) *Climate change impacts on fire-weather in south-east Australia, Report C/1061*. CSIRO Marine and Atmospheric Research, Bushfire CRC and Australian Bureau of Meteorology. CSIRO, Aspendale, Victoria.

Hennessy K, Whetton P, Smith I, Bathols J, Hutchinson M and Sharples J (2003) *The impact of climate change on snow conditions in mainland Australia*. CSIRO Atmospheric Research, Aspendale, Victoria.

Hewitt CL, Campbell ML, Thresher RE, Martin RB, Mays N, Ross DJ, Boyd S, Gomon MF, Lockett MM, O'Hara TD, Poore GCB, Storey MJ, Wilson RS, Cohen BF, Currie DR, McArthur MA, Keough MJ, Lewis JA and Watson JE (2004) Introduced and cryptogenic species in Port Phillip Bay, Victoria. *Australian Marine Biology* **144**: 183–202.

Hewitt G and Nicholls RA (2005) Genetic and evolutionary impacts of climate change. In *Climate change and biodiversity*. (Eds TE Lovejoy and L Hannah) pp. 176–192. Yale University Press, New Haven, CT.

Higgins SI, Nathan RC and Cain ML (2003) Are long-distance dispersal events in plants usually caused by nonstandard means of dispersal? *Ecology* **84**: 1945–1956.

Hilbert DW, Ostendorf B and Hopkins M (2001) Sensitivity of tropical forests to climate change in the humid tropics of north Queensland. *Austral Ecology* **26**: 590–603.

Hilbert DW, Graham AW and Parker TA (2001) Tall open forest and woodland habitats in the Wet Tropics: responses to climate and implications for the northern bettong (*Bettongia tropica*). *Tropical Forest Research Series* **1**: 1–46.

Hilbert DW, Graham A and Hopkins MS (2007a) Glacial and interglacial refugia within a long-term rainforest refugium: the Wet Tropics Bioregion of NE Queensland, Australia. *Palaeogeography, Palaeoclimatology, Palaeoecology* **251**: 104–118.

Hilbert DW, Hughes L, Johnson J, Lough JM, Low T, Pearson RG, Sutherst RW and Whittaker S (Eds) (2007b) *Biodiversity conservation research in a changing climate. Workshop report: research needs and information gaps for the implementation of key objectives of the National Biodiversity and Climate Change Action Plan.* A report produced by CSIRO for the Department of Environment and Heritage, Canberra.

Hobbs RJ and Cramer VA (2008) Restoration ecology: interventionist approaches for restoring and maintaining ecosystem function in the face of rapid environmental change. *Annual Review of Environment and Resources* **33**: 39–61.

Hobbs RJ and McIntyre S (2005) Categorizing Australian landscapes as an aid to assessing the generality of landscape management guidelines. *Global Ecology and Biogeography* **14**: 1–15.

Hobbs RJ and Suding KN (Eds) (2009) *New models for ecosystem dynamics and restoration.* Island Press, Washington, DC.

Hobbs RJ and Yates CJ (2003) Turner Review. No. 7. Impacts of ecosystem fragmentation on plant populations: generalising the idiosyncratic. *Australian Journal of Botany* **51**: 471–488.

Hobbs RJ, Arico S, Aronson J, Baron JS, Bridgewater P, Cramer VA, Epstein PR, Ewel JJ, Klink CA, Lugo AE, Norton D, Ojima D, Richardson DM, Sanderson EW, Valladares F, VilÃ M, Zamora R and Zobel M (2006) Novel ecosystems: theoretical and management aspects of the new ecological world order. *Global Ecology and Biogeography* **15**: 1–7.

Hobday AJ, Okey TA, Poloczanska ES, Kunz TJ and Richardson AJ (2006) *Impacts of climate change on Australian marine life - Part B: Technical Report.* Report to the Australian Greenhouse Office, Department of the Environment and Heritage. CSIRO Marine and Atmospheric Research, Canberra.

Hobday AJ, Poloczanska ES and Matear RJ (Eds) (2007) *Implications of climate change for Australian fisheries and aquaculture: a preliminary assessment.* Report to the Department of Climate Change, Canberra.

Hoegh-Guldberg O (1999) Climate change, coral bleaching and the future of the world's coral reefs. *Marine and Freshwater Research* **50**: 839–866.

Hoegh-Guldberg O, Hughes L, McIntyre S, Lindenmayer DB, Parmesan C, Possingham HP and Thomas CD (2008) Assisted colonization and rapid climate change. *Science* **321**: 345–346.

Hoegh-Guldberg O, Mumby PJ, Hooten AJ, Steneck RS, Greenfield P, Gomez E, Harvell CD, Sale PF, Edwards AJ, Caldeira K, Knowlton N, Eakin CM, Iglesias-Prieto R, Muthiga N, Bradbury RH, Dubi A and Hatziolos ME (2007) Coral reefs under rapid climate change and ocean acidification. *Science* **318**: 1737–1742.

Hoffman N and Brown A (1998) *Orchids of south-west Australia.* University of Western Australia Press, Nedlands.

Hoffmann AA and Weeks AR (2007) Climatic selection on genes and traits after a 100 year-old invasion: a critical look at the temperate-tropical clines in *Drosophila melanogaster* from eastern Australia. *Genetica* **129**: 133–147.

Hoffmann BD and O'Connor S (2004) Eradication of two exotic ants from Kakadu National Park. *Ecological Management and Restoration* **5**: 98–105.

Holling CS (Ed.) (1978) *Adaptive environmental assessment and management.* International Series on Applied Systems Analysis. John Wiley and Sons, Toronto.

Holmes J (1997) Diversity and change in Australia's rangeland regions: translating resource values into regional benefits. *The Rangeland Journal* **19**: 3–25.

Holmes J (2006) Impulses towards a multifunctional transition in rural Australia: gaps in the research agenda. *Journal of Rural Studies* **22**: 142–160.

Holmes J (2009) Comment: Agency in facilitating the transition to a multifunctional countryside. *Journal of Rural Studies* **25**: 248–249.

Homan RN, Windmiller BS and Reed JM (2004) Critical thresholds associated with habitat loss for two vernal pool-breeding amphibians. *Ecological Applications* **14**: 1547–1553.

Hopper S and Gioia P (2004) The southwest Australian floristic region: evolution and conservation of a global hot spot of biodiversity. *Annual Review of Ecology, Evolution, and Systematics* **35**: 623–650.

Hopper SD, Harvey MS, Chappill JA, Main AR and Main BY (1996) The Western Australian biota as Gondwanan heritage – a review. In *Gondwanan heritage*. (Eds SD Hopper, JA Chappill, MS Harvey and AS George) pp. 1–46. Surrey Beatty & Sons, Chipping Norton, New South Wales.

Horwitz P and Sommer B (2005) Water quality responses to fire, with particular reference to organic-rich wetlands and the Swan Coastal Plain: a review. *Journal of the Royal Society of Western Australia* **88**: 121–128.

Howden SM, Reyenga PJ and Gorman JT (1999) *Current evidence of global change and its impacts: Australian forests and other ecosystems*. Working paper 99/01. CSIRO Wildlife and Ecology, Canberra.

Hughes L (2000) Biological consequences of global warming: is the signal already apparent? *Trends in Ecology & Evolution* **15**: 56–61.

Hughes L (2003) Climate change and Australia: trends, projections and impacts. *Austral Ecology* **28**: 423–443.

Hughes L and Westoby M (1994) Climate change and conservation policies in Australia: coping with change that is far away and not yet certain. *Pacific Conservation Biology* **1**: 308–318.

Hughes L, Westoby M and Cawsey M (1996) Climatic range sizes of Eucalyptus species in relation to future climate change. *Global Ecology and Biogeography Letters* **5**: 23–29.

Humphries SE, Groves RH and Mitchell DS (1991) *Plant invasions of Australian ecosystems: a status review and management directions*. Kowari 2. Australian National Parks and Wildlife Service, Canberra.

Hunter M (2007) Climate change and moving species: furthering the debate on assisted colonization. *Conservation Biology* **21**: 1356–1358.

Intergovernmental Panel on Climate (IPCC) (2000) *Special report on emissions scenarios*. (Eds N Nakicenovic and R Swart). Cambridge University Press, Cambridge, UK.

Intergovernmental Panel on Climate (IPCC) (2001) *Climate change 2001: the scientific basis. Contribution of Working Group I to the Third Assessment Report of the Intergovernmental Panel on Climate Change*. (Eds J Houghton, Y Ding, DJ Griggs, M Noguer, PJ van der Winden and X Dai). Cambridge University Press, Cambridge, UK.

Intergovernmental Panel on Climate (IPCC) (2007a) *Climate change 2007: the physical science basis. Contribution of Working Group I to the Fourth Assessment Report of the Intergovernmental Panel on Climate Change*. (Eds S Solomon, D Quin, M Manning, M Marquis, K Averyt, MMB Tignor, HL Miller and Z Chen). Cambridge University Press, Cambridge, UK.

Intergovernmental Panel on Climate (IPCC) (2007b) *Climate change 2007: impacts, adaptation and vulnerability. Contribution of Working Group II to the Fourth Assessment Report of the Intergovernmental Panel on Climate Change*. (Eds M Parry, O Canziani, J Palutikof, P van der Linden and C Hanson). Cambridge University Press, Cambridge, UK.

Intergovernmental Panel on Climate (IPCC) (2007c) *Climate change 2007: mitigation of climate change. Summary for policymakers. Contribution of Working Group III to the Fourth Assessment Report of the Intergovernmental Panel on Climate Change*. (Eds B Metz, O Davidson, P Bosch, R Dave and L Meyer). Cambridge University Press, Cambridge, UK.

James CD, Landsberg J and Morton SR (1999) Provision of watering points in the Australian arid zone: a review of the effects on biota. *Journal of Arid Environments* **41**: 87–121.

James CD and Shine R (2000) Why are there so many coexisting species of lizards in Australian deserts? *Oecologia* **125**: 127–141.

Jarman PJ (1986) The red fox – an exotic, large predator. In *The ecology of exotic animals and plants: some Australian case histories*. (Ed. RL Kitching) pp. 43–61. Wiley, Brisbane.

Johns CV and Hughes L (2002) Interactive effects of elevated CO_2 and temperature on the leaf-miner *Dialectica scalariella* Zeller (Lepidoptera: Gracillariidae) in Paterson's Curse, *Echium plantagineum* (Boraginaceae). *Global Change Biology* **8**: 142–152.

Johnson B (1998) Consequences for the Macquarie Marshes. In *Climate change scenarios and managing the scarce water resources of the Macquarie River*. pp.

61–68. Hassall and Associates, and the Australian Greenhouse Office, Australia.

Johnson C (2007) *Australia's mammal extinctions: a 50,000 year history.* Cambridge University Press, Melbourne.

Johnson JE and Marshall PA (Eds) (2007) *Climate change and the Great Barrier Reef: a vulnerability assessment.* Great Barrier Reef Marine Park Authority and Australian Greenhouse Office, Australia.

Jones RN, Preston B, Brooke C, Aryal S, Benyon R, Blackmore J, Chiew F, Kirby M, Maheepala S, Oliver R, Polglase P, Prosser I, Walker G, Young B and Young M (2008) *Climate change and Australian water resources: a preliminary risk assessment.* Report to the Australian Greenhouse Office and the National Water Commission, Canberra, Australia.

Jones RE and Crome FH (1990) The biological web: plant–animal interactions in the rainforest. In *Australian tropical rainforests: science – values – meaning.* (Eds LJ Webb and J Kikkawa) pp. 74–87. CSIRO, Australia.

Jones RN and Pittock B (1997) Assessing the impacts of climate change: the challenge for ecology. In *Frontiers of ecology.* (Eds N Klomp and I Lunt) pp. 311–322. Elsiever Science, Oxford.

Jordon A (2000) The politics of multilevel environmental governance: subsidiarity and environmental policy in the European Union. *Environment and Planning A* **32**: 1307–1324.

Kanowski JJ (2001) Effects of elevated CO_2 on the foliar chemistry of seedlings of two rainforest trees from north-east Australia: implications for folivorous marsupials. *Austral Ecology* **26**: 165–172.

Kathiresan K and Rajendran N (2006) Coastal mangrove forests mitigated tsunami. *Estuarine, Coastal and Shelf Science* **65**: 601–606.

Keast A, Crocker RL, and Christian CS (Eds) (1959) *Biogeography and ecology in Australia.* Dr. W Junk, The Haage, Netherlands.

Keogh K, Chant D and Frazer B (2006) *Review of arrangements for regional delivery of natural resource management programmes.* Report by the Ministerial Reference Group for Future NRM Programme Delivery, Canberra.

King RW, Pate JS and Johnston J (1995) Ecotypic differences in the flowering of *Pimelea ferruginea* (Thymelaeaceae) in response to cool temperatures. *Australian Journal of Botany* **44**: 47–55.

Kingsford RT and Auld KM (2005) Waterbird breeding and environmental flow management in the Macquarie Marshes, arid Australia. *River Research and Applications* **21**: 187–200.

Kingsford RT and Norman FI (2002) Australian waterbirds – products of the continent's ecology. *Emu* **102**: 47–69.

Kinnear JE, Sumner NR and Onus ML (2002) The red fox in Australia – an exotic predator turned biocontrol agent. *Biological Conservation* **108**: 335–359.

Kirschbaum MUF (1999a) CenW, a forest growth model with linked carbon, energy, nutrient and water cycles. I. Model description. *Ecological Modelling* **118**: 17–59.

Kirschbaum MUF (Ed.) (1999b) The impacts of climate change on Australia's forests and forest industries. In *Impacts of global change on Australian temperate forests.* (Eds SM Howden and JT Gorman) pp. 56-61. Working Paper Series 99/08. CSIRO Wildlife and Ecology, Canberra.

Kleypas JA, McManus JW and Meñez LAB (1999) Environmental limits to coral reef development: where do we draw the line? *American Zoologist* **39**: 146–159.

Kothavala Z (1999) The duration and severity of drought over eastern Australia simulated by a coupled ocean–atmosphere GCM with a transient increase in CO_2. *Environmental Modelling and Software* **14**: 243–252.

Krebs C (2008) *The ecological world view.* CSIRO Publishing, Melbourne.

Kriticos DJ, Sutherst RW, Brown JR, Adkins SW and Maywald GF (2003a) Climate change and the potential distribution of an invasive alien plant: *Acacia nilotica* ssp. *indica* in Australia. *Australian Journal of Applied Ecology* **40**: 111–124.

Kriticos DJ, Sutherst RW, Brown JR, Adkins SW and Maywald GF (2003b) Climate change and biotic invasions: a case history of a tropical woody vine. *Biological Invasions* **5**: 147–165.

Kuleshov YA (2003) *Tropical cyclone climatology for the southern hemisphere. Part I: spatial and tempo-*

ral profiles of tropical cyclones in the southern hemisphere. National Climate Centre, Australian Bureau of Meteorology.

La Marca E, Lips KR, Lötters S, Puschendorf R, Ibáñez R, Rueda-Almonacid JV, Schulte R, Marty C, Castro F, Manzanilla-Puppo J, García-Pérez JE, Bolaños F, Chaves G, Pounds JA, Toral E and Young BE (2005) Catastrophic population declines and extinctions in neotropical harlequin frogs (Bufonidae: *Atelopus*). *Biotropica* **37**: 190–201.

Landsberg J, James CD, Morton SR, Hobbs TJ, Stol J, Drew A and Tongway H (1999) *The effects of artificial sources of water on rangeland biodiversity.* Final report to the biodiversity convention and strategy section of the biodiversity group, Environment Australia. Biodiversity Technical Paper No. 3, Environment Australia Biodiversity Group and CSIRO Division of Wildlife and Ecology, Canberra.

Landsberg J and Stafford Smith DM (1992) A functional scheme for predicting the outbreak potential of herbivorous insects under global atmospheric change. *Australian Journal of Botany* **40**: 565–577.

Landsberg J and Wylie FR (1983) Water stress, leaf nutrients and defoliation: a model of dieback of rural eucalypts. *Australian Journal of Ecology* **8**: 27–41.

Larcombe J and McLaughlin K (Eds) (2007) *Fishery status reports 2006: status of fish stocks managed by the Australian Government.* Bureau of Rural Sciences, Canberra.

Laurance WF (2008) Global warming and amphibian extinctions in eastern Australia. *Austral Ecology* **33**: 1–9.

Lehmann CER, Prior LD, Williams RJ and Bowman DMJS (2008) Spatio-temporal trends in tree cover of a tropical mesic savanna are driven by landscape disturbance. *Journal of Applied Ecology* **45**: 1304–1311.

Lenton TM, Held H, Kriegler E, Hall JW, Lucht W, Rahmstorf S and Schnellnhuber HJ (2008) Tipping elements in the Earth's climate system. *Proceedings of the National Academy of Sciences of the United States of America* **105**: 1786–1793.

Leslie LM, Karoly DJ, Leplastrier M and Buckley BW (2007) Variability of tropical cyclones over the southwest Pacific Ocean using a high-resolution climate model. *Meteorology and Atmospheric Physics* **97**: 171–180.

Levey DJ, Bolker BM, Tewksbury JJ, Sargent S and Haddad NM (2005) Effects of landscape corridors on seed dispersal by birds. *Science* **309**: 146–148.

Lindenmayer DB (2007a) *On borrowed time: Australia's environmental crisis and what we must do about it.* CSIRO Publishing and Penguin, Melbourne, <http://www.publish.csiro.au/pid/5691.htm>.

Lindenmayer DB (2007b) *The variable retention harvest system and its implications for biodiversity in the mountain ash forests of the central highlands of Victoria.* ANU Fenner School of Environment and Society Occasional Paper No. 2. Department of Primary Industries, Victoria.

Lindenmayer DB and Franklin JF (2002) *Conserving forest biodiversity: a comprehensive multiscaled approach.* Island Press, Washington, DC.

Lindenmayer DB and Burgman MA (2005) *Practical conservation biology.* CSIRO Publishing, Melbourne.

Lindenmayer DB and Fischer J (2006) *Habitat fragmentation and landscape change: an ecological and conservation synthesis.* Island Press, Washington, DC.

Lindenmayer DB and Fischer J (2007) Tackling the habitat fragmentation panchreston. *Trends in Ecology & Evolution* **22**: 127–132.

Lindenmayer DB and Hobbs RJ (Eds) (2007) *Managing and designing landscapes for conservation: moving from perspectives to principles.* Blackwell Publishing, Oxford.

Lindenmayer DB, Cunningham RB, Donnelly CF, Tanton MT and Nix HA (1993) The abundance and development of cavities in Eucalyptus trees: a case study in the montane forests of Victoria, southeastern Australia. *Forest Ecology and Management* **60**: 77–104.

Lindenmayer DB, Cunningham RB, MacGregor C and Incoll RD (2003) A survey design for monitoring the abundance of arboreal marsupials in the Central Highlands of Victoria. *Biological Conservation* **110**: 161–167.

Lindenmayer DB, Fischer J and Cunningham RB (2005) Native vegetation cover thresholds associated with species responses. *Biological Conservation* **124**: 311–316.

Lindenmayer DB, Franklin JF and Fischer J (2006) General management principles and a checklist of strategies to guide forest biodiversity conservation. *Biological Conservation* **131**: 433–445.

Lindenmayer DB, Fischer J, Felton A, Crane M, Michael D, Macgregor C, Montague-Drake R, Manning A and Hobbs RJ (2008a) Novel ecosystems resulting from landscape transformation create dilemmas for modern conservation practice. *Conservation Letters* **1**: 129–135.

Lindenmayer DB, Burton P and Franklin J (2008b) *Salvage logging and its ecological consequences.* Island Press, Washington, DC.

Lindgren E (2000) *The new environmental context for disease transmission, with case studies on climate change and tick-borne encephalitis.* Natural Resources Management, Department of Systems Ecology. Sweden, Stockholm University, Sweden.

Lindsay AM (1985) Are Australian soils different? *Proceedings of the Ecological Society of Australia* **14**: 83–97.

Ling SD, Johnson CR, Frusher S and King CK (2008) Reproductive potential of a marine ecosystem engineer at the edge of a newly expanded range. *Global Change Biology* **14**: 907–915.

Ling SD, Johnson CR, Ridgway K, Hobday AJ and Haddon M (2009) Climate-driven range extension of a sea urchin: inferring future trends by analysis of recent population dynamics. *Global Change Biology* **15**: 719–731.

Linnell JDC and Strand O (2000) Interference interactions, co-existence and conservation of mammalian carnivores. *Diversity and Distributions* **6**: 169–176.

Liverpool City Council and Ecological Australia (2003) *Biodiversity strategy.* <http://www.liverpool.nsw.gov.au/biodiversitystrategy.htm>.

Logan JA, Régnière J and Powell JA (2003) Assessing the impacts of global warming on forest pest dynamics. *Frontiers in Ecology and Environment* **1**: 130–137.

Lonsdale WM (1994) Inviting trouble: introduced pasture species in northern Australia. *Australian Journal of Ecology* **19**: 345–354.

Lonsdale WM, Miller IL and Forno IW (1989) The biology of Australian weeds. 20. *Mimosa pigra* L. *Plant Protection Quarterly* **4**: 119–131.

Lorius C, Jouzel J, Raynaud D, Hansen J and Le Treut H (1990) The ice-core record: climate sensitivity and future greenhouse warming. *Nature* **347**: 139–145.

Low T (1994) Invasion of the savage honeyeaters. *Australian Natural History* **24**: 27–34.

Low T (1999) *Feral future: the untold story of Australia's exotic invaders.* Penguin, Melbourne.

Low T (2002) *The new nature: winners and losers in wild Australia.* Penguin, Melbourne.

Low T (2007) *Climate change and Brisbane biodiversity.* Report for Brisbane City Council, Brisbane.

Low T (2008) *Climate change and invasive species: a review of interactions.* Department of the Environment and Water Resources, Canberra.

Lucas C, Hennessey K, Mills G and Bathols J (2007) *Bushfire weather in southeast Australia: recent trends and projected climate change impacts.* Consultancy report prepared for the Climate Institute of Australia.

Lucas R and Kirschbaum MUF (Eds) (1999) Australia's forest carbon sink: threats and opportunities from climate change. In *Impacts of global change on Australian temperate forests.* (Eds SM Howden and JT Gorman) pp. 118–124. Working Paper Series 99/08. CSIRO Wildlife and Ecology, Canberra.

Luck GW, Ricketts TH, Daily GC and Imhoff M (2004) Alleviating spatial conflict between people and biodiversity. *Proceedings of the National Academy of Sciences of the United States of America* **101**: 182–186.

MA (Millennium Ecosystem Assessment) (2005) *Ecosystems and human well-being: synthesis.* Island Press, Washington, DC.

Mackey B, Lindenmayer D, Gill M, McCarthy M and Lindesay J (2002) *Wildlife, fire and future climate: a forest ecosystem analysis.* CSIRO Publishing, Melbourne.

Mackey BG, Keith H, Berry S and Lindenmayer DB (2008) *Green carbon: the role of natural forests in carbon storage.* The Australian National University, EPress, <http://epress.anu.edu.au/green_carbon_citation.html>.

Mann ME, Bradley RS and Hughes MK (1999) Northern hemisphere temperatures during the past millennium: inferences, uncertainties, and limitations. *Geophysical Research Letters* **26**: 759–762.

Mann ME, Ammann CM, Bradley RS, Briffa KR, Crowley TJ, Hughes MK, Jones PD, Oppenheimer M, Osborn TJ, Overpeck JT, Rutherford S, Trenberth KE and Wigley TML (2003) On past temperatures and anomalous late-20th century warmth. *Eos* **84**: 256–258.

Manning AD, Fischer J, Felton A, Newell B, Steffen W and Lindenmayer DB (2009) Landscape fluidity – a unifying perspective for understanding and adapting to global change. *Journal of Biogeography* **36**: 193–199.

Manning AD, Fischer J and Lindenmayer DB (2006) Scattered trees are keystone structures. *Biological Conservation* **132**: 311–321.

Mansergh I and Cheal D (2007) Protected area planning and management for eastern Australian temperate forests and woodland ecosystems under climate change – a landscape approach. In *Protected areas: buffering nature against climate change*. (Eds M Taylor and P Figgis) pp. 58–72. Proceedings of a WWF and IUCN World Commission on Protected Areas Symposium, 18–19 June, Canberra. WWF-Australia, Sydney.

Mansergh I, Cheal D and Fitzsimons JA (2008) Future landscapes in south-eastern Australia: the role of protected areas and biolinks in adaptation to climate change. *Biodiversity* **9**: 59–70.

Mansergh I, Lau A and Anderson R (2007) Adaptation to climate change in Victorian "agricultural" landscapes – nature conservation. In *Place and purpose* – conference, 30–31 May 2007, Bendigo. Department of Primary Industries, Melbourne.

Margules CR and Pressey RL (2000) Systematic conservation planning. *Nature* **405**: 243–253.

Marine Biodiversity Decline Working Group (2008) *A national approach to addressing marine biodiversity decline*. Natural Resource Management Ministerial Council, Canberra.

Marsh LM (1993) The occurrence and growth of *Acropora* in extra-tropical waters off Perth, Western Australia. In *Proceedings of the 7th International Coral Reef Symposium, Guam 2*. (Ed. RH Richmond), pp. 1233–1238. University of Guam Press, UOG Station, Guam.

Marshall GR (2005) *Economics for collaborative environmental management: renegotiating the commons*. Earthscan, London.

Marshall GR (2008) Nesting, subsidiarity, and community-based environmental governance beyond the local level. *International Journal of the Commons* **2**: 75–97.

Martin P (1991) Environmental care in agricultural catchments: toward the communicative catchment. *Environmental Management* **15**: 773–783.

Mason R (2005) Potential impacts of climate change on threatened plants in NSW. Honours thesis, Macquarie University, Sydney.

McAllan I, Cooper D and Curtis B (2007) Changes in ranges: an historical perspective. Supplement to *Wingspan* **17**: 12–13.

McAlpine CA, Syktus J, Deo RC, Lawrence PJ, McGowan HA, Watterson IG and Phinn SR (2007) Modeling the impact of historical land cover change on Australia's regional climate. *Geophysical Research Letters* **34**: L22711.1–L22711.6.

McCarty JP (2001) Ecological consequences of recent climate change. *Conservation Biology* **15**: 320–331.

McDougall KL, Morgan JW, Walsh NG and Williams RJ (2005) Plant invasions in treeless vegetation of the Australian Alps. *Perspectives in Plant Ecology, Evolution and Systematics* **7**: 159–171.

McIntyre S, McIvor JC and MacLeod ND (2000) Principles for sustainable grazing in eucalypt woodlands: landscape-scale indicators and the search for thresholds. In *Management for sustainable ecosystems*. (Eds P Hale, A Petrie, D Moloney and P Sattler) pp. 92–100. University of Queensland, Brisbane.

McKean MA (2002) *Nesting institutions for complex common-pool resource systems*. In *Landscape futures II. Social and institutional dimensions. Proceedings of the 2nd International Symposium on Landscape Futures*. 4–6 December 2001, Armidale. (Eds J Graham, I Reeve and D Brunkhorst). Institute for Rural Futures, University of New England, Armidale.

McKellar RJ, Abbott I, Coates D, Keighery G, McKenzie NL, Williams M and Yates CJ (2007) *A review of climate change-biodiversity modelling, done in 1999 by Dr Odile Pouliquan-Young and Professor Peter Newman.* Department of Environment and Conservation, Perth.

McKenzie NL, Burbidge AA, Baynes A, Brereton RN, Dickman CR, Gordon G, Gibson LA, Menkhorst PW, Robinson AC, Williams MR and Woinarski JCZ (2007) Analysis of factors implicated in the recent decline of Australia's mammal fauna. *Journal of Biogeography* **34**: 597–611.

McKergow LA, Prosser IP, Hughes AO and Brodie J (2004) Regional scale nutrient modelling: exports to the Great Barrier Reef World Heritage Area. *Marine Pollution Bulletin* **51**: 186–199.

McKinney ML (1997) Extinction vulnerability and selectivity: combining ecological and paleontological views. *Annual Reviews of Ecology and Systematics* **28**: 495–516.

McKinnon GE, Jordan GJ, Vaillancourt RE, Steane DA and Potts BM (2004) Glacial refugia and reticulate evolution: the case of the Tasmanian eucalypts. *Philosophical Transactions of the Royal Society of London B Biological Sciences* **359**: 275–284.

Meynecke JO (2004) Effects of global climate change on geographic distributions of vertebrates in North Queensland. *Ecological Modelling* **174**: 347–357.

Michael DR and Lindenmayer DB (2008) Records of the inland carpet python, *Morelia spilota metcalfei* (Serpentes: Pythoniade) from the south-western slopes of New South Wales. *Proceedings of the Linnean Society of New South Wales* **129**: 253–261.

Millar AJK (2003a) *Vanvoorstia bennettiana.* In *2006 IUCN Red List of Threatened Species.* IUCN, viewed 12 January 2007, <http://www.iucnredlist.org>.

Millar AJK (2003b) The world's first recorded extinction of a seaweed. In *Proceedings of the XVIIth International Seaweed Symposium.* (Eds ARO Chapman, RJ Anderson, VJ Vreeland and IR Davison) pp. 313–318, Cape Town. Oxford University Press, New York.

Mittermeier R, Gil PR, Hoffman M, Pilgrim J, Brooks T, Mittermeier CG, Lamoreaux J and da Fonseca GAB (2005) *Hotspots revisited: Earth's biologically richest and most endangered terrestrial ecoregions.* Conservation International, Arlington, VA.

Mittermeier RA, Myers N, Thomsen JG, da Fonseca GA and Oliveri S (1998) Biodiversity hotspots and major tropical wilderness areas: approaches to setting conservation priorities. *Conservation Biology* **12**: 516–520.

Mittermeier RA, Myers N and Mittermeier CG (Eds) (1999) *Hotspots: Earth's biologically richest and most endangered terrestrial ecoregions*, Cemex and Conservation International, Mexico City.

Molony S and Vanderwoude C (2008) Red imported fire ants: a threat to eastern Australian wildlife? *Ecological Management & Restoration* **3**: 167–175.

Monteith GB (1985) Altitudinal transect studies at Cape Tribulation, north Queensland. VII. Coleoptera and Hemiptera (Insecta). *Queensland Naturalist* **26**: 70–78.

Monteith GB (1995) *Distribution and altitudinal zonation of low vagility insects of the Queensland wet tropics* (Part 4, p. 120). Queensland Museum, Brisbane.

Monteith GB and Davies VT (1991) Preliminary account of a survey of arthropods (insects and spiders) along an altitudinal transect in tropical Queensland. In *The rainforest legacy* (Eds G Werren and P Kershaw) vol. 2, pp. 345–362. Australian Government Publishing Service, Canberra.

Monte-Luna P del, Lluch-Belda D, Serviere-Zaragoza E, Carmona R, Reyes-Bonilla H, Aurioles-Gamboa D, Castro-Aguirre JL, Guzmán del Próo SA, Trujillo-Millán O and Brook BW (2007) Marine extinctions revisited. *Fish and Fisheries* **8**: 107–122.

Mooney H (2007) The costs of losing and restoring ecosystem services. In *Managing and designing landscapes for conservation.* (Eds DB Lindenmayer and RJ Hobbs) pp. 365–375. Blackwell Press, Oxford.

Moritz C, Patton JL, Schneider CJ and Smith TB (2000) Diversification of rainforest faunas: an integrated molecular approach. *Annual Review of Ecology and Systematics* **31**: 533–563.

Morton SR, Hoegh-Guldberg O, Lindenmayer DB, Olson MH, Hughes L, McCulloch MT, McIntyre S, Nix HA, Prober SM, Saunders DA, Andersen AN, Burgman MA, Lefroy EC, Lonsdale WM, Lowe I, McMichael AJ, Parslow JS, Steffen W, Williams JE and Woinarski JCZ (2009) The big ecological questions inhibiting effective environmental management in Australia. *Austral Ecology* **34**: 1–9.

Morton SR and James CD (1988) The diversity and abundance of lizards in arid Australia: a new hypothesis. *American Naturalist* **132**: 237–256.

Mudd GM (2008) Mining. In *Ten commitments: reshaping the lucky country's environment.* (Eds DB Lindenmayer, S Dovers, MH Olsen and S Morton) pp. 113–118. CSIRO Publishing, Melbourne.

Munday PL, Leis JM, Lough JM, Paris CB, Kingsford MJ, Berumen ML and Lambrechts J (2009) Climate change and coral reef connectivity. *Coral Reefs* **28**. doi: 10.1007/s00338-008-0461-9.

Murray C and Marmorek DR (2004) Adaptive management: a spoonful of rigour helps the uncertainty go down. In *16th International Annual Meeting of the Society for Ecological Restoration.* 23–27 August, Victoria, British Columbia, Canada.

Murray-Darling Basin Commission (2003) *Native Fish Strategy for the Murray-Darling Basin 2003–2013.* Murray-Darling Basin Commission, Canberra.

Musgrave W and Kingma O (2003) Economic policy and sustainable use of natural resources. In *Managing Australia's environment.* (Eds S Dovers and S Wild River) chap. 21. The Federation Press, Annandale, Sydney.

Myers N and Kent J (2001) *Perverse subsidies: how tax dollars can undercut the environment and the economy.* Island Press, Washington, DC.

Myers N, Mittermeier RA, Mittermeier CG, da Fonseca GAB and Kent J (2000) Biodiversity hotspots for conservation priorities. *Nature* **403**: 853–858.

National Biodiversity Strategy Review Task Group (2009) *Australia's Biodiversity Conservation Strategy 2010–2020.* Department of the Environment, Water, Heritage and the Arts, Canberra.

National Tidal Centre (2006) The Australian Baseline Sea level Monitoring Project Monthly Data Report August 2006. Bureau of Meteorology, Kent Town, South Australia, <http://www.environment.gov.au/soe/2006/publications/drs/pubs/366/co/co_03_aust_mean_sea_level_survey_2003.pdf>.

Newsome AE (2001) The biology and ecology of the dingo. In *A symposium on the dingo.* (Eds CR Dickman and D Lunney) pp. 20–33. Royal Zoological Society, Mosman, New South Wales.

Nicholls N (2004) The changing nature of Australian droughts. *Climatic Change* **63**: 323–336.

Nicholls N (2005) Climate variability, climate change and the Australian snow season. *Australian Meteorological Magazine* **54**: 177–185.

Nicholls N (2006) Detecting and attributing Australian climate change: a review. *Australian Meteorological Magazine* **55**: 199–211.

Nicholls N (2008) *Australian climate and weather extremes: past, present and future.* Department of Climate Change, Canberra.

Nicholls N, Lavery B, Frederiksen C, Drosdowsky W and Torok S (1996) Recent apparent changes in relationships between the El Niño–Southern Oscillation and Australian rainfall and temperature. *Geophysical Research Letters* **23**: 3357–3360.

Noble IR and Slatyer RO (1980) The use of vital attributes to predict successional changes in plant-communities subject to recurrent disturbances. *Vegetatio* **43**: 5–21.

Norgaard RB (2004) Learning and knowing collectively. *Ecological Economics* **49**: 231–241.

Norgaard RB and Baer P (2005) Collectively seeing complex systems: the nature of the problem. *Bioscience* **55**: 953–960.

Notaro M, Vavrus S and Liu ZY (2007) Global vegetation and climate change due to future increases in CO_2 as projected by a fully coupled model with dynamic vegetation. *Journal of Climate* **20**: 70–90.

Oakerson RJ (1999) *Governing local public economies: creating the civic metropolis.* ICS Press, Oakland, CA.

O'Connor MI, Bruno JF, Gaines SD, Halpern BS, Lester SE, Kinlan BP and Weiss JM (2007) Temperature control of larval dispersal and the implications for marine ecology, evolution, and conservation. *Proceedings of the National Academy of Sciences of the United States of America* **104**: 1266–1271.

O'Dowd DJ and Lake PS (1989) Red crabs in rainforest, Christmas Island: removal and relocation of leaf fall. *Journal of Tropical Ecology* **5**: 337–348.

O'Dowd DJ and Lake PS (1990) Red crabs in rainforest, Christmas Island: differential herbivory of seedlings. *Oikos* **58**: 289–292.

O'Dwyer TW, Buttemer WA and Priddel DM (2000) Inadvertent translocation of amphibians in the shipment of agricultural produce into New South

Wales: its extent and conservation implications. *Pacific Conservation Biology* **6**: 40–45.

Olsen P (2007) The state of Australia's birds 2007: birds in a changing climate. Supplement to *Wingspan* **14**: 1–32.

Olsen P, Weston MA, Cunningham R and Silcocks A (2003) The state of Australia's birds 2003. Supplement to *Wingspan* **13**.

Opdam P, Pouwels R, van Rooij S, Steingröver E and Vos CC (2008) Setting biodiversity targets in participatory regional planning: introducing ecoprofiles. *Ecology and Society* **13**: 20.

Opdam P, Steingröver E and van Rooij S (2006) Ecological networks: a spatial concept for multi-actor planning of sustainable landscapes. *Landscape and Urban Planning* **75**: 322–332.

Opdam P and Wascher D (2004) Climate change meets habitat fragmentation: linking landscape and biogeographical scale levels in research and conservation. *Biological Conservation* **117**: 285–297.

Oppenheimer M, O'Neill BC, Webster M and Agrawala S (2007) Climate change: the limits of consensus. *Science* **317**: 1505–1506.

Orell P and Morris KD (1994) *Chuditch recovery plan 1992–2001. Western Australian Wildlife Management Program No. 13.* Department of Conservation and Land Management, Como, Western Australia.

Orians GH and Milewski AV (2007) Ecology of Australia: the effects of nutrient-poor soils and intense fires. *Biological Reviews* **82**: 393–423.

Osborne WS, Davis MS and Green K (1998) Temporal and spatial variation in snow cover. In *Snow: a natural history, an uncertain future.* (Ed. K Green) pp. 56–68. Australian Alps Liaison Committee, Surrey Beattie and Sons, Sydney.

Ostrom E (1990) *Governing the commons: the evolution of institutions for collective action.* Cambridge University Press, Cambridge.

Ostrom E (2005) *Understanding institutional diversity.* Princeton University Press, Princeton.

Ostrom E, Burger J, Field CB, Norgaard RB and Policansky D (1999) Revisiting the commons: local lessons, global challenges. *Science* **284**: 278–282.

Ostrom V, Tiebout CM and Warren R (1999) [1961]. The organization of government in metropolitan areas: a theoretical inquiry. In *Polycentricity and local public economies: readings from the Workshop in Political Theory and Policy Analysis.* (Ed. MD McGinnis). University of Michigan Press, Ann Arbor, MI.

Pardon LG, Brook BW, Griffiths AD and Braithwaite RW (2003) Determinants of survival for the northern brown bandicoot under a landscape-scale fire experiment. *Journal of Animal Ecology* **72**: 106–115.

Parker M and MacNally R (2002) Habitat loss and the habitat fragmentation threshold: an experimental evaluation of impacts on richness and total abundances using grassland invertebrates. *Biological Conservation* **105**: 217–229.

Parmesan C (2006) Ecological and evolutionary responses to recent climate change. *Annual Review of Ecology, Evolution, and Systematics* **37**: 637–669.

Parmesan C and Yohe G (2003) A globally coherent fingerprint of climate change impacts across natural systems. *Nature* **421**: 37–42.

Parmesan C, Root TL and Willig MR (2000) Impacts of extreme weather and climate on terrestrial biota. *Bulletin of the American Meteorological Society* **81**: 443–450.

Parry ML, Canziani OF, Palutikof JP and PJH van der Linden PJH (Eds) (2007) *Climate change 2007: impacts, adaptation and vulnerability. Contribution of Working Group II to the Fourth Assessment Report.* Cambridge University Press, Cambridge, UK.

Peterson AT, Tian H, Martínez-Meyer E, Soberón J, Sánchez-Cordero V and Huntley B (2006) Modeling distributional shifts of individual species and biomes. In *Climate change and biodiversity.* (Eds TE Lovejoy and L Hannah) pp. 211–228. Yale University Press, New Haven, CT.

Peterson CH and Bishop MJ (2005) Assessing the environmental impacts of beach nourishment. *Bioscience* **55**: 887–896.

Petit JR, Jouzel J, Raynaud D, Barkov NI, Barnola J-M, Basile I, Bender M, Chappellaz J, Davis M, Delaygue G, Delmotte M, Kotlyakov VM, Legrand M, Lipenkov VY, Lorius C, Pépin L, Ritz C, Saltzman E and Stievenard M (1999) Climate and atmospheric history of the past 420,000 years from the Vostok ice core, Antarctica. *Nature* **399**: 429–436.

Petty AM, Werner PA, Lehmann CER, Riley JE, Banfai DS and Elliott LP (2007) Savanna responses to feral buffalo in Kakadu National Park, Australia. *Ecological Monographs* **77**: 441–463.

Phillips BL, Brown GP and Shine R (2003) Assessing the potential impact of cane toads on Australian snakes. *Conservation Biology* **17**: 1738–1747.

Phillips BL and Shine R (2004) Adapting to an invasive species: toxic cane toads induce morphological change in Australian snakes. *Proceedings of the National Academy of Sciences of the United States of America* **101**: 17150–17155.

Phillips OL, Baker TR, Arroyo L, Higuchi N, Killeen TJ, Laurance WF, Lewis SL, Lloyd J, Malhi Y, Monteagudo A, Neill DA, Núñez Vargas P, Silva JNM, Terborgh J, Vásquez Martínez R, Alexiades M, Almeida S, Brown S, Chave J, Comiskey JA, Czimczik CI, Di Fiore A, Erwin T, Kuebler C, Laurance SG, Nascimento HEM, Olivier J, Palacios W, Patiño S, Pitman NCA, Quesada CA, Saldias M, Torres Lezama A and Vinceti B (2004) Pattern and process in Amazon tree turnover, 1976–2001. *Philosophical Transactions of the Royal Society of London Series B-Biological Sciences* **359**: 381–407.

Phillips OL, Martinez RV, Arroyo L, Baker TR, Killeen T, Lewis SL, Malhi Y, Mendoza AM, Neill D, Vargas PN, Alexiades M, Ceroon C, Di Flore A, Erwin T, Jardim A, Palacios W, Saldlas M and Vinceti B (2002) Increasing dominance of large lianas in Amazonian forests. *Nature* **418**: 770–774

Pickering C, Good R and Green K (2004) *Potential effects of global warming on the biota of the Australian Alps*. Australian Greenhouse Office, Canberra.

Pitman AJ, Narisma GT and McAneney J (2007) The impact of climate change on the risk of forest and grassland fires in Australia. *Climatic Change* **84**: 383–401.

Pitman AJ, Narisma GT, Pielke RA Sr. and Holbrook NJ (2004) Impact of land cover change on the climate of southwest Western Australia. *Journal of Geophysical Research Atmospheres* **109**: D18109.

Pitt N (2008) *Climate-driven changes in Tasmanian intertidal fauna: 1950s to 2000s*. School of Zoology, University of Tasmania, Hobart.

Pittock AB (2003) *Climate change: an Australian guide to the science and potential impacts*. Australian Greenhouse Office, Canberra.

Pittock AB (2005) *Climate change: turning up the heat*. CSIRO Publishing, Melbourne.

Plummer R (2006) Managing for resource sustainability: the potential of civic science. *Environmental Sciences* **3**: 5–13.

Poloczanska ES, Babcock RC, Bulter A, Hobday AJ, Hoegh-Guldberg O, Kunz TJ, Matear R, Milton DA, Okey TA and Richardson AJ (2007) Climate change and Australian marine life. *Oceanography and Marine Biology: An Annual Review* **45**: 407–478.

Poloczanska ES, Hobday AJ and Richardson AJ (2008) In hot water: preparing for climate change in Australia's coastal and marine systems. Proceedings of conference held in Brisbane, 12–14 November 2007. CSIRO Marine and Atmospheric Research, Hobart.

Pounds JA, Bustamante MR, Coloma LA, Consuegra JA, Fogden MPL, Foster PN, La Marca E, Masters KL, Merino-Viteri A, Puschendorf R, Ron SR, Sánchez-Azofeifa GA, Still CJ and Young BE (2006) Widespread amphibian extinctions from epidemic disease driven by global warming. *Nature* **439**: 161–167.

Power S, Haylock M, Colman R and Wang XD (2006) The predictability of interdecadal changes in ENSO activity and ENSO teleconnections. *Journal of Climate* **19**: 4755–4771.

Pratchett MS, Wilson SK and Baird AH (2006) Long-term monitoring of the Great Barrier Reef. *Journal of Fish Biology* **69**: 1269–1280.

Pressey RL (1995) Conservation reserves in NSW: Crown jewels or leftovers? *Search* **26**: 47–51.

Preston BL and Jones RN (2006) *Climate change impacts on Australia and the benefits of early action to reduce global greenhouse gas emissions: a consultancy report for the Australian Business Roundtable on Climate Change*. CSIRO, Canberra.

Primack R (2001) Causes of extinction. In *Encyclopedia of biodiversity, Vol. 2*. (Eds S Levin *et al.*) pp. 697–714. Academic Press, San Diego, CA.

Productivity Commission (2004) *Impacts of native vegetation and biodiversity regulations*. Report No. 29. Productivity Commission, Melbourne.

Pyke G (1990) Apiarists vs scientists: a bittersweet case. *Australian Natural History* **23**: 386–392.

Radford JQ, Bennett AF and Cheers GJ (2005) Landscape-level thresholds of habitat cover for woodland-dependent birds: an introduction to ecological thresholds. *Biological Conservation* **124**: 317–337.

Rahmstorf S (2007) A semi-empirical approach to projecting future sea-level rise. *Science* **315**: 368–370.

Rahmstorf S, Cazenave A, Church JA, Hansen JE, Keeling RF, Parker DE and Somerville RCJ (2007) Recent climate observations compared to projections. *Science* **316**: 709.

Raupach MR, Marland G, Ciais P, Le Quéré C, Canadell JG and Field CB (2007) Global and regional drivers of accelerating CO_2 emissions. *Proceedings of the National Academy of Sciences of the United States of America* **104**: 10288–10293

Rees WE (1992) Ecological footprints and appropriated carrying capacity: what urban economics leaves out. *Environment and Urbanization* **4**: 121–130.

Reid PC, Johns DG, Edwards M, Starr M, Poulin M and Snoeijs P (2007) A biological consequence of reducing Arctic ice cover: arrival of the Pacific diatom *Neodenticula seminae* in the North Atlantic for the first time in 800 000 years. *Global Change Biology* **13**: 1910–1921.

Ribot JC (2004) *Waiting for democracy: the politics of choice in natural resource decentralization.* World Resources Institute, Washington, DC.

Ridgway KR (2007) Long-term trend and decadal variability of the southward penetration of the East Australian Current. *Geophysical Research Letters* **34**: L13613.1–L13613.5.

Ringold PL, Alegria J, Czaplewksi RL, Mulder BS, Tolle T and Burnett K (1996) Adaptive monitoring design for ecosystem management. *Ecological Applications* **6**: 745–747.

Ritchie EG, Martin JK, Krockenberger AK, Garnett S and Johnson CN (2008) Large-herbivore distribution and abundance: intra- and interspecific niche variation in the tropics. *Ecological Monographs* **78**: 105–122.

Rolls EC (1969) *They all ran wild.* Angus & Robertson, Sydney.

Root TL, Price JT, Hall KR, Schneider SH, Rosenzweig C and Pounds JA (2003) Fingerprints of global warming on wild animals and plants. *Nature* **421**: 57–60.

Root TL, MacMynowski DP, Mastrandrea MD and Schneider SH (2005) Human modified temperatures induce species changes: joint attribution. *Proceedings of the National Academy of Sciences of the United States of America* **102**: 7465–7469.

Rose DB (Ed.) (1995) *Country in flames. Proceedings of the 1994 Symposium on Biodiversity and Fires in North Australia.* Biodiversity Unit, Department of the Environment, Sports and Territories, Canberra.

Roshier DA, Whetton PH, Allan RJ and Robertson AI (2001) Distribution and persistence of temporary wetland habitats in arid Australia in relation to climate. *Austral Ecology* **26**: 371–384.

Rossiter NA, Setterfield SA, Douglas MM and Hutley LB (2003) Testing the grass-fire cycle: alien grass invasion in the tropical savannas of northern Australia. *Diversity and Distributions* **9**: 169–176.

Rouse WR, Douglas MSV, Hecky RE, Hershey AE, Kling GW, Lesack L, Marsh P, McDonald M, Nicholson BJ, Roulet NT and Smol JP (1997) Effects of climate change on the freshwaters of arctic and subarctic North America. *Hydrological Processes* **11**: 873–902.

Sattler PS and Taylor MFJ (2008) *Building nature's safety net 2008. Progress on the directions for the national reserve system.* WWF-Australia, Sydney.

Saunders DA, Smith GT, Ingram JA and Forrester RI (2003) Changes in a remnant of salmon gum *Eucalyptus salmonophloia* and York gum *E. loxophleba* woodland, 1978 to 1997. Implications for woodland conservation in the wheat-sheep regions of Australia. *Biological Conservation* **110**: 245–256.

Schneider C and C Moritz (1999) Rainforest refugia and evolution in Australia's wet tropics. *Proceedings of the Royal Society of London Series B* **266**: 191–196.

Schneider CJ and Williams SE (2005) Effects of quaternary climate change and rainforest diversity: insights from spatial analyses of species and genes in Australia's Wet Tropics. In *Tropical rainforests: past, present & future.* (Eds E Bermingham, CW Dick and C Moritz) pp. 401–424. Chicago University Press, Chicago.

Schreider SY, Jakeman AJ, Whetton PH and Pittock AB (1997) Estimation of climate impact on water availability and extreme events for snow-free and snow-affected catchments of the Murray–Darling Basin. *Australian Journal of Water Resources* **2**: 35–46.

Seastedt TR, Hobbs RJ and Suding KN (2008) Management of novel ecosystems: are novel approaches required? *Frontiers in Ecology and the Environment* **6**: 547–553.

Sekercioglu CH, Schneider SH, Fay JP and Loarie SR (2008) Climate change, elevational range shifts, and bird extinctions. *Conservation Biology* **22**: 140–150.

Semeniuk V (1994) Predicting the effect of sea-level rise on mangroves in northwestern Australia. *Journal of Coastal Research* **10**: 1050–1076.

Severinghaus JP, Sowers T, Brook EJ, Alley RB and Bender ML (1998) Timing of abrupt climate change at the end of the Younger Dryas interval from thermally fractionated gases in polar ice. *Nature* **391**: 141–146.

Seymour E, Pannell D, Ridley A, Marsh S and Wilkinson R (2007) *Capacity needs for technical analysis and decision making within Australian catchment management organisations.* CVCB Project No. UWA-92A, <http://cyllene.uwa.edu.au/~dpannell/sif30702.pdf>.

Shanks AL, Grantham BA and Carr MH (2003) Propagule dispersal distance and the size and spacing of marine reserves. *Ecological Applications* **13**: S159–S169.

Sharp BR and Bowman DMJS (2004) Patterns of long-term woody vegetation change in a sandstone-plateau savanna woodland, Northern Territory, Australia. *Journal of Tropical Ecology* **20**: 259–270.

Sharp BR and Whittaker RJ (2003) The irreversible cattle-driven transformation of a seasonally flooded Australian savanna. *Journal of Biogeography* **30**: 783–802.

Shearer BL, Crane CE and Cochrane A (2004) Quantification of the susceptibility of the native flora of the South-West Botanical Province, Western Australia to *Phytophthora cinnamomi. Australian Journal of Botany* 52: 435–443.

Shoo LP, Williams SE and Hero J-M (2005a) Potential decoupling of trends in distribution area and population size of species with climate change. *Global Change Biology* **11**: 1469–1476.

Shoo LP, Williams SE and Hero J-M (2005b) Climate warming and the rainforest birds of the Australian Wet Tropics: using abundance data as a sensitive predictor of change in total population size. *Biological Conservation* **125**: 335–343.

Short J and Smith A (1994) Mammal decline and recovery in Australia. *Journal of Mammalogy* **75**: 288–297.

Skeat AJ, East TJ and Corbett LK (1996) Impact of feral water buffalo. In *Landscape and vegetation ecology of the Kakadu region, northern Australia.* (Eds CM Finlayson and I von Oertzen) pp. 155–177. Kluwer Academic Publishers, Dordrecht.

Smajgl A, Leitch A *et al.* (2009 forthcoming) *Outback institutions: an application of the Institutional Analysis and Development (IAD) framework to four case studies in Australia's outback.* Report 31. Desert Knowledge CRC, Alice Springs.

Smit B, Burton I, Klein RJT and Wandel J (2000) An anatomy of adaptation to climate change and variability. *Climatic Change* **45**: 223–251.

Smith AP and Quin DG (1996) Patterns and causes of extinction and decline in Australian conilurine rodents. *Biological Conservation* **77**: 243–267.

Smithers BV, Peck DR, Krockenberger AK and Congdon BC (2003) Elevated sea-surface temperature, reduced provisioning and reproductive failure of wedge-tailed shearwaters (*Puffinus pacificus*) in the southern Great Barrier Reef, Australia. *Marine and Freshwater Research* **54**: 973–977.

Specht RL and Specht A (1999) *Australian plant communities: dynamics of structure, growth and biodiversity.* Oxford University Press, Melbourne.

Stafford Smith DM and Pickup G (1990) Patterns and production in arid lands. *Proceedings of the Ecological Society of Australia* **16**: 195–200.

Stafford Smith M and McAllister RRJ (2008) Managing arid zone natural resources in Australia for spatial and temporal variability – an approach from first principles. *The Rangeland Journal* **30**: 15–27.

Stankowski S, Byrne M, *et al.* (in review) Limited pollen mediated gene flow between fragmented populations of *Calothamnus* sp. Whicher, a narrow endemic bird-pollinated shrub. *Conservation Genetics.*

Steffen W, Sanderson A, Tyson PD, Jäger J, Matson P, Moore III B, Oldfield F, Richardson K, Schellnhuber HJ, Turner II BL and Wasson RJ (2004) *Global change and the Earth system: a planet under pressure*. IGBP Book Series, Springer-Verlag, Berlin Heidelberg.

Sunnucks P, Blacket MJ, Taylor JM, Sands CJ, Ciavaglia SA, Garrick RC, Tait NN, Rowell DM and Pavlova A (2006) A tale of two flatties: different responses of two terrestrial flatworms to past environmental climatic fluctuations at Tallaganda in montane southeastern Australia. *Molecular Ecology* **15**: 4513–4531.

Suppiah R, Hennessy K, Whetton PH, McInnes K, Macadam I, Bathols J, Ricketts J and Page CM (2007) Australian climate change projections derived from simulations performed for the IPCC 4th Assessment Report. *Australian Meteorological Magazine* **56**: 131–152.

Taminiau C and Haarsma RJ (2007) Projected changes in precipitation and the occurrence of severe rainfall deficits in central Australia caused by global warming. *Australian Meteorological Magazine* **56**: 167–175.

Thackway R and Creswell ID (1995) *An interim biogeographic regionalisation for Australia: a framework for setting priorities in the National Reserves System Cooperative Program. Version 4.0*. Australian Nature Conservation Agency, Canberra.

Thomas CD, Cameron A, Green RE, Bakkenes M, Beaumont LJ, Collingham YC, Erasmus BF, De Siqueira MF, Grainger A, Hannah L, Hughes L, Huntley B, Van Jaarsveld JS, Midgley GF, Miles L, Ortega-Huerta MA, Peterson AT, Phillips OL and Williams SE (2004) Extinction risk from climate change. *Nature* **427**: 145–148.

Thoms M and Sheldon F (2000) Water resource development and hydrological change in a large dryland river: the Barwon–Darling River, Australia. *Journal of Hydrology* **228**: 10–21.

Threatened Species Scientific Committee (2005) Loss of biodiversity and ecosystem integrity following invasion by the yellow crazy ant (*Anoplolepis gracilipes*) on Christmas Island, Indian Ocean, <http://www.environment.gov.au/biodiversity/threatened/ktp/christmas-island-crazy-ants.html>.

Thresher R, Proctor C, Ruiz G, Gurney R, MacKinnon C, Walton W, Rodriguez L and Bax N (2003) Invasion dynamics of the European shore crab, *Carcinus maenas*, in Australia. *Marine Biology* **142**: 867–876.

Thresher RR, Koslow JA, Morison AK and Smith DC (2007) Depth-mediated reversal of the effects of climate change on long-term growth rates of exploited marine fish. *Proceedings of the National Academy of Sciences of the United States of America* **104**: 7461–7465.

Thuiller W, Lavorel S and Araújo MB (2005) Niche properties and geographical extent as predictors of species sensitivity to climate change. *Global Ecology and Biogeography* **14**: 347–357.

Tidemann CR (1999) Biology and management of the grey-headed flying fox, *Pteropus poliocephalus*. *Acta Chiropterologica* **1**: 151–164.

Trowbridge CD (1996) Introduced versus native subspecies of *Codium fragile*: how distinctive is the invasive subspecies *tomentosoides*? *Marine Biology* **126**: 193–204.

Ujvari B and Madsen T (2008) Invasion of cane toads associates with a significant increase in mortality in a naïve Australian varanid lizard. *Proceedings Cane Toad Control Research Forum*, 13 June 2008, Darwin, <http://www.invasiveanimals.com/downloads/Cane-Toad-Proceedings.pdf>.

Umina PA, Weeks AR, Kearney MR, McKechnie SW and Hoffmann AA (2005) A rapid shift in a classic clinal pattern in *Drosophila* reflecting climate change. *Science* **308**: 691–693.

van der Wal R, Truscott A-M, Pearce ISK, Cole L, Harris MP and Wanless S (2008) Multiple anthropogenic changes cause biodiversity loss through plant invasion. *Global Change Biology* **14**: 1428–1436.

van Dyke S and Strahan R (Eds) (2008) *The mammals of Australia*. 3rd edn. Reed New Holland, Chatswood, Australia.

van Vliet A and Leemans R (2006) Rapid species responses to changes in climate require stringent climate protection requirements. In *Avoiding dangerous climate change*. (Eds HJ Schellnhuber, W Cramer, N Nakícénovic, T Wigley and G Yohe) pp. 135–143. Cambridge University Press, Cambridge, UK.

Venn SE and Morgan JW (2007) Phytomass and phenology of three alpine snowpatch species across a natural snowmelt gradient. *Australian Journal of Botany* **55**: 450–456.

Veron JEN (2008) *A reef in time: the Great Barrier Reef from beginning to end*. Belknap Press, Cambridge, MA.

Visser ME and Both C (2005) Shifts in phenology due to global climate change: the need for a yardstick. *Proceedings of the Royal Society of London B Biological Sciences* **272**: 2561–2569.

Wackernagel M and Rees WE (1996) *Our ecological footprint: reducing human impact on the Earth*. New Society Press, Gabriola Island, BC, Canada.

Walker B and Salt D (2006) *Resilience thinking*. Island Press, Washington, DC.

Walker B, Steffen W and Langridge J (1999) Interactive and integrated effects of global change on terrestrial ecosystems. In *The terrestrial biosphere and global change: implications for natural and managed ecosystems*. (Eds B Walker, W Steffen, J Canadell and JS Ingram) pp. 329–375. Cambridge University Press, Cambridge, UK.

Walker BH, Holling CS, Carpenter SR and Kinzig AR (2004) Resilience, adaptability and transformability in social-ecological systems. *Ecology and Society* **9**: 5.

Wallace KJ (2007) Classification of ecosystem services: problems and solutions. *Biological Conservation* **139**: 235–246.

Walsh K, Hennessy K, Jones R, McInnes KL, Page CM, Pittock, AB, Suppiah R and Whetton P (2001) *Climate change in Queensland under enhanced greenhouse conditions: third annual report, 1999–2000*. CSIRO consultancy report for the Queensland Government. CSIRO, Aspendale, Victoria.

Walsh KJE, Nguyen K-C and McGregor JL (2004) Fine-resolution regional climate model simulations of the impact of climate change on tropical cyclones near Australia. *Climate Dynamics* **22**: 47–56.

Walther G-R, Post E, Convey P, Menzel A, Parmesan C, Beebee TJC, Fromentin J-M, Hoegh-Guldberg O and Bairlein F (2002) Ecological responses to recent climate change. *Nature* **416**: 389–395.

Walker B, Holling CS, Carpenter SR and Kinzig A (2004) Resilience, adaptability and transformability in social–ecological systems. *Ecology and Society* **9**: 5.

Ward P (2004) The father of all mass extinctions. *Conservation Magazine* **5**: 12–19.

Ward TM (2000) Factors affecting the catch rates and relative abundance of sea snakes in the by-catch of trawlers targeting tiger and endeavour prawns on the northern Australian continental shelf. *Marine and Freshwater Research* **51**: 155–164.

Wardle P (1988) Effects of glacial climates on floristic distribution in New Zealand: a review of the evidence. *New Zealand Journal of Botany* **26**: 541–555.

Waterhouse DF and Sands DPA (2001) *Classical biological control of arthropods in Australia*. Australian Centre for International Agricultural Research, Canberra.

Watkins D (1993) *A national plan for shorebird conservation in Australia*. RAOU Report No. 90. Royal Australasian Ornithologists Union, Moonee Ponds, Victoria.

Wearne LJ and Morgan JW (2001) Recent forest encroachment into subalpine grasslands near Mount Hotham, Victoria, Australia. *Arctic, Antarctic, and Alpine Research* **33**: 369–377.

Weimerskirch H, Inchausti P, Guinet C and Barbraud C (2003) Trends in bird and seal populations as indicators of a system shift in the Southern Ocean. *Antarctic Science* **15**: 249–256.

Welbergen JA, Klose SM, Markus N and Eby P (2007) Climate change and the effects of temperature extremes on Australian flying-foxes. *Proceedings of the Royal Society B-Biological Sciences* **275**: 419–425.

Werner PA (1976) Ecology of plant populations in successional environments. *Systematic Botany* **1**: 246–268.

Werner PA (1977) Colonization success of a 'biennial' plant species: experimental field studies of species cohabitation and replacement. *Ecology* **58**: 840–849.

Werner PA (1979) Competition and coexistence of similar species. In *Topics in plant population biology*. (Eds OT Solbrig, S Jain, GB Johnson and P Raven) pp. 287–310. Columbia University Press, New York.

Werner PA and Wigston DL (1994) Biomes. In *The encyclopedia of the environment.* (Eds RA Eblen and WR Eblen) pp. 63–67. Houghton Mifflin Co., Boston.

Westoby M, Falster DS, Moles AT, Vesk PA and Wright IJ (2002) Plant ecological strategies: some leading dimensions of variation between species. *Annual Review of Ecology and Systematics* **33**: 125–159.

Wethered R and Lawes MJ (2005) Nestedness of bird assemblages in fragmented afromontane forest: the effect of plantation forestry in the matrix. *Biological Conservation* **123**: 125–137.

Whelan RJ (1995) *The ecology of fire.* Cambridge University Press, Cambridge, UK.

Whinam J and Copson G (2006) Sphagnum moss: an indicator of climate change in the sub-Antarctic. *Polar Record* **42**: 43–49.

White TCR (2005) *Why does the world stay green? Nutrition and survival of plant-eaters.* CSIRO Publishing, Melbourne.

Williams AAJ, Karoly DJ and Tapper N (2001) The sensitivity of Australian fire danger to climate change. *Climatic Change* **49**: 171–191.

Williams JW and Jackson ST (2007) Novel climates, no-analog communities, and ecological surprises. *Frontiers in Ecology and the Environment* **5**: 475–482.

Williams JW, Jackson ST and Kutzback JE (2007) Projected distributions of novel and disappearing climates by 2100 AD. *Proceedings of the National Academy of Sciences of the United States of America* **104**: 5738–5742.

Williams MAJ (1984) Cenozoic evolution of arid Australia. In *Arid Australia.* (Eds HG Cogger and EE Cameron) pp. 59–78. Australian Museum, Sydney.

Williams RJ, Bradstock RA, Cary GJ, Enright NJ, Gill AM, Liedloff A, Lucas C, Whelan RJ, Andersen AN, Bowman DMJS, Clarke PJ, Cook GD, Hennessy K and York A (2009) *Interactions between Climate Change, Fire Regimes and Biodiversity in Australia - A Preliminary Assessment.* Report to Department of Climate Change and Department of the Environment, Water, Heritage and the Arts, Canberra.

Williams RJ, Cook GD, Braithwaite RW, Andersen AN and Corbett LK (1995) Australia's wet-dry tropics: identifying the sensitive zones. In *Impacts of climate change on ecosystems and species: terrestrial ecosystems.* (Eds JC Pernetta, R Leemans, D Elder and S Humphrey) pp. 39–65. IUCN, Gland, Switzerland.

Williams RJ, Cook GD, Gill AM and Moore PHR (1999) Fire regimes, fire intensity and tree survival in a tropical savanna in northern Australia. *Australian Journal of Ecology* **24**: 50–59.

Williams RJ and Costin AB (1994) Alpine and subalpine vegetation. In *Australian Vegetation.* (Ed. RH Groves) pp. 467–500. Cambridge University Press, Cambridge.

Williams RJ, Wahren C-H, Bradstock RA and Müller WJ (2006) Does alpine grazing reduce blazing? A landscape test of a widely-held hypothesis. *Austral Ecology* **31**: 925–936.

Williams SE (1997) Patterns of mammalian species richness in the Australian tropical rainforests: are extinctions during historical contractions of the rainforest the primary determinants of current regional patterns in biodiversity? *Wildlife Research* **24**: 513–530.

Williams SE, Bolitho EE and Fox S (2003) Climate change in Australian tropical rainforests: an impending environmental catastrophe. *Proceedings of the Royal Society of London. Biological Sciences* **270**: 1887–1892.

Williams SE and Hero J-M (2001) Multiple determinants of Australian tropical frog biodiversity. *Biological Conservation* **98**: 1–10.

Williams SE and Middleton J (2008) Climatic seasonality, resource bottlenecks, and abundance of rainforest birds: implications for global climate change. *Diversity and Distributions* **14**: 69–77.

Williams SE and Pearson RG (1997) Historical rainforest contractions, localized extinctions and patterns of vertebrate endemism in the rainforests of Australia's Wet Tropics. *Proceedings of the Royal Society of London–Biological Sciences* **264**: 709–716.

Williams SE, Pearson RG and Walsh PJ (1996) Distributions and biodiversity of the terrestrial vertebrates of Australia's Wet Tropics: a review of current knowledge. *Pacific Conservation Biology* **2**: 327–362.

Williams SE, Shoo LP, Isaac J, Hoffmann AA and Langham G (2008) Toward an integrated framework for assessing the vulnerability of species to climate change. *PLOS Biology* **6**: 2621–2626.

Williamson I (1999) Competition between the larvae of the introduced cane toad *Bufo marinus* (Anura: Bufonidae) and native anurans from the Darling Downs area of southern Queensland. *Australian Journal of Ecology* **24**: 636–643.

Wilson RJ, Gutierrez D, Gutierrez JE, Martinez D, Agudo R and Monserrat VJ (2005) Changes to the elevational limits and extent of species ranges associated with climate change. *Ecology Letters* **8**: 1138–1146.

Winn KO, Saynor MJ, Eliot MJ and Eliot I (2006) Saltwater intrusion and morphological change at the mouth of the East Alligator River, Northern Territory. *Journal of Coastal Research* **22**: 137–149.

Winter JW (1997) Responses of non-volant mammals to late Quaternary climatic changes in the Wet Tropics region of north-eastern Australia. *Wildlife Research* **24**: 493–511.

Woehler EJ, Auman HJ and Riddle MJ (2002) Long-term population increase of black-browed albatrosses *Thalassarche melanophrys* at Heard Island, 1947/1948 – 2000/2001. *Polar Biology* **25**: 921–927.

Woinarski JCZ (Ed.) (1999) Fire and Australian birds: a review. In *Australia's biodiversity: responses to fire. Plants, birds and invertebrates.* (Eds AM Gill, JCZ Woinarski and A York) pp. 55–112. Department of the Environment and Heritage, Canberra.

Woldendorp G, Hill MJ, Doran R and Ball MC (2008) Frost in a future climate: modelling interactive effects of warmer temperatures and rising atmospheric [CO_2] on the incidence and severity of frost damage in a temperate evergreen (*Eucalyptus pauciflora*). *Global Change Biology* **14**: 294–308.

Woodroffe CD and Mulrennan ME (1993) *Geomorphology of the Lower Mary River Plains, Northern Territory.* Australian National University and the Conservation Commission of the Northern Territory, Australia.

Woodroffe CD, Thom BG, Chappell J, Wallensky E, Grindrod J and Head J (1987) Relative sea level in the South Alligator River Region, North Australia, during the Holocene. *Search* **18**: 198–200.

Wright HE (1984) Sensitivity and response time of natural systems to climate change in the late Quaternary. *Quaternary Science Reviews* **3**: 91–131.

Yates CJ, Elliott C, Byrne M, Coates DJ and Fairman R (2007) Seed production, germinability and seedling growth for a bird-pollinated shrub in fragments of kwongan in south-west Australia. *Biological Conservation* **136**: 306–314.

Yates CJ, Ladd PG, Coates DJ and McArthur S (2007) Hierarchies of cause: understanding rarity in an endemic shrub *Verticordia staminosa* (Myrtaceae) with a highly restricted distribution. *Australian Journal of Botany* **55**: 194–205.

Yeates DK, Bouchard P and Monteith GB (2002) Patterns and levels of endemism in the Australian Wet Tropics rainforest: evidence from flightless insects. *Invertebrate Systematics* **16**: 605–619.

Zeidberg LD and Robison BH (2007) Invasive range expansion by the Humboldt squid, *Dosidicus gigas*, in the eastern North Pacific. *Proceedings of the National Academy of Sciences of the United States of America* **104**: 12948–12950.

Zhang J, Jørgensen E, Beklioglu M and Ince O (2003) Hysteresis in vegetation shift: Lake Mogan prognoses. *Ecological Modelling* **164**: 227–238.

Index

C

O

P

R